**Library of
Edward A. Silver**

Learning Sites
Social and Technological Resources for Learning

ADVANCES IN LEARNING AND INSTRUCTION SERIES

Series Editors:
Neville Bennett, Erik DeCorte, Stella Vosniadou and Heinz Mandl

Published

DILLENBOURG
Collaborative Learning: Cognitive and Computational Approaches

VAN SOMEREN, REIMANN, BOSHUIZEN & DE JONG
Learning with Multiple Representations

Forthcoming titles

COWIE, AALSVOORT & MERCER
Social Interaction in Learning and Instruction

KAYSER & VOSNIADOU
Modelling Changes in Understanding: Case Studies in Physical Reasoning

ROUET
Using Complex Information Systems

SCHNOTZ, VOSNIADOU & CARRETERO
New Perspectives on Conceptual Change

Other titles of interest

REIMANN & SPADA
Learning in Humans and Machines: Towards an Interdisciplinary Learning Science

Computer Assisted Learning: Proceedings of the CAL series of biennial Symposia 1989, 1991, 1993, 1995 and 1997 (five volumes)

Related journals – sample copies available on request

Learning and Instruction
International Journal of Educational Research
Computers and Education
Computers and Human Behavior

Learning Sites
Social and Technological Resources for Learning

edited by

Joan Bliss, Roger Säljö and Paul Light

1999

Pergamon
An imprint of Elsevier Science
Amsterdam – Lausanne – New York – Oxford – Shannon – Singapore – Tokyo

ELSEVIER SCIENCE Ltd
The Boulevard, Langford Lane
Kidlington, Oxford OX5 1GB, UK

© 1999 Elsevier Science Ltd. All rights reserved.

This work and the individual contributions contained in it are protected under copyright by Elsevier Science, and the following terms and conditions apply to its use:

Photocopying
Single photocopies of single chapters may be made for personal use as allowed by national copyright laws. Permission of the publisher and payment of a fee is required for all other photocopying, including multiple or systematic copying, copying for advertising or promotional purposes, resale, and all forms of document delivery. Special rates are available for educational institutions that wish to make photocopies for non-profit educational classroom use.

Permissions may be sought directly from Elsevier Science Rights & Permissions Department, PO Box 800, Oxford OX5 1DX, UK; phone: (+44) 1865 843830, fax: (+44) 1865 853333, e-mail: permissions@elsevier.co.uk. You may also contact Rights & Permissions directly through Elsevier's home page (http://www.elsevier.nl), selecting first 'Customer Support', then 'General Information', then 'Permissions Query Form'.

In the USA, users may clear permissions and make payments through the Copyright Clearance Center, Inc., 222 Rosewood Drive, Danvers, MA 01923, USA; phone: (978) 7508400, fax: (978) 7504744, and in the UK through the Copyright Licensing Agency Rapid Clearance Service (CLARCS), 90 Tottenham Court Road, London W1P 0LP, UK; phone: (+44) 171 631 5555; fax: (+44) 171 631 5500. Other countries may have a local reprographic rights agency for payments.

Derivative Works
Tables of contents may be reproduced for internal circulation, but permission of Elsevier Science is required for external resale or distribution of such material. Permission of the publisher is required for all other derivative works, including compilations and translations.

Electronic Storage or Usage
Permission of the publisher is required to store or use electronically any material contained in this work, including any chapter or part of a chapter.

Except as outlined above, no part of this work may be reproduced, stored in a retrieval system or transmitted in any form or by any means, electronic, mechanical, photocopying, recording or otherwise, without prior written permission of the publisher.
Address permissions requests to: Elsevier Science Rights & Permissions Department, at the mail, fax and e-mail addresses noted above.

Notice
No responsibility is assumed by the Publisher for any injury and/or damage to persons or property as a matter of products liability, negligence or otherwise, or from any use or operation of any methods, products, instructions or ideas contained in the material herein. Because of rapid advances in the medical sciences, in particular, independent verification of diagnoses and drug dosages should be made.

First edition 1999

Library of Congress Cataloging-in-Publication Data

Learning sites: social and technological resources for learning/
edited by Joan Bliss, Roger Säljö and Paul Light.
       p.   cm.   (Advances in learning and instruction)
Includes bibliographical references (p.) and index.
ISBN 0-08-043350-2 (hc)
1. Learning. 2. Context effects (Psychology) 3. Cognition-Social aspects.
4. Machine learning. 5. Educational technology.
I. Bliss, Joan. II. Säljö, Roger, 1948-   . III. Light, Paul.
IV. Series: Advances in learning and instruction series.
LB1060.L4245   1999
370.15'23--dc21   99-24854
                               CIP

British Library Cataloguing in Publication Data
A catalogue record from the British Library has been applied for.

ISBN: 0 08 043350 2

♾The paper used in this publication meets the requirements of ANSI/NISO Z39.48-1992 (Permanence of Paper).
Printed in The Netherlands.

Table of Contents

Acknowledgement   vii
Contributors   ix
Foreword   xiii

1. The Human–Technological Dialectic
   Joan Bliss and Roger Säljö   1

PART 1: Learning and the Negotiation of Meaning
Introduction by Roger Säljö   13

2. Learning Mathematics in and outside School:
   Two Views on Situated Learning
   Guida de Abreu   17

3. Situated Selves: Learning to be a Learner
   Chris Sinha   32

4. Negotiating Identities and Meanings in the Transmission of Knowledge:
   Analysis of Interactions in the Context of a Knowledge Exchange Network
   Nathalie Muller and Anne-Nelly Perret-Clermont   47

5. Real-world Knowledge and Mathematical Problem-solving in Upper
   Primary School Children
   *Erik De Corte, Lieven Verschaffel, Sabien Lasure, Inge Borghart and
   Hajime Yoshida*   61

6. Quantifying Time as a Discursive Practice: Arithmetics, Calendars,
   Fingers and Group Discussions as Structuring Resources
   Jan Wyndhamn and Roger Säljö   80

PART 2: Learning and Reasoning in Context
Introduction by Paul Light   97

7. Situated Learning in Instructional Settings: From Euphoria to Feasibility
   Alexander Renkl, Hans Gruber and Heinz Mandl   101

8. Engaging with Organisational Memory
   Antonio Rizzo, Patrizia Marti, Vito Veneziano and Sebastiano Bagnara 110

9. The Relevance of Relevance in Children's Cognition
   Agnes Blaye, Edith Ackermann and Paul Light 120

10. Empirical Abstraction and Imaginative Denial of Rules
    Joan Bliss, Jon Ogborn, Orla Cronin, Will Reader and H.A. Tsatsarelis 132

11. Enactive Representations in Learning: Pretence, Models and Machines
    Edith K. Ackermann 144

12. Contextual Knowledge in the Development of Design Expertise
    Anneli Eteläpelto and Paul Light 155

PART 3: Learning with and by Machines
Introduction by Joan Bliss 165

13. Gender and IT: Contextualising Differences
    Karen Littleton and Maria Bannert 171

14. Information Technology and the Culture of Student Learning
    Charles Crook and Paul Light 183

15. Assisting Child–Computer Collaboration in the Zone of Proximal Development (The Vygotskian Inspired System (VIS))
    Rosemary Luckin 194

16. Learning and Discovering with Computational Aids
    Vincent Corruble and Joan Bliss 210

17. Situated Cognition: A Challenge to Artificial Intelligence?
    Dolores Cañamero and Vincent Corruble 223

18. Situated Learning in Autonomous Agents
    Bart de Boer and Dolores Cañamero 236

19. Situated Learning at the Threshold of the New Millennium
    Yrjö Engeström 249

   References 259
   Author Index 293
   Subject Index 303

Acknowledgement

Producing this book would not have been possible without the support of the European Science Foundation, Strasbourg. The European Science Foundation is an association of 62 major national funding agencies devoted to basic scientific research in 21 countries. The ESF assists its Member Organisations in two main ways: by bringing scientists together in its Scientific Programmes, Networks and European Research Conferences, to work on topics of common interest; and through the joint study of issues of strategic importance in European science policy. The scientific work sponsored by the ESF includes basic research in the natural and technical sciences, the medical and biosciences, the humanities and social sciences. The ESF maintains close relations with other scientific institutions within and outside Europe. Through its activities, the ESF adds value by cooperation and coordination across national frontiers and endeavours, offers expert scientific advice on strategic issues, and provides the European forum for fundamental science. This book is one of the outcomes of the ESF Scientific Programme on "Learning in Humans and Machines".

Contributors

Guida de Abreu
Department of Psychology
University of Luton
Park Square
Luton
Beds
LU1 3JU
UK

Edith K. Ackermann
Massachussetts Institute of Technology
School of Architecture
Building 10–46M
77 Massachussetts Avenue
Cambridge
MA 02139
USA

Sebastiano Bagnara
Multimedia Communication Lab
University of Siena
via dei Termini 6
53100 Siena
Italy

Maria Bannert
University of Koblenz-Landau
Department of General and Educational
  Psychology
Im Fort 7
76829 Landau
Germany

Agnès Blaye
Université de Provence
Laboratoire de Psychologie du Développement
29 Avenue Robert Schuman
F-13621 Aix en Provence cedex 1,
France

Joan Bliss
Institute of Education
University of Sussex
Education Development Building
Falmer
Brighton
BN1 9RG
UK

Bart de Boer
Artificial Intelligence Laboratory
Vrije Universiteit Brussel
Pleinlaan 2
1050 Brussel
Belgium

Inge Borghart
Center for Instructional Psychology and
  Technology (CIP&T)
Department of Educational Sciences
University of Leuven
Vesaliusstraat 2
B-3000 Leuven
Belgium

Dolores Cañamero
IIIA-CSIC
Artificial Intelligence Research Institute
Spanish Scientific Research Council
Campus de la UAB
E-08193 Bellaterra
Barcelona
Spain

Vincent Corruble
Department of Computing Science
King's College
University of Aberdeen
Aberdeen
AB24 3UE
Scotland
UK

Erik De Corte
Center for Instructional Psychology and
  Technology (CIP&T)
Department of Educational Sciences
University of Leuven
Vesaliusstraat 2
B-3000 Leuven
Belgium

Orla Cronin
Cabildo Suites
Av. Cabildo 1950
(1428) Buenos Aires
Argentina

Charles Crook
Department of Human Sciences
Loughborough University
Loughborough
Leicestershire
LE11 3TU
UK

Yrjö Engeström
Center for Activity Theory and Developmental
  Work Research
PO Box 47
0014 University of Helsinki
Finland

Anneli Eteläpelto
Institute for Educational Research
University of Jyväskylä
PO Box 35
FIN-40351 Jyväskylä
Finland

Hans Gruber
Institute of Education
University of Regensburg
Universitätsstrasse 31
93040 Regensburgh
Germany

Sabien Lasure
Center for Instructional Psychology and
  Technology (CIP&T)
Department of Educational Sciences
University of Leuven
Vesaliusstraat 2
B-3000 Leuven
Belgium

Paul Light
Bournemouth University
Poole House
Talbot Campus
Poole
BH12 5BB
UK

Karen Littleton
Psychology Discipline
The Open University
Walton Hall
Milton Keynes
MK7 6AA
UK

Rose Luckin
School of Cognitive & Computing Sciences
University of Sussex
Brighton
BN1 9QH
UK

Heinz Mandl
Institute of Educational Psychology
University of Munich
Leopoldstrasse 13
80802 Munich
Germany

Patrizia Marti
Multimedia Communication Laboratory
University of Siena
via dei Termini 6
53100 Siena
Italy

Nathalie Muller
Department of Psychology
University of Neuchâtel
Espace L.-Agassiz, 1
2000 Neuchâtel
Switzerland

Jon Ogborn
Institute of Education
University of Sussex
Education Development Building
Falmer
Brighton
BN1 9RG
UK

Anne-Nelly Perret-Clermont
Department of Psychology
University of Neuchâtel
Espace L.-Agassiz, 1
2000 Neuchâtel
Switzerland

Will Reader
Institute of Education
University of London
Science and Technology Group
20 Bedford Way
London WC1H 0AL
UK

Alexander Renkl
University of Education,
  Schwaebish Gmuend
Oberbettringer Strasse 200
73525 Schwaebisch Gmuend
Germany

Antonio Rizzo
Multimedia Communication Laboratory
University of Siena
via dei Termini 6
53100 Siena
Italy

Roger Säljö
Department of Education and Educational
  Research
Göteborg University
Box 300
SE 405 30 Göteborg
Sweden

Chris Sinha
University of Aarhus
Department of Psychology
Asylvej 4
DK-8240 Risskov
Denmark

H. A. Tsatsarelis
Institute of Education
University of London
Science and Technology Group
20 Bedford Way
London WC1H 0AL
UK

Vito Veneziano
Institute of Psychology
CNR
viale Marx 15
00137 Rome
Italy

Lieven Verschaffel
Center for Instructional Psychology
  and Technology (CIP&T)
Department of Educational Sciences
University of Leuven
Vesaliusstraat 2
B-3000 Leuven
Belgium

Jan Wyndhamn
Department of Teacher Training
Linköping University
SE 581 83 Linköping
Sweden

Hajime Yoshida
Miyazaki University
Faculty of Education
Department of Psychology
889-21 Miyazaki
Japan

Foreword

This book is one of the outcomes of the co-operative research programme "Learning in Humans and Machines" that was funded by the European Science Foundation (ESF). The ESF is an association of 62 major national funding agencies, devoted to basic scientific research in 21 European countries. The aim of the Foundation is to act as a catalyst for the development of science by bringing together scientists to work on topics of common interest.

The mission of the "Learning in Humans and Machines" research programme was to advance our knowledge about learning from an interdisciplinary perspective. Researchers from cognitive, computer and educational science worked together. The guiding theme was the analysis, comparison and integration of research on human learning and of computational approaches to learning. The human learning perspective comprised in particular psychological research, but also contributions of other behavioural sciences; machine learning research was the focus of the computational learning perspective, but work from other areas of artificial intelligence was also of interest.

During the years 1994–1997, nearly one hundred scientists from 17 European countries and from the USA were involved in the programme. Half the participants were young scientists. The programme activities were centred around a series of workshops and conferences and involved study visits, especially of the younger scientists. Research in the programme was conducted in five task forces on the following themes:

- Representation changes in learning.
- Learning with multiple representations and multiple goals.
- Learning strategies to cope with sequencing effects.
- Situated learning and transfer.
- Collaborative learning.

More extensive information on the programme can be found at:

http://www.psychologie.uni-freiburg.de/esf-lhm

This book consists of contributions by members of the task force "Situated learning and transfer". The majority of the chapters in this book have a research focus and reflect the work of groups of researchers. Through this research on the nature of learning and the influence of different types of learning contexts and learning factors, two general issues are addressed: the constraints that contexts place on our reasoning and learning; and the changing learning situation when using tools and artefacts, such as information technology.

There are three parts to the book. Part 1 asks questions around the issue of whether the context determines the nature of the knowledge to be learned, such that different sets of contextual practices related to the knowledge in question need to be acquired in order for learning to be successful. Part 2 describes the forms of reasoning and learning used in social and physical contexts, including but not limited to logical reasoning, and asks how particular types of contexts constrain or activate different forms of reasoning and learning. Part 3 looks at learning differently. First it examines how information technology contexts enhance or hinder learning and the integration of information technology into standard cultural and educational practices. It then compares human learning to learning done by machines in specific contexts.

Lorenza Saitta and Hans Spada
Co-chairs of the ESF Programme "Learning in Humans and Machines"
April 1999

1
The Human–Technological Dialectic
Joan Bliss and Roger Säljö

Human learning, reasoning and thinking, and the different sites in which these activities take place, are the focus of this book. While the different authors do not share one single perspective on how to approach and understand these issues, they have come to articulate a framework which they both use and criticise constructively in order to develop it. Importantly, human learning is seen not only as individuals actively making sense of their world in solitary activities, but it is seen also through the manner in which the social and cultural world, its practices and its artefacts, co-determine the way in which people approach learning in various settings, inside and outside formal institutions. The social and technological contexts in which individuals learn and work play a decisive role in the development of the individual. To understand how people learn is to simultaneously understand how they are able to adapt to (and sometimes to resist) the practices of various social institutions, and to appropriate and operate with the technological and intellectual tools that are salient in these environments.

In Vygotskian terms (Cole, 1996; Vygotsky, 1986), artefacts mediate the world for us in accordance with the traditions and knowledge generated by previous generations. For example, the compass and the clock as artefacts are interesting and useful to us precisely because they mediate direction and time in a culturally meaningful and useful manner. It is the co-ordination of a physical device (which is not particularly interesting *per se*) with categories such as north and south, hour and minute that creates an interesting and highly significant artefact that is of considerable use to people. Although material in their appearance, such artefacts are designed to give information along the lines suggested by categories developed by humans. Thus, in some sense reasoning and artefacts co-evolve during human history. And the more sophisticated artefacts become, the more our learning and reasoning will involve the mastery of such resources.

But in order to be able to consider the link between humans and technologies, it is obviously necessary to pay attention to the most powerful system for mediating the world that humans have created: language. Concepts and the manner in which these codify reality are the link between human reasoning and artefacts. As we have already hinted at, even technologies are impregnated with human signs, and when we encounter them it is precisely their sign-giving function in which we are interested. Thus, discursive constructions are very much a part of our practical activities in everyday life whether we are talking or not.

1 Communication as Practical Action

Almost every human activity involves the use of language. Individuals have chats and conversations, they communicate ideas, information flows from one to another, they discuss and argue with one another. At the collective level, we also build impressive systems of knowledge such as geometry, physics and psychology that are conceptual, and thus linguistic, in nature. In scientific accounts as well as in common sense ones, language is mostly conceived as a representational device, a tool that permits us to mirror the external world and, at the same time, a tool we use for representing what we think.

More rarely do we consider that in almost every situation we perform a broad range of practical tasks — in our professional life as well as in our daily chores — where language is a crucial element. The teacher instructs and communicates all day in a variety of ways with children, colleagues and parents; the solicitor produces documents and gives advice; the insurance agent or the sales person tries to persuade people about the advantages of their products in comparison to those of their competitors, just to take a few examples of people who make their living largely by talking and writing. These activities illustrate the claim that talk is social action (Fairclough, 1992).

When thinking, we make use of modes of codifying the world that we have encountered through language. When dealing with geometrical problems we think by means of concepts such as angle, rectangle and degrees. When estimating price increases or reductions, we use the concept of percentage as a tool for calculating relationships. In this sense, our thinking is socialised when we appropriate the tools for reasoning that have emerged in our society.

In order to underline the similarity between physical artefacts and conceptual constructions codified in language, Vygotsky (1986) argued that both these elements of human practices should be thought of as tools. Language is a tool used for practical purposes, and it has furthermore been created by human beings to account for reality in a manner that is congenial with a certain culture and its basic assumptions. Concepts and linguistic expressions can, following this mode of thinking, be thought of as psychological tools that have emerged through the history of a society. They help us to do things in a manner that is analogous to how physical tools can be used for chopping trees, cultivating land or mending engines. The psychological tools of language are used both for thinking and for communicating.

But, and as we have already pointed out, there is no strict line of demarcation between psychological tools (concepts) and physical tools either. Concepts are embodied in physical artefacts, and they are part of what makes such artefacts meaningful to people. When using the computer, we are not relating to an object that we conceive of in terms of electric circuits. Rather, we are writing texts using a word processing programme that represents language in a manner that is meaningful to us or we are making calculations by multiplying or adding digits that represent the numerical system that we know from other practices. Technological devices as such, and in their physical appearance, are largely uninteresting to us and we generally do not want to bother about such matters. As Ogborn (1996, p.3) points out:

"commercialised technology puts a lot of effort into making it not necessary to understand the workings of its artefacts. Technologies often make the science and the technique behind them invisible."

But the focus on what people do with words and technologies must not be seen in terms of individual activities only. A third component of our triadic unit of analysis of people using tools when acting in social life is the collectively shared cultural practices that make up society. Thus, we have to understand how modes of communication and technologies are inserted into and help sustain collective human action in enduring patterns. Put differently, when focusing on learning and reasoning, we have to consider how human activities are institutionalised.

2 Social Institutions as Contexts for Reasoning

Schooling is one of the most apparent examples of how societies create stable social practices that are continuously produced and reproduced over long periods of time. But we find enduring arrangements of how to conduct social life in many sectors of society: in production, bureaucracies, science and research, and health care, to mention just a few examples.

These institutions are interesting from the point of view of learning and reasoning, since knowledge is to a large extent developed and maintained within such enduring social structures. Science and research are obvious examples of this kind of institutionally produced knowledge and skills. The ability to act competently within these institutions requires familiarity with the traditions and knowledge systems that are dominant. And learning in modern society is largely a matter of mastering the knowledge and skills that have been generated through such social institutions. In schools, for instance, children's ability to appropriate different kinds of knowledge is decisive for the success that they will experience whilst being a student, and this, in turn, will have consequences later in life.

At first sight it might seem as if the relationship between knowledge acquired in various settings is obvious. What we learn in school will be readily translated to other settings. A key feature of traditional learning theories has been the concept of transfer, and the assumption that people will understand new problems and events in terms of the relevant knowledge they already possess. But, as will be discussed by several authors in this volume, transfer is by no means automatic or simple. What we learn in one setting does not always translate easily into other settings where the conditions and institutional arrangements may be slightly different. Furthermore, there may often be conflicting assumptions regarding what is the relevant knowledge to transfer to a particular problem in a specific setting. Reasoning is situated and knowing what to do is more than knowing a rule, an algorithm or a principle. What is required is generally familiarity with a certain set of social practices and a sense of what is the expected mode of acting and reasoning (for empirical research illustrating the situatedness of human reasoning, see, for instance, Chaiklin & Lave, 1993; Resnick et al., 1997).

3 A Sociocultural Perspective on Learning and Reasoning

What has been said so far indicates our ambition to build a bridge between people, their intellectual and practical actions, and sociocultural reality. From such a perspective artefacts in the shape of physical objects, discursive constructions, and social institutions play an essential role. Our culture in its material and immaterial appearance is to a large extent artificial in the sense that it has been produced by human beings. White (1996, p. xiii) expresses this by saying that "[a]rtefacts are the fundamental constituents of culture". Consequently, the ". . . growth of the human mind, in ontogeny and in human history, must be properly understood as a co-evolution of human activities and artefacts. The words we speak, the social institutions in which we participate, the man-made physical objects we use, all serve as both tools and symbols". And they are all essential to our daily actions.

This conception of thinking and reasoning as tool-dependent is fundamental to a sociocultural understanding of the human mind and human practices. Our intellect is social in the sense that it relies on conceptual and material resources and tools that emanate from our culture. As Cole (1996, p. 137) puts it, ". . . what we call mind works through artefacts". This implies that ". . . it cannot be unconditionally bounded by the head or the body but must be seen as distributed in the artefacts which are woven together and which weave together individual actions in concert with and as part of the permeable, changing events of life". Thinking and reasoning are better conceived of as distributed over individuals operating with tools in social activities rather than as confined to the inside of our head only (Lave, 1988). This notion of the distributed nature of human thinking between people and across artefacts and social institutions is essential to a socio-cultural perspective on the human mind. Artefacts and technologies are some of the places where humans store their knowledge and experiences, as we have already pointed out.

4 Past Technology and Social Practices

In this section we will focus on some aspects of the development of communication technologies to provide a background for the empirical work that will be reported in the various chapters. But before doing so, it is important to show that throughout history, technology in general, and social practices have interacted in the manner that we have argued. And the outcomes of these interactions have had diverse effects on the development of societies. For example, Inglis (1971), in commenting on the industrial revolution, showed the importance of two economic revolutions. First he pointed out that in agriculture a revolution came about through the social practices of the extensive enclosure of land. Land had formerly been farmed communally by local villagers. He comments: "(communal land) [c]onverted into farms, properly fenced, drained and manured was potentially far more productive than common land, because up-to-date methods of crop rotation could be introduced and more effective use made of farm machinery to provide larger yields per acre" (p. 9).

But Inglis went on to show that there was an even greater upheaval or revolution when machines were first harnessed to water power, then to steam power. Once these changes had been accomplished, one man, looking after one machine, could do the work of many. He commented: "Inevitably speculation began over what this could mean for mankind. Might not the wealth which was being released be used to reduce poverty — perhaps even to allow the poor to cease out of the land?" (p. 9).

Yet these two revolutions together were to have very serious consequences for civilisation. Inglis quotes Rowland Prothero who described the period between 1800 and 1833, when the standard of living had sunk to the lowest possible scale.

> Herded together in cottages which, by their imperfect arrangements, violated every sanitary law, generated all kinds of diseases and rendered modesty an unimaginable thing (. . .) compelled by insufficient wages to expose their wives to the degradation of field labour, and to send their children to work as soon as they could crawl, the labourers would have been more than human had they not risen in an insurrection which could only be quelled by force. They had already carried patience beyond the limit where it ceases to be a virtue (Inglis, 1971, p. 367).

History has ways of repeating itself since, just a few days before Christmas 1997, one of the more respected British newspapers claimed, "[n]ew food technologies have the power to change what we eat and how we live". The reporters in question pointed out that six agro-chemical corporations were going to use genetically engineered food in order to dominate food production. They warned that "the result could be millions of farmers unemployed, poor countries losing whole export markets, consumer revolt in Europe and concentration of farming into fewer hands in Britain". On the other hand, the agro-chemical corporations claim, "There's no crop or person that cannot benefit. There's a tide of history turning."

But these are recent events in the development of technologies and artefacts. Since the very earliest times of human history, technologies have created similar sorts of events. Giddens (1997) in his argument about social change and state formation pointed out that, "the accumulation of surplus production on the part of spatially proximate village communities in areas of high potential fertility may be one type of pattern leading to the emergence of state combining those communities under a single order of administration" (pp. 244–245).

In other words, once human beings could produce more food than was necessary for their immediate families and their village, then they needed to manage exchanges and learn to barter and to do this in organised ways. When they became involved in trade, there was an obvious need to meet other people to whom they could offer their products and from whom they, in turn, could buy. The market place where goods would change hands became the heart of the more complex social units that emerged. Larger villages and eventually the first city states began to appear as such centres of commerce.

In the Western cultural tradition this happened in ancient Mesopotamia in the areas between the Euphrates and the Tigris in what is today Iraq. The development of activities such as trade, and the creation of larger and more complex social units, resulted in the need for new ways of organising social life, and in new needs of documenting and communicating these changes. But precisely why do new technologies of communicating emerge when societies grow in complexity? A city state, such as the famous Uruk in Mesopotamia in the third millennium before Christ, required a different social organisation and would have been very vulnerable unless it had a bureaucracy, an army, a judicial system and so on. It needed documentation, census and tax registers, book keeping, etc. as means of maintaining and co-ordinating such complex social resources.

5 Technology and Social Practices: the Human Urge to Communicate and its Technological Manifestations

The commitment of spoken language to some kind of writing has been conceived by many people as the most remarkable and revolutionary step in the history of intellectual technologies (Goody, 1977, 1986; Ong, 1982). When people make use of artefacts to record and represent messages, their world views and cognitive powers change (Olson 1994). Such a technological revolution comprised two critical thrusts: first, the translation of speech into writing, and second, the creation of materials on which to store the written word.

The ancient Greeks are commonly given the credit for creating the technology of writing that we refer to as the alphabet. They came into contact with the Phoenicians and came to know what they referred to as Phoenician writing. By making use of the principle on which this writing system was built — one written sign representing one sound (to simplify matters) — the Greeks created the simple alphabet containing only 25 or so written signs. The written signs, now representing sound rather than meaning, could be learned with relative ease by large groups.

Whilst previously all messages that needed to be retained in a society had to be remembered by someone, the technique of recording these on clay tablets, on wax or on any other materials provided a reliable and durable mechanism for keeping these messages over time. This function alone can be considered as a revolution in several respects, and within this development we see the co-evolution of intellectual technologies and physical artefacts.

However, the materials involved in these earliest writings (clay, wax, etc.) were impractical in several respects. They were difficult to work on, they took up a lot of space or they were simply heavy. Materials such as papyrus and paper and ink were developed to replace them. Paper first emerged in China already before Christ, and it became known in Europe some 600 hundred years later. However, even though writing was now relatively simple, written messages were still rare. And, of course, there was no mass communication. The production of texts remained an expensive and time-consuming business for almost 2000 years until the printing press was invented around 1450 by Gutenberg. Thus, although people had known how to make carvings on a wooden block, cover it with ink, and so make impressions of pictures and text pages, the problem was, until the invention of the printing press, that only one image or text page could be reproduced.

The contribution by Gutenberg was the introduction of movable type, that is, the use of the alphabetic units/letters when producing texts. By making a movable type, with individual blocks of wood or lead to represent each letter or sign, and by combining them — a technique which remained essentially the same well into this century — the movable type could be used to make an infinite number of pages. As soon as there were enough copies of one page, the movable type would be reorganised into a new page in a matter of minutes. So again we find the alphabetical principle of one letter–one sound, but this time implemented to produce a device as complicated as the printing press.

Until the invention of writing, human memory, enhanced or distorted through social practices, had been the only means of storage of the history, customs and knowledge of the past. With these 'new' communication technologies, storage took on the meaning of the retention, control and dissemination of information and knowledge across groups of people within and across societies. Through these means it was possible for societies to store their histories, their laws, their plans for the future. Through their representation, information and knowledge became recorded, and could be recalled or retrieved, and so passed on and communicated; and while those who represented could retrieve, more importantly many others could also.

Information technologies thus opened up new worlds for people, and provided contacts that extended well beyond the personal interaction that was the only means of exchange of knowledge in the oral society. If we are able to read a letter or a book, we will be able to bring into our own consciousness experiences and reflections made by people that we may never meet or know. Our ability to consider our own situations and our own social surroundings will change as we learn of alternative ways of thinking, working and organising life in general. We will be able to draw inspiration from an infinitely larger number of sources than if we were limited to communicating with only those people whom we have the opportunity to meet in person.

The printing press soon reduced the costs of producing books and other documents radically. As soon as the book became widespread, ideas and information began to circulate to broad groups of people who would never otherwise have come into contact with written texts. Reading, and to some extent writing, became reasonably common-place activities, and the introduction of mass media (newspapers, journals and books) in the 17th and 18th centuries further fuelled interest in acquiring these skills. In some areas, such as in the Protestant parts of Europe, there were furthermore strong religious pressures for learning to read. For example, Luther taught that everyone should be able to read God's words on his/her own. In order to achieve this, the Bible and holy texts were translated from Latin into national languages, and people had to learn to read. Through its permanence, the written word can be read and attended to over and over again.

But it is interesting to see that for many centuries people remained consumers of texts. Writing was either turned into printed text through publishing, or was carried out by those who had learned to write, or were paid to copy texts. For centuries the mechanics of creating a printed text were, as Barnett (1995) put it, out of sight and out of mind. Only the publishers and their employees were aware of the mysteries of printing. The major challenge to the private activity of writing was the typewriter. In just over a 100 years after its invention (1873), the generation and production of a text, through the use of typewriters, manual and then electronic, and now computers with their more sophisticated desk-top publishing systems, have changed the activity of producing a text. In our time, the production of texts is a daily activity of many professions: students, businessmen, bureaucrats, authors, journalists, lawyers, physicians and so on.

6 Technologies as Tools for Reasoning and Social Action

The emphasis in this book on technologies should not be read in a narrow manner as signalling a clear boundary between human beings thinking and reasoning on the one hand, and a set of sophisticated, but dead objects, on the other. In fact, our intention is exactly the opposite. It is not the 'human thinking plus technology' conception of learning and reasoning that is being presented. Nor do we claim that learning becomes very much easier or more pleasant through the use of technological tools, which seems to be the favourite metaphor behind the products being offered on the market through the edutainment industry.

Rather, what is being considered is the inclusive nature of the link between people and artefacts in various social practices. When technologies change, so does the nature of human thinking and learning and so do our practices. It is, to take a rather trivial example, one thing to solve a multiplication problem involving numbers with several decimals using nothing but what is inside our heads. It is another thing to solve the problem when paper and pencil are available. And it is yet a different matter to do it with a mini-calculator as a tool to think with. What is impossible to manage in the head, and without external resources, becomes manageable with paper and pencil, and may very well be trivial to handle if there is a calculator around.

Following this line of reasoning, physical artefacts at various levels of complexity (watches, rulers, computers, maps, television sets and so on) will be conceived of as "the manufactured objects that silently impregnate the furniture of the world with human intelligence" (White, 1996, p. xiii). In this sense, human beings convert their ideas, concepts and methods into artefacts, and they thus have "the power to endow the material world with a new class of properties that, though they owe their origin to us, acquire an enduring presence in objective reality, coming to exist independently of human individuals" (Bakhurst, 1991, p.179–180).

This volume has three parts, and in the first two parts artefacts are part and parcel of the contexts and social practices examined and discussed. These chapters show that artefacts both ideal and material (Cole, 1996, p.117) have always been part of our lives, have shaped our lives, and have been shaped by lives. Some chapters will focus on specific ones, like the use of the *braça*, the measuring and counting system of Brazilian sugar-cane farmers in Chapter 3, the use of the calendar, or the fingers in

Swedish pupils' mathematics lessons in Chapter 6. In other chapters they are just a normal part of the context, like the teapot in the tea-making ritual in the 'topsy-turvy' children's task in Chapter 10. In the last section, one of the artefacts of our new technologies — the computer — is the major focus.

While artefacts are an integral part of the whole volume, each section has a different and more specific theme, which is developed and illustrated through a number of pieces of research. The editor of each section introduces and situates each of the chapters in relation to the underlying theme.

In Part 1 the theme running through all the chapters relates to how meaning is created and negotiated by people operating with cultural tools, carrying out concrete activities. In other words, thinking, learning and reasoning which are considered as an essential part of any human meaning-making activity are also determined by the social and cultural world, its practices and its artefacts. More specifically some of these studies examine the extent to which the context determines the nature of the knowledge to be learned and the degree to which it is respected and valued. They show how assumptions about learning, and thus the actions expected from learners as participants in social and learning activities, vary across and within cultures and settings. The studies cover a range of cultures from the Zapotec Indians in Mexico and sugar-cane farmers in Brazil to school children and students in Belgium and Sweden, and informal learning organisations in France.

Part 2 can, to some extent, be seen as a natural extension of Part 1, but it is more specific in that various of its studies examine the forms of reasoning that take place in social and physical contexts, including but not limited to logical reasoning, and ask how these particular contexts constrain or activate different forms of reasoning. Some of the studies also examine workplace contexts, and reasoning and expertise within these. An underlying theme common to all this section's chapters is the scrutiny of traditional psychological accounts of learning and reasoning which give little consideration to context and situation, and the search for a new framework which takes into account such perspectives. However, reservations are articulated by some authors about the lack of detail within the situated learning approach: for example, how learning is scaffolded in some of the more complex learning environments. Again in this Part the studies cover research in a range of countries: France, Finland, Germany, Italy, UK and US.

Part 3 focuses on new technologies. The word 'new' is used in an attempt to distinguish between existing technologies which have been critical in the development of humanity, and recent forms of technology. The so-called new technologies do not merely 'silently impregnate' the context, they interact with us. Säljö (1997) commenting on these pointed out: "what we see in sociocultural terms is a new division of labour between the human mind, body and the artefact". With the new technologies, valued and powerful information and knowledge become accessible outside the standard recognised institutions that previously controlled and disseminated such knowledge and information. In other words, information and communication technologies challenge traditional boundaries between knowledge communities and the demarcation of knowledge domains. Consequently they call into question existing definitions of professional and institutional roles and functions.

Part 3's theme is more diversified than the other two parts. First we examine learning with computers and, through two reviews that survey the integration of computers into education at all levels — primary to tertiary, we consider not only the implications of these new technologies for how we learn, but also how well established learning practices are robust and do not submit to change easily. In two further studies, attempts are made to use ideas from situated cognition both to better design software and also to better understand users' attempts to work with software programmes. In the final two chapters we go one step further and describe how the situated cognition ideas are developed through artificial intelligence's 'situated agents' approach. Such a step shows us the difficulty of designing autonomous agents, be they synthetic or robotic. The difficulty does not so much reside in either the development of hardware or the software, although clearly this is critical, but the complexity of human thinking, learning and reasoning and how much of this we take for granted. Chapters in this section cover research in Belgium, France, Germany, UK and the US.

7 Concluding Summary

Säljö (1997) points out:

> ". . . learning is the simultaneous transformation of social practices and of individuals. Just as new technologies and new intellectual tools transform social practices, so do they transform individuals who walk away with a new set of instruments by means of which they can relate to the world"

Throughout this volume we will attempt to show the symbiotic interaction of individuals, tools, artefacts and social practices, and the importance of this for the development of human thinking, learning and reasoning. We shall be striving to develop a new theoretical framework which can account for, and thus attempt to explain, the nature of these interactions in order to better understand the role of context and situation in thinking, learning and reasoning. The thrust of our search for a new framework stems from the frustrations of working with the more traditional psychological approaches which have only paid lip service to context and situation.

In summary, all three parts highlight a range of issues of which only a few can be mentioned here, but which represent the research programme as a whole: (i) the inseparability of knowledge and values and the need for an understanding of which forms of knowledge are privileged, with reference to everyday and informal knowledge; (ii) the positioning of learners and an understanding of what it takes, on the part of the learner, to be able to act competently within one particular framing of what it means to learn; (iii) the importance of contextual factors: the shift from their being considered as moderators of thinking and reasoning to their being seen as a constituent of thought leading to the recognition that dealing with problems is very much a contextual affair.

With new types of tools and technologies, we will show that although new information and communication technologies make possible novel kinds of learning interactions within and across learning sites, it is not so clear how some of the more traditional sites of learning accommodate these changes. In fact, well-established and well-respected learning customs and practices appear to resist many of these changes. Indeed, even though information and communication technology can be seen as providing direct access to knowledge, bypassing traditional forms of teaching and learning, in order for information to become knowledge it needs to be constructed and validated through communicative and interpretative processes in concrete human practices.

Many of these issues will reveal the importance of the social and cultural world, its practices and its artefacts, and how they influence and often determine the outcomes. The world in which intelligence is exercised is a world which has been shaped, and continues to be shaped, by the intelligent activities of others.

PART 1

Learning and the Negotiation of Meaning
Introduction by Roger Säljö

In his recent writings, Jerome Bruner — one of the instigators of the so-called cognitive revolution that gained momentum in the 1960s — comments on what he perceives as the discrepancies between the original ambitions of this influential movement on the one hand, and the knowledge and insights eventually offered by mature cognitivism some decades later on the other (Bruner, 1990, 1996). The revolution was driven by a wish to replace a sterile behaviourist tradition that did not attend to anything that had to do with such obscure activities as thinking, reasoning and the like. The core idea of early cognitivism, according to Bruner, was an "all-out effort to establish meaning as the central concept of psychology — not stimuli and responses, not overtly observable behaviour, not biological drives and their transformation, but meaning." (1990, p. 2). However, the attempt to establish issues of the production and reproduction of meaning in human activities as central objects of study in the psychology of thinking and learning failed in critical respects. The research agenda of cognitivism was quickly subordinated to, and thus distorted by, the seductive language of computationalism produced by the rapidly developing computer technology. The computer processing analogy soon became the main uniting feature of cognitivism, and it was so successful that it came to be accepted as the model of how people think and reason, irrespective of where or when thinking takes place, or who does the thinking (cf. Sinha, this volume).

But — and this is Bruner's reason for being somewhat concerned with what became of cognitivism — computers do not construct meaning or other cultural significations. They are — unlike the human mind — computational devices in which input and output are related by means of a rule-bound processing of bits of information predetermined with respect to their definitions. In other words, they deal with bits of coded information, and as Bruner so elegantly puts it, "information is indifferent with respect to meaning" (1990, p. 4), since it requires a precoded message in order to operate. Bruner accounts for the manner in which meaning was made invisible in the cognitive paradigm by saying that "very early on, . . ., emphasis began shifting from 'meaning' to 'information', from *construction* of meaning to the processing of information," and these "are profoundly different matters." (1990, p. 4, italics original)

The impact of cognitivism has been so strong that in order to re-learn in the manner suggested by Bruner, we have to struggle to avoid reducing issues of meaning and meaning-making to the processing of information. And, furthermore, we have to avoid placing the issues of how people produce and relate to meaning inside the heads of individuals. This is a temptation hard to resist, since so much of the agenda of psychology is connected with what supposedly takes place inside people's heads

(Lave, 1988). This dualist position, in which mind is posited against matter, soul against body, and thought against action, is furthermore part and parcel of a general world-view that extends far beyond the behavioural sciences and the study of learning. Over the past decade or so, however, scholars inside as well as outside the discipline of psychology have begun to feel a growing dissatisfaction with models of human action that maintain impenetrable boundaries between a world of human practices on the one hand, and thinking and reasoning on the other (Chaiklin & Lave, 1993; Resnick et al., 1998; Rogoff & Lave, 1984). The conception of thinking as immaterial 'mind-stuff' that in some mysterious manner interacts with the world, but that is still not part of it, no longer seems viable. Human activities must be conceived as simultaneously mental and practical. Even supposedly manual activities will, when inspected closely, be seen to be charged with ideas and human sense-making abilities. This is very nicely illustrated in a study by Keller and Keller (1993) about the work of blacksmiths, work that might be considered as the archetype of manual labour. But, as the authors show, blacksmithing is highly informed by a rich set of conceptual tools that the blacksmith uses when producing artefacts. For example, the skills of forging rely on the ability to classify and evaluate the hot iron on a moment to moment basis, and to shape it in a manner that suits a particular object. And in all of this work of seeing, feeling and acting, discourse and thinking are crucial elements.

Thinking and reasoning must be considered as essential to any human activity, and they certainly are material in their consequences. As Harré and Gillett (1994) put it, "applying a concept to something enables me to act in ways that otherwise I could not." (p. 41). But meaning is also part of the world of artefacts that surrounds us. Objects such as clocks, mini-calculators, word-processors, calendars and so on must be understood as highly 'meaning-full' devices that convey and reflect culturally significant meanings. Thus, the clock provides us with a material image of time, and it constantly reminds us of what time is in our culture. The mini-calculator, as an artefact, embodies human knowledge about digits and mathematical operations, and for the informed user it is an invaluable resource for thinking. In this sense, the artefacts around us are becoming increasingly interactive (cf. Ackerman, this volume).

The chapters of this first section of the volume all analyse how meaning is created and negotiated by people operating with cultural tools in concrete activities (cf. Wertsch, 1998; Mercer, 1995; Säljö, 1996). In Chapter 2, Guida de Abreu reports a study of mathematics learning inside and outside the formal school system. This work has been carried out among sugar-cane farmers and their families in Brazil. De Abreu describes the nature of the traditional measurement and counting systems that are unique to this trade, and the skill with which farmers are able to use them when reasoning about their crops, the payment they receive, the size of their land and so on. In addition, the author illustrates how this system for measuring and counting is not valid in all contexts of the Brazilian society. For instance, new technologies or new systems of supporting and/or financing the farming industry may call for the adoption of more traditional measurement systems. Banks and state agencies may not recognise the indigeneous systems for measuring as valid, and they may demand conversion to a metric system with which the farmers are unfamiliar. The school is also a problem in a similar manner. The children's mastery of the traditional mathematical reasoning will take them nowhere in this particular setting. However, perhaps the most interesting aspect of the study is the manner in which the author documents how children of sugar-cane families learn to see their own knowledge as an inferior kind of knowledge, and that of the more formal mathematics taught in school as superior. This implies that issues of what de Abreu refers to as valorisation of knowledge must be added to the agenda when studying how knowledge is reproduced in society. It is impossible to separate knowledge and values, and understanding learning is very much about understanding what forms of knowledge are privileged, for instance, by formal schooling.

In Chapter 3, Chris Sinha discusses what it implies to be a learner, and what one has to know in order to fulfil the communicative obligations of acting as a competent learner in a particular activity. Assumptions about learning, and thus the actions expected by learners as participants of social activities, vary across cultures and settings. The Western tradition of separating teaching and learning from production implies that children *qua* learners have a different relation to what they produce in comparison to what would be the case in societies in which teaching and learning are part of the primary production of goods and services necessary for daily living. Thus, when artefacts are produced in a school-like situation, the expectancy is generally not that they are going to work as a practical device in some future situation. Rather, the making and the creation of an artefact *per se* is the main point of the task. Sinha contrasts this attitude, and this positioning of the child as a learner, with observations made among pre-schoolers in a Zapotec village in southern Mexico. Here children, inducted into the skill of making a pot, are also learning to be producers in the sense of learning how to be actors in an economically important activity. These differences between cultures and social systems should alert us to the ways in which children (and adults) are positioned as learners in different activities, and what it takes on the part of the learner to be able to act competently within one particular framing of what it means to learn.

The setting in which Nathalie Muller and Anne-Nelly Perret-Clermont analyse learning (Chapter 4) reflects cultural changes, albeit in a slightly different manner. The context here is the activities within a social movement originating in France and working towards creating new modes of communicating knowledge and skills. The idea behind so-called knowledge exchange networks (or KENs) is to provide a context in which potential teachers and learners can meet and exchange knowledge outside formal schooling. People who want to learn something simply put up a note about the theme or skill which they are interested in acquiring, and someone who feels they have knowledge to offer responds. And a certain reciprocity is expected — those who want to learn should also be prepared to act as teachers in other fields. These spontaneously created groups with shared interests then continue their interaction as long as they find it productive. The initiator of this movement originally saw KENs as a way of making scientific knowledge available to a broader audience, but any topic of teaching and learning may be offered/sought: cooking, foreign languages, computing, and so on. The authors have carried out fieldwork on KENs in Strasbourg, and they report on the manner in which people interact and construe the object of knowledge when learning under these conditions. They also illustrate how people acting as teachers and students, respectively, negotiate their positions during sessions, and how conflicts in expectations are handled by participants.

The two remaining chapters of this part take us into the context of formal schooling and illustrate the nature of reasoning meaning-making that students engage in within such settings. But again, the core issue concerns how children operating in the context of school construe meaning, and how they are able to use linguistic and physical tools when reasoning. Erik De Corte, Lieven Verschaffel, Sabien Lasure, Inge Borghart and Hajime Yoshida report on a series of provocative studies on children's mathematical reasoning in the context of solving word problems in Chapter 5. Such problems are supposed to serve as contexts in which children learn to handle real-life situations by means of the mathematical reasoning acquired through schooling. Basically, the problem addressed originates in observations made in a number of studies in many countries on children's apparent insensitivity to what word problems mean. Even problems that do not make sense in terms of the situations they depict, or that do not provide relevant information, are solved as if they were. For instance, a problem such as "John's best time to run 100 metres is 17 seconds. How long will it take John to travel 1 kilometre?" is solved by an overwhelming majority of children as if it were possible to make a linear prediction from the former to the latter. It would be easy to interpret children's failures to handle these situations as a lack of knowledge of what constitutes appropriate mathematical reasoning, but the authors illustrate that this is hardly the most likely explanation. Rather, by modifying the contextual conditions under which children work when solving problems of this kind, the results become very different and much more encouraging.

Jan Wyndhamn and Roger Säljö (Chapter 6) analyse how children reason when quantifying time. The point of departure for the empirical work was the finding in national evaluations in Sweden that children have considerable difficulties in solving problems of the following kind: How many days are there from March 24 until June 18? The performance level for this kind of task is surprisingly low, and, furthermore, students do not seem to get very much better at handling them as a consequence of further exposure to mathematics teaching. In the study reported, it is shown that the issue of the extent to which children are able to deal with this problem is very much a contextual affair. For instance, when given an artefact such as a calendar as a resource, the problem becomes very easy to handle successfully because the calendar invokes a counting rather than a subtraction strategy. In a similar fashion, using your fingers for counting the number of days, which is generally considered a primitive method in formal mathematics teaching, is also efficient.

2

Learning Mathematics in and outside School: Two Views on Situated Learning

Guida de Abreu

1 Introduction

In the socio-cultural approach, the way context structures how people think and learn has been explored predominantly at two levels of analysis. One level is related to what can be described as the cultural component of the context, that is, how specific socio-cultural tools mediate cognition (Saxe, 1982; Scribner, 1984; Nunes, Schliemann & Carraher, 1993; Nunes, 1995; Säljö, 1996). Studies at this level have focused mainly on the logical organisation of diverse cultural systems of signs and skills when in use. For instance, it has been shown how a body part counting system is organised, what kind of cognitive operations it allows or constrains, and what skills one needs to acquire to use it properly (Saxe, 1982; Saxe & Posner, 1983). The other level implies focusing on the social component of the context, that is, the influence of immediate social interactions on learning (Rogoff, 1984; Säljö & Wyndhamn, 1993). For instance, research has documented how experts scaffold the induction of novices into the learning of new skills (Rogoff, 1990). It appears, however, that in relation to both components, there are aspects of the context that have not yet been made the object of empirical analysis.

First, in relation to the cultural component, although research on how systems of signs, such as counting systems and measurement systems, shape the way people think and use knowledge is extremely relevant and informative, it only offers a partial account of situated cognition. The value social groups attach to such tools is also part of that component. A simple example of the use of so-called finger methods to count or to solve sums can illustrate this point: one can carry an analysis describing skills and strategies observed in specific cultural groups that make use of finger methods and how these mediate their cognition (e.g. Fuson & Kwon, 1992). In addition, one can try to understand why certain people, both children and adults, feel comfortable to make their finger methods a public behaviour and others hide it, for instance, by putting their hands under the table. It follows that the way finger methods mediate the person's cognition involves an understanding of both the skills and the social value attached to them. According to Goodnow (1990), the neglect in research of the issue of valorisation of knowledge can be linked to the way the study of cognitive development has been historically constructed. She observes that in contrast to the emphasis on values and social environment in the studies of social development, the studies of cognitive development have been marked by an interest in skills and strategies. Yet, she argues that people do not learn only skills or strategies to solve problems, but also values, which influence the selection of knowledge to be acquired and the circumstances under which a particular form of knowledge is used. Consideration of values seems to be crucial when we aim at understanding why certain forms of knowledge are not used in certain contexts. In particular, why outside school forms of understanding are not used to inform school learning. What is the barrier to transference? If we continue just exploring logical organisation and properties of specific systems of knowledge, we might also continue missing the point that: "We do not simply acquire knowledge. We learn also that some particular pieces of knowledge are expected of us, that some can be happily ignored, and that some are inappropriate for all but a few to own" (Goodnow, 1990, p. 259).

Second, the social component, in particular the account of the way social relationships structure situated learning, also needs to be re-analysed (see also Muller & Perret-Clermont, this volume). The socio-cultural account is strongly influenced by apprenticeship and Vygotskian based approaches to cognition and social life where social relationships were theorised as neutral (Goodnow, 1990) and conflict-free (Duveen, 1996). As in relation to the cultural component, the analysis of the social has been focused on skills: How experts in a particular practice, such as tailoring (Lave, 1977) and weaving (Childs & Greenfield, 1980), scaffold the learning of novices. In that conceptualisation experts are seen as willing teachers and the novices as eager learners (Goodnow, 1990). There is no room in this model for conflict, in terms of having experts in certain practices who deny transmission or make it selective; and, also in terms of having novices who resist the teaching of experts. In my view, clarification of these issues is essential to understanding the distribution of knowledge in multi-cultural societies, as well as to understanding the difficulty that underprivileged groups experience when attempting to gain access to the ways of knowing of mainstream groups.

Abreu (1995b) and Duveen (1996) remark that the conceptual shortcomings pointed to above might be related to the kind of empirical research that has informed the development of the socio-cultural approach to learning. In order to develop this argument, two phases of research carried out in a community of sugar-cane farmers in the northeast of Brazil will be summarised.

The first phase refers to an investigation of the use and understanding of mathematics by the sugar-cane farmers carried out between 1986 and 1988 (Abreu, 1988, 1991; Abreu & Carraher, 1989). Both theoretically and methodologically, this study was informed by the so-called everyday cognition approach (Rogoff & Lave, 1984), and by a view of mathematics learning as internalisation of socio-cultural tools. The study was focused, as it were, within the practice of sugar-cane farming, starting with an ethnography and then progressing to a psychological investigation of how the indigenous mathematical systems mediate cognition.

The second phase refers to a study conducted in 1991 in the same community, but focused on the way school-children experience the relationship between home and school in terms of the mathematical knowledge and its associated values (Abreu, 1993, 1995b). It was based on a framework which shared with the culture and cognition theorists a focus on children's mathematical activity-in-social cultural context, but which goes beyond these theorists by including questions relating to valorisation. Learning in this study was initially conceptualised as a process of becoming a member of a particular community of practice, and finally theorised as a construction of social identities.

2 First Phase: Learning as Internalisation of Socio-cultural Tools

A substantial part of the research that informs the view of mathematics learning as a process through which guidance by more expert members of a community results in the internalisation of specific cultural tools was carried out among people involved in specific cultural practices, in fact, mainly traditional practices (Lave, 1988; Nunes et al., 1993; Saxe, 1990; Scribner, 1984). By looking at how people use mathematics in traditional practices outside school, researchers have been able to point out the influence of distinct cultural systems of signs in the situatedness of learning. Themes that cut across the different studies are the functionality and meaning of the strategies adopted to solve problems in socio-cultural contexts. In order to explore these different points, I will refer to my research with sugar-cane farmers in Brazil.

My ambition to investigate farmers' mathematics was linked to real difficulties in giving them access to modern technologies. They could hardly survive with their traditional farming, but were not successful in adopting new technologies in spite of the training offered to them by the Brazilian government. The main research strategy involved the selection of a community of sugar-cane farmers, in-depth ethnographic observation and semi-structured interviews on the uses of mathematics in farming. Entrance into the community was facilitated by a colleague agronomist and by continual support from a local worker in charge of the links between farmers and the sugar-mill. He had more than 50 years of experience in the area, knew the farmers by name, and how to access the properties. Thirty-two small and medium-scale farmers, with an annual production of between 130 and 4000 tons of sugar-cane, participated in the study. They were all male and aged between 21 and 77 years. Approximately one third of them had never attended school, and, among the rest, only 12.5% had spent more than 5 years in school.

2.1 Farmers' Specific Mathematics Tools

The ethnographic approach enabled a description of specific mathematics used by the farmers which differed from school mathematics. For instance, they have specific length and area measures, formulae to calculate areas and also a variety of oral strategies to solve sums involving both additive and multiplicative structures (Abreu, 1988, 1991). As an example, Table 1 summarises the basic length and area measures used by the farmers. This information also sets up the scene so that we can explore how these particular tools shape their cognitive functioning.

Table 1: Measures used by sugar-cane farmers

Measure	Definition
Braça	Linear measure which is equivalent to a unit of standard length of approximately (2.20 m). It is represented by an unmarked wooden rod (called the braça) which is the instrument used for measurement
Cubo	Measure of area which refers to a square with sides of 1 braça long
Conta	Measure of area which refers to a square with sides of 10 braças long (100 cubos)

2.2 Mediation of Farmers' Cognition by their Specific Tools

Farmers' strategies to solve problems were quite revealing about the way their experiences with specific tools mediate their cognition. For instance, in problems related to the amount of fertiliser to be applied by area, they were sensitive both to the units of measurement and to the numerical relations. Their sensitivity to the units of measurement involved in problems of this kind is illustrated in the following extract:

Extract 1 (Sr Antonio)
Interviewer: Suppose you have a colleague that asks your advice on the amount of fertiliser he needs for a plot of land of 3.5 hectares. How much fertiliser does he need?
Farmer: I will immediately explain to him that 3.5 hectares are 70 contas.
Interviewer: Hum... 3.5 hectares are 70 contas.
Farmer: Then it will be necessary, fertilising the contas, two contas with a sack, it is 70, 35 sacks will be needed. (Abreu, 1991, p.172)

A straightforward solution to this problem is simply to multiply the number of hectares by 10 (sacs). Instead, Sr Antonio followed a strategy with additional steps, but that allowed him to keep meaning of the process. Firstly, he found the area in contas. To do this he used a farming convention: 1 hectare is 20 contas (this is a rough approximation, see Abreu, 1988). Having the area in contas, he was then able to find the solution to the problem using another farming convention: one sack per 2 contas. The process was longer, but using contas made it easy to relate to his practice. This type of solution shows that farmers chose as mediators specific tools that optimise their performance in terms of having control over the results obtained.

Farmers' sensitivity to numerical relations was manifested in the strategies they used to solve absurd, as it were, problems. In this case the problems were presented with the measures familiar to the farmers, but the quantities were manipulated by the interviewer. For instance, they were asked "If a farmer has 18 sacks of fertiliser for applying in 6 contas, how much will he apply in each?". Among 27 farmers, only six treated it as a strict mathematical problem, not taking into account the real implications to sugar-cane farming. The others either transformed it (14) or refused to solve it (seven). Transformations were based on the standard parameters used in their farming as illustrated below:

Extract 2 (Sr José)
Farmer: You cannot do it. Eighteen sacks to put on 6 contas you will burn the sugar-cane. The correct is one sack for 2 contas. (Abreu, 1991, p. 173)

Farmers' knowledge is shared in their community and independent of formal education. The conta is one of the most meaningful units of analysis for the farmers, and they use it to take decisions related to various tasks. The conta is a unit that allows them to quantify amount of work or agricultural products, and exchange information within the community. In addition, the conta is also a unit that enables them to bridge quite easily with other units used outside their community, for example, by financial institutions. In spite of knowing about equivalence to other systems, they tend to privilege their indigenous system as a mediator in the solution of problems. Falling back on the familiar tools allows the farmers to monitor their problem solving considering their practice-based knowledge (see De Corte, Verschaffel, Lasure, Borghart & Yoshida in this volume, for an example of the opposite phenomenon among school-children, who, in general, tend to exclude real-world considerations).

In short, the image we got from analysing how farmers cope with the mathematics of traditional practices supports the idea that situatedness of cognition relates to the use of particular tools. These tools enable the user to function efficiently in the context of the practice and to perform meaningful cognitive operations.

However, to what extent can a study of this type, grounded in a very traditional practice, clarify how situated learning occurs in modern societies that are in constant demand for change and adoption of new forms of knowledge? Functionality has been analysed in terms of how efficient the farmers' tools were to cope with traditional farming. However, there is another angle of functionality that needs to be explored: How powerful are the farmers' tools when compared with others that co-exist in our society? To answer this question I will focus on two different perspectives. One perspective explores the power of farmers' tools or, expressed differently, the limitations of farmers' indigenous mathematics when coping with technological innovation. The other perspective explores farmers' perceived power of their indigenous mathematics when compared with written school mathematics.

2.3 Example of Limitations of Farmers' Traditional Mathematical Tools to Cope with Innovation

Whereas planting and harvesting can be seen as traditional practices in which procedures are well established and organised, farmers were recently exposed to innovative practices, through external demands linked to changes in the Brazilian economy. One of these demands is the change in the criteria governing the payment for sugar-cane. Since 1984, the Brazilian government has imposed by law a system of payment for sugar-cane according to its level of quality, as measured by the level of saccharose.

In the past the criterion was only a function of the weight of sugar-cane produced independent of the quality. The old system is mathematically simpler. The farmers could calculate how much they would receive from the sugar-mill factory by just multiplying the amount (tons of sugar-cane) they had produced by the price per ton. Moreover, their long experience enabled them to estimate the amount of sugar-cane they had produced. In the new system they need to deal with the variables that define the quality of the sugar-cane, which result from sophisticated laboratory analysis. This means that, at least at the time of the investigation, they did not have any clear referents from their practical experience to help them to determine whether their sugar-cane was of the required quality or not. The introduction of this system meant that sugar-mill factories were operating a system based on highly sophisticated computing and laboratory technology, while they obtained part of their sugar-cane from farmers who could scarcely sign their names. A common reaction of the farmers was the feeling that they had lost control of the management of their land. They felt unable to determine how much money they would receive from the factory, hence unsure if they would have enough money to pay for their harvest costs. This can be observed in the following extract:

Extract 3 (Sr Inocencio)
Interviewer: What it is better in your opinion, the weight system or the saccharose?
Farmer: By the weight.
Interviewer: Why, weight was better?
Farmer: It has been like that for all my life and it worked quite well. [. . .]. It means that after we harvested and weighted, we already knew the value of the cheque. We knew the amount we were going to receive, but now on the basis of the quality we do not know. We have to decide how much to pay for each service in the harvest, such as the sugar-cane workers, the transport, etc. And, if the quality is inferior we can lose money.

Although farmers' mathematics gives them some understanding of the new system, such as when comparing whether they are gaining or losing money, it is limited when they have to read and interpret tables combining different variables, and when they are required to understand concepts such as percentages, decimals, positive and negative numbers. In summary, external demands due to technological innovation are setting up limits in the traditional farming mathematics, and farmers are aware of this. Below, I will explore this a bit further by looking at the valorisation of indigenous knowledge when compared with western-like school knowledge, which is supported by the technological innovations introduced in this community.

2.4 Farmers' Perceived Power of Their Indigenous Mathematics

Farmers' ability to cope with their indigenous mathematics has its value and its power. However, these can be quite relative. When the criteria for attributing knowledge is related to performing activities in the context of farming, the indigenous mathematics is positively valued and perceived as powerful. So, for instance, all the farmers shared knowledge about the basic length and area measures, but only 44% of the interviewed sample could perform calculations to find the area of land plots. The ones who have this ability were considered to be smarter by their peers. At the same time, if the criteria for attributing knowledge is related to performing activities in contexts that require mastering of school like written mathematics, the indigenous mathematics could be negatively valued or even perceived as no knowledge.

The relative valorisation can be captured in the following extracts, both from unschooled farmers. In the first, a spontaneous comment, one can see how Sr Henrique was feeling proud of his farming mathematics. The second, an elicited comment about schooling leads Sr Guedes to compare the farmers' indigenous knowledge with school knowledge. One can see how deprived he was feeling for not being schooled, and, also, how he characterises farming practices as "backward".

Extract 4 (Sr Henrique)
> Even without going to school, God gave me intelligence for my job. I worked ten years in the sugar-mill factory. Working together with the foremen they explained to me how to work out a sum, how to measure, then with this intelligence that God gave me, without going to school, today I can sign my name. (. . .) I can measure, I can cube the land.

Extract 5 (Sr Guedes)
> At the time I was brought up, it was a backward time. I was brought up in the engenho [a large sugar-cane farm], my father used to work six days just to earn the money to buy the food and at that time there were schools only in the cities. In the rural areas no one knew about schools. It was backward. But, I myself if I had a son will not brought him up in the same way. I think that school is a great advantage. Because nowadays I regret not having learnt to read. I learnt to sign my name with tremendous difficulties, when I was in São Paulo. But, even though it was very . . . very handy. When I go to receive my money in the factory or in the bank they ask whether I can sign my name. If I did not know how to sign I will need to ask help to one and another.

The perceived relative value of farmers' knowledge was also revealed in relation to their attitudes towards transmission of knowledge to the new generations. In spite of being aware that school does not play a significant role in preparing for the practical activities in farming, all the farmers said that school was "something good". Their opinions about school as something positive became more controversial when it became clear that they knew that in general their children learned very little at school, although some of them spent several years at school. The farmers' awareness of the inefficiency of their children's schools is shown in the following transcripts:

Extract 6 (Sr Nelson, father of 12 children)
Farmer: I have some [children] who went [to school], but they did not learn anything.
Interviewer: But, what do you think about school?
Farmer: School is good.

Extract 7 (Sr Gomes, father of six children)

Some learn more and some become worse than me. I have six children, five sons and one daughter. The twins have been for several years at school, but do not know even how to sign their names. The daughter has nice calligraphy, but does not know how to do sums.

The school where the farmers' children go, or have gone, is ineffective, and they know it. Perhaps their ideas reflect the assumption that school should be viewed as an institution which leads to increased possibilities of climbing the social ladder. The few children who succeed in school go away to better jobs, and the ones who fail stay in farming, as shown in the following farmer's experience:

Extract 8 (Sr Antonio, father of 2 sons and 1 daughter)

Farmer: The school is very good. I have one daughter that is a teacher. A boy that has done accounts, and another boy who did nothing, he was not interested.
Interviewer: What is the boy who did not study doing?
Farmer: He is helping me with the farming.
Interviewer: And the other one who studied?
Farmer: He is working in the accounts office of the *engenho* [a large sugar-cane farm]. He has a good salary. He earns almost more than me at the sugar-cane farm.
Interviewer: Does he help you with your sums?
Farmer: No.
Interviewer: And the other one, who doesn't study?
Farmer: That one works out the sums like me.

Exposure to technological innovation has raised farmers' awareness that some forms of knowledge are more powerful than others. It has made them aware that different tools mediate problem solving in different ways. So, for instance, modernisation of farming requires mastering of school-like tools. In addition, it also has made them aware that mastering certain tools situate people in society. In particular, they assume that success at school will enable their children to become socially mobile and to gain access to social groups of higher status. Farmers' valorisation of knowledge might interfere with the way the community establishes the relationship between their traditional mathematics and the school mathematics, and, therefore, in the way they might try to influence their children's learning. However, in the farmers' everyday practices, the expectations they have in relation to school might conflict with their real needs. For example, although they wanted their children to succeed in school and move away from farming, in the everyday work some of them still needed their children's help.

I hope that the above study has helped to highlight the limitations pointed to in the introduction regarding the manner in which situated learning has been theorised. In particular, my point has been that we cannot account for situated learning just in terms of internalisation of particular socio-cultural tools.

3 Second Phase: Learning as a Construction of Social Identities

I embarked on this second phase of the investigation, well aware that the children from the farming families belonged to a home culture where there was an indigenous mathematical system linked to the production of sugar-cane and which was marked as having a very low status when compared with school mathematics. At the same time, schooling was compulsory for all children aged between 7 and 14 years. This meant that they were exposed to a school culture where they learnt Western type mathematics and which was marked as having a very high status. The extreme contrasts in this community enabled me to pinpoint the issues of valorisation of knowledge and conflict, which, as I said above, have not yet been explored in the sociocultural approach. I used this context to explore the idea that situated use of knowledge cannot be accounted for solely in terms of a progressive acquisition of tools and skills.

The consideration of valorisation of knowledge and consequent conflicts, leading to emergence of diversity among children belonging to the same home culture, was the impetus behind a theoretical development which implied moving beyond the culture and cognition framework towards a conceptualisation of learning mathematics as a construction of social identities (Abreu, 1993, 1995b). Before presenting some of the empirical findings that supported this move, let me try to explain what is entailed in this view of learning. Firstly, mastering of specific tools will be related to becoming a member of a specific community of practice. Secondly, groups can allow varied forms of membership. Thirdly, forms of membership can be constructed over time. This construction has been referred to by Lave and Wenger (1991) in terms of a move from legitimate peripheral to central participation and motivated by the value of participation and by the desire to become full member of a community. In short, learning in this perspective is part of a process of construction of social identities, which can be conceptualised in terms of: (1) the knowledge and experiences the person acquires are associated with varying degrees of membership in particular communities of practice; and, (2) the values, beliefs and attitudes associated with the positioning of the person in relation to that membership (cf. Tajfel, 1978).

Next, I locate some of the issues discussed above in the experiences of the children of the farming community. The study aimed at exploring how they experienced the relationship between their home and school mathematics. It followed a multi-method approach combining the use of ethnographic and clinical interview techniques. I lived in this community for 2 months and collected information in two primary schools by interviewing the children, by means of interviews and informal contacts with the classroom teachers, through classroom observation, videotapes, and notes from school documents and pupils' files. A sample of 20 school-children, aged between 8 and 16 and enrolled in third to sixth grade classes, was selected as case studies, balancing successful and unsuccessful pupils and also boys and girls. In addition, 41 other children from the same classes participated at different stages of the research, for instance as partners in group work.

A complete description of the findings is provided in Abreu (1993). Here I will focus on findings that illustrate the child's participation and valorisation of the home and school mathematics.

3.1 Children's Participation in the School and Farming Practices

Official statistics provided the evidence of some diversity among the children with respect to their participation in the school practices. The farming community studied was just one example of a great number of rural villages in the north-east of Brazil where: (a) a large number of children drop out from school in the primary grades; (b) a large number of children have to repeat the same school grade several times because they fail to learn; (c) a minority of children succeed in school; (d) children who fail in school succeed in learning mathematics in their home cultural practices (Nunes et al. 1993). Data collected in the two schools corroborated this official information. What was not known was the extent to which the children were also diverse in relation to their participation in home practices. In order to understand the children's participation in their home culture, I will focus on two questions. Could we treat children from the sugar-cane farming community as a homogeneous group? Did the children know their home mathematics?

Although all the children lived in the community, not all the families or children engaged in sugar-cane farming. Among the twenty target children, three different levels of engagement were observed: (1) neither the child nor the family worked directly in farming (four children); (2) the head of the family had a job linked to farming, but the child did not work in it (nine children); (3) the child had personal experience of farming either as a part-time job or helping a parent (seven children).

With regard to their knowledge of sugar-cane farming mathematics each child was questioned about: What is a *braça*? What are the conta and cubo? How are people paid when harvesting sugar-cane? The results revealed that the proportion of children who knew about sugar-cane farming mathematics increased with their level of engagement in the practice. For instance, all the children who had been directly engaged in farming activities knew about the *braça*, while only four (out of nine) that had no direct personal experience did so (although they were living in families engaged in farming). Only one (of four) of the group where neither the child nor the family worked in farming knew about the braça. These results appear to reflect their specific responsibilities in sugar-cane farming, that is, helping in the harvest. Fewer children could identify what a conta was. None of the children could describe the other unit commonly used by the farmers, the cubo. In fact, both the conta and the cubo are mainly used for tasks related to the fertilising and clearing of land, which the children usually did not engage in.

The level of engagement appears to be the most important factor in the children's appropriation of sugar-cane farming mathematics, since among the seven children with direct experience in farming there were younger and older children, boys and girls, third and fifth/sixth graders, successful and unsuccessful students.

The influence of the level of engagement on the knowledge of sugar-cane farming mathematics was also observed in some extra interviews with five boys, student-workers, in the evening classes. All the boys knew the farming measures, including the cubo and conta. For instance, José, a 16 year-old boy who had worked in sugar-cane farming from the age of 11, who started school at 14, and who was in the first/second grade of the primary school in the evening session, demonstrated his knowledge of the units and its relationships exemplifying with the calculation of the area of a rectangular piece of land 10 *braças* width:

Extract 9 (José)
> 15 braça (length) like this gives 150 cubo. A square like this, here is 15, 10, 15 and 10 here. (. . .) if we have 20 here and 10 here (20 length and 10 width) it gives 200 cubos, 2 conta.

This complementary information obtained from a young boy of the same age range as the target pupils, although at a less advanced level of schooling, seemed to reinforce the idea that sugar-cane farming mathematics was learned gradually and was related to the level of engagement in the activity. None of the target pupils revealed sugar-cane farming knowledge at a similar level of understanding as José. He was less advanced in schooling, but more knowledgeable of farming mathematics.

These findings show that it is misleading to treat children from the same community as an homogeneous group. They were engaged in their home practices in different ways, and this affected the mathematical knowledge they acquired. The most striking fact, however, was the repeated observation that some of the children who demonstrated more knowledge about home mathematics were among those judged by the teachers as unsuccessful in school mathematics. For instance, it was found that the majority of children who had direct experience in sugar-cane farming were part of the group of unsuccessful pupils, that is, from the seven children who were directly involved, only one was part of the group who were successful in school.

From a cognitive point of view, there was no reason why a child who knew more home mathematics should fail at school. On the contrary, the mastery of one system of representation could have enhanced the learning of a new system in a positive way (Saxe, 1990). But, in considering learning as a process of construction of social identities, it became obvious that there was another level of analysis that needed to be taken into account. That was the value and emotional significance which was attached to membership in the home farming community. What were the positions the children assumed in relation to their home and school practices? Did the children attribute the same value to their home and school mathematics?

3.2 Children's Valorisation of the School and Farming Mathematical Practices

To access the value children attribute to each mathematical practice, we started by exploring children's beliefs about situations where people use mathematics. They were asked to separate pictures which involved people doing things in farming, schools, markets and offices into two groups: in one group they were to place the pictures in which they considered that the people use mathematics, and in the other group the pictures they considered as illustrating situations in which people do not use mathematics. After choosing, the children were asked to explain why they had divided the pictures in the way they had. The results are displayed in Table 2, and it was evident that farming was the situation where children least believed that mathematics was used. In contrast, office jobs, which were rare in this community, were chosen by the majority as situations where people use mathematics.

Similar findings were obtained when they were asked to respond to the following questions on the basis of the same set of pictures: Who do you think was the best pupil in school mathematics? Who do you think was the worst pupil in school mathematics? Are there any of them who might not have been to school? The children's answers (Table 3) again revealed that the majority chose people in an office as the ones who performed best in school mathematics. The sugar-cane farming people were chosen by the majority as the ones who would have given the worst performance and also as the ones who might be unschooled.

Some examples of the justifications given by the children for their choices are condensed in Table 4, where it can be observed that children believe that those who were successful at school got a good job (office), and that work in sugar-cane was considered a low value job, and therefore associated with failure.

Table 2: Frequency and percentage of children who believed that mathematics was used in each specific situation ($n = 26$)

Description of the picture	Number of children who believed that people use mathematics
Farming	
A man cutting sugar-cane	8 (31%)
A man loading sugar-cane	8 (31%)
A man planting sugar-cane	6 (23%)
A man operating a tractor	7 (28%)
School	
A girl measuring	18 (69%)
Children playing 'hop scotch'	17 (65%)
Children in a classroom	20 (77%)
Office	
A woman in an office	25 (96%)
A man in an office	25 (96%)
A man operating a computer	24 (92%)
Market	
A woman buying fruit	22 (85%)
A woman buying cheese	20 (77%)
A boy weighing flour	19 (73%)

(Adapted from Abreu, 1995a; p. 90)

Table 3: The children's choices about who they assumed would be the best and the worst pupil in school mathematics and which people were likely to be unschooled ($n = 26$)

Situation	The best	The worst	Unschooled
Farming	0	19	20
School	2	3	0
Office	21	0	0
Market	3	3	2
Do not know	0	1	4

Adapted from Abreu, 1995a, p. 91

The most striking point to observe was that some of the children who were personally engaged in sugar-cane farming did not choose any of the farming pictures as situations where people use mathematics. For several children, farming mathematics was not mathematics at all because it was a kind of knowledge used by people who had low status jobs. They could not dissociate the type of knowledge from the social status of the group who used it.

To sum up, children's beliefs suggested that: (a) they have constructed an understanding of the status of different kinds of mathematics and how these are situated among people in particular social groups; (b) that several children experienced the relationship between the two practices as disruptive – farming mathematics was associated with belonging to low status groups and negative social identity, and school mathematics with high status groups and positive social identity.

4 Conclusions

The two phases of empirical research described in this chapter provided three analytical scenarios: (1) uses of mathematics in a traditional sugar-cane farming practice; (2) encounters with new mathematics in the process of social change or innovation of farming technology and management; (3) learning of mathematics by the new generation, the children, exposed to the coexisting traditional and innovative practices. These scenarios were distinct in terms of the salient features of the socio-cultural context and were instrumental to my reflection on situated learning.

In the first scenario, the focus on farmers' traditional practices enabled me to document the existence of mathematical tools specific to sugar-cane farming. Furthermore, following on the already established methodologies in the field of culture and mathematical cognition (see for instance, Nunes et al., 1993; Scribner, 1984), I described the ways in which these specific cultural tools mediate cognition. The analysis at this level reinforces the importance of the internalisation of socio-cultural tools and handling of cultural artefacts as part of situated learning (see also Säljö, 1996; Wyndhamn and Säljö, this volume).

Table 4: Examples of the children's justifications for their choices of the best and the worst pupil in school mathematics and the unschooled people

The choice	Examples of justifications
The best: Person working in an office	I see that he is in a good job. I think he was good at mathematics, because he needs a lot of mathematics to do that.
	He is working with a computer, he has studied a lot of mathematics, he has done very well in that subject.
The worst: Person working in sugar-cane farming	Because he is in a bad job. Because he does not know.
	Because if he had studied more he could get a better job.
	Because he did not learn properly he did not look for a better job.
	Because if he was good in mathematics he would not be working in the field. He would be in another job.
Unschooled: People working in sugar-cane farming	If they had studied they would be working in a bank; repairing cars and other things. Because they are working in that difficult job, they work a lot and earn very little.
	Because I know, he cuts sugar-cane.
	If they had been to school they would not be there working in that place. That is an example of someone who has never been to school, like my father.

The second scenario was marked by farmers' struggle to cope with changes imposed by the macro-social structures and which involved varied degrees of exposure to new mathematical tools. Although this scenario could be analysed from the perspective described above, focusing on the internalisation of the new tools and on the limits on cognition imposed by not mastering the new mediating tools, this will be a partial account of farmers' experiences only. As highlighted previously in this chapter, exposure to new forms of knowledge served to make them aware of the relative value and status of their farming mathematics. They developed ideas about the valorisation attached to specific tools and also about how mastering of specific tools situated people in social groups.

While in the two first scenarios the actors were an adult generation, full participants in a practice, in the third scenario there was a shift to young actors, schoolchildren growing up in the farming community. This made it possible to turn participation in the practices and understanding of valorisation into objects of empirical analysis. The results were quite complex. Children were diverse in their participation in home farming practices. Like the farmers they have also developed an awareness of the valorisation of the different tools and how mastering of specific tools situated the person socially. Although there was a clear link between degree of participation and knowledge of tools, there was no simple relationship between participation and valorisation. In several instances, the relationship appeared to involve some amount of conflict. This means that a simple view of situated learning as internalisation of socio-cultural tools does not account for the complexity of the phenomena observed. An alternative is to expand the concept of valorisation. One way of doing it is to view the experiencing of valorisation through the social identity lens, which involves trying to subdivide this experiencing into the understanding of the social markers of particular tools (social rules about where use of each system can be expected, the status of users in a particular society) and into the positioning each particular individual assumes towards it.

From this perspective, situated learning and uses of mathematics are mediated by social identities. This view is not incompatible with Vygotskian-based propositions of cognition as a process that is socio-culturally mediated. However, it pushes the analysis a step further; it requires an understanding of how the valorisations which social groups attach to different forms of knowledge are mediators of situated learning: (a) as part of the way specific cultural tools are represented in society at particular historical times; (b) as part of the structuring of social relationships; (c) as informing the individual positioning towards participation in specific practices. Both from a social and from an individual perspective what one learns and how one uses knowledge locates him or her in particular social groups, and, thus, defines specific social identities. This appears to be equally true of farming, and oral, knowledge as it is of access to more recent information technology tools and, consequently, this stresses the importance of a continuation of empirical research on these issues.

3

Situated Selves: Learning to be a Learner

Chris Sinha

1 Introduction

To view learning and cognition as situated is usually understood to involve rethinking cognitive and learning processes in terms of their framing by context, communication and social practice — in contradistinction to traditional views which focus upon the "isolated, individual subject" in confrontation with a cognitive or learning task. There is, however, a danger in counterposing, in simplistic fashion, the learning situation to the individual learner. The danger is that we either lose sight of the learner altogether, or we fail to challenge the very notion of what it is "to be a learner", which, explicitly or implicitly, underlies traditional views of learning. In this chapter I try to explore how we may better conceptualize the subjectivity of the learner as not only a crucial perspective in the constitution of learning situations, but also as a perspective which itself is discursively constructed.

Re-thinking "the learner" from a situated perspective requires, first and foremost, the abandonment of the universalistic presupposition that learners, in all essentials, are the same in all times and places, and that their transactions with the human and natural environment are (at some suitable level of abstraction) equally universal. This challenge to universalism is not new; it was explicit in many formulations in the 1970s of "contextual" approaches to cognitive development, which drew inspiration from the cultural–historical approach of L.S. Vygotsky. To give one example, Valerie Walkerdine and I (Walkerdine & Sinha, 1978) critically cited the following passage from Piaget:

> Whether we study children in Geneva, Paris, New York or Moscow, in the mountains of Iran or the heart of Africa, or on an island in the Pacific, we observe everywhere certain ways of conducting social exchanges between children, or between children and adults, which act through their functioning alone, regardless of the context of information handed down through education. In all environments, individuals ask questions, work together, discuss, oppose things and so on; and this constant exchange between individuals takes place throughout the whole of development according to a process of socialisation which involves the social life of children among themselves as much as their relationships with older children or adults of all ages. (Piaget, 1972: 35)

Walkerdine and I entitled our article 'The Internal Triangle: language, reasoning and the social context'. In it we criticized the assumption of the universality of learning processes, and the neglect of the socio-culturally specific dynamics of communication, exemplified in the quotation from Piaget. We wanted to break away from what we saw as the dominant tradition of Western psychological thinking about learning and development: the posing of learning in terms of the confrontation between the individual learner, and an 'environment' of people and objects abstracted from their cultural context. We criticized existing accounts for presupposing 'a Western rationalist paradigm as the model of a universal intellect' (op. cit. p. 174). Our plea was for an approach to learning and development according due weight to the importance of context, at both the macro-level of culture and society, and the micro-level of settings of learning and settings of cognitive experimentation. We were far from alone, at the time, in this critical re-evaluation of developmental psychology; or in deriving inspiration from the work of Vygotsky in trying to formulate an alternative approach. (See, to cite just a few examples, Bates, 1976; Bruner, 1975; Cole, Hood & McDermott, 1978; Lock, 1978, 1980; McGarrigle & Donaldson, 1975; Rommetveit, 1978; Wertsch, 1985).

Since then, "situated learning" has blossomed from a critique to a multi-faceted and multi-disciplinary approach. It is not my intention here to review the extensive literature on situated learning. Rather, as I have indicated, I want to pose some largely unasked questions about the discursive construction of the central character in the language game which we call learning: namely, the learner himself or herself.

In using terms like "discursive construction", and "language game", I am emphasizing the affinity of at least some versions of the 'situated' approach, with the social constructionist movement in contemporary psychology (Gergen, 1994; Shotter, 1993); which in turn owes its inspiration to diverse strands within the "linguistic turn" in philosophy and the human sciences, including the later work of Wittgenstein, post-structuralism and post-modernism. Social constructionism, with its emphasis on both the primacy of "conversational reality" in everyday life, and the rhetorical regimentation of language which defines scientific disciplines, is usually (and rightly) perceived as being fundamentally in opposition to the Cartesian and mechanistic vision which traditionally dominates "classical" cognitive science. I will leave to later in this chapter the question of whether an understanding of learners and learning situations as being fundamentally discursively constructed can fruitfully be combined with some more recent approaches in cognitive science, as is argued for example by Harré and Gillett (1994).

My main, and initial, focus, however, will be on the following question: How is the developing human being constructed and positioned in particular and specific kinds of discursive and non-discursive practices, in such a way that he or she becomes a learning subject, or self, of the kind required by the culture within which teaching-learning situations and opportunities are situated? In posing this question, I am questioning a (usually unstated) presupposition of most theories of learning. This presupposition, or assumption, is that the goal of an adequate theory of learning is to characterize the capacities, processes and mechanisms which are naturally brought to bear by developing human beings in their encounters with specific domains of cognition and understanding. That is, the learner, in most theories of learning, is taken for granted as a learner. The task of the theory of learning is seen as being one of accurately characterizing this already existing learner — with the further, practical goal of optimizing the learning situation so as to achieve a "best fit" with the learner's capacities. This predominant naturalistic view of learning is so deeply ingrained in our scientific culture that we rarely stop to examine it. What, we might reply, if asked about it, could be more natural than a child's learning?

Confronting, for example, the "miracle" of language acquisition, we turn "naturally" to naturalistic explanations: we attribute to the child the possession of an innate language acquisition device, or we model the "input" to the learner in constructing a connectionist model of the acquisition of inflectional morphology. Or, in confronting cases in which learning fails or goes wrong, we seek explanations either in a deficit in the natural (neurobiological) mechanisms which we suppose to underpin the learner's abilities and strategies, or in an inadequacy of the arrangement of the learning situation with respect to the learner's natural capacities. In either case, the learner is constructed, in our discourses of development and learning, as the already given term in a two-term relation with an input which is structured by formal or informal teaching.

Now, my purpose in questioning the naturalistic assumption is in no way to deny the relevance and importance of human biology, and particularly neurobiology, for the study of human cognition and learning. I will go further, and say that the acknowledgement of the neural embodiment of learning processes in connectionist artificial neural networks, in contrast with previous computational approaches to modelling human cognition, is a major advance in "developmentalizing" cognitive science (Elman et al., 1996; Plunkett & Sinha, 1992). Equally, current advances in developmental cognitive neuroscience (Johnson, 1997) are already revolutionizing our understanding of the biological foundations of human learning and development. The naturalistic stance is the hallmark of scientific understanding, and we do ourselves no service if we neglect or reject the fruits of its application.

However, when, in the sciences of mind, the naturalistic stance is adopted by itself, exclusively, unqualified and unsupplemented by an equally rigorous attention to the specifically socio-cultural construction of the human mind and human subjectivity, it turns easily into a limiting and impoverishing scientism. The reason for this, as has been brilliantly argued by Bruner (1990), is that the naturalistic stance offers no purchase on meaning. Human beings are meaning-creating, the practices in which their cognitive and communicative activities are embedded are meaning-saturated, and the "grounding" of the human mind is to be sought as much in our participation in acts of meaning, as in its embodiment in neurobiology. It is my contention, and here perhaps I part company as much with radical social constructionism as with scientific reductionism, that a truly adequate science of learning and cognition can only emerge from viewing the human subject bi-perspectivally, through a double prism. A fully adequate approach to human learning and understanding requires the recognition of the complementarity, not opposition, of the objectivizing stance of naturalism, and the reflexive stance of the sciences of meaning. This bi-perspectivism, or perspectival complementarity, I have called a "socio-naturalistic" approach to human development (Sinha, 1988).

My purpose in this chapter is to sketch some implications of socio-naturalistic perspectival complementarity for our conceptualization of human learning. I suggested above that the uni-perspectival and univocal stance of naturalism leads to a view of the learner as "already given", a biological ensemble of cognitive processes without a subject, which is "plugged in", as it were, to a surround of input. Learning and development are then definable in terms of changes in the observable and measurable outputs of this system, and/or changes in the complexity of 'inputs' which the system can appropriately process. I emphasize that this is not a caricature: most cognitive accounts of learning and development are based, implicitly or explicitly, on such a definition of learning, with the goal of explaining behavioural changes in terms of changes and development in processing systems.

The perspective upon the learner which views him or her as an already existing processing system can, undoubtedly, lead to interesting theoretical insights into developmental processes. It can also be illuminating for the purposes of optimizing or remediating learning, as for example may be the case for specific language impairments, literacy and reading disabilities. However, in its neglect of what it means to be a learner, and what learning settings mean for learners, this perspective is also limiting and distorting.

2 Romancing the Subject

The naturalistic stance, despite the insights it is able to deliver into learning processes, rests upon what is essentially a fiction. That fiction is that there is nothing more to the learner than the already-existing learning processes: the learner as a subject has dropped out of the picture. But does this matter? Advocates of the exclusively naturalistic stance in psychology typically dismiss talk of subjectivity, self and identity as being irrelevant to their more 'scientific' concerns. My thesis, on the contrary, is that the exclusive adoption of the naturalistic stance involves a more than merely 'innocent' scientific fiction or fabulation. Because it fails to recognize the grounding of learning in the discursive and non-discursive practices which frame and construct the learning subject, it cannot grasp the full complexity of what we loosely call match and mis-match between learner and learning situation.

To illustrate this point, I offer an observation of a learner in a learning situation of the kind which is often referred to as anecdotal, inasmuch as the observation emerges from an unplanned occurrence in a setting which has not been previously selected as a site for observation, and whose nature is such that it does not lend itself readily to artificial reproduction for the purposes of further experimental manipulation and study. To the extent that the observation occurred in a setting which came into being as a result of doing fieldwork, it could perhaps charitably be interpreted as a contribution to the "ethnography of learning", but no recording was made of what transpired. I am not therefore going to pretend that I can offer an exhaustive analysis of what occurred in what I narrate below. Nevertheless I would maintain that the observation can serve as symptomatic evidence of the importance of understanding the construction of the learner as a learning subject, in attempting to explain a cultural "mismatch" between learner and learning situation.

The setting for the observation was provided by my participation in fieldwork in a non-Western setting, in early 1996.[1] The research involved working with my student and colleague, Kristine Jensen de López, who was investigating language and cognitive development in pre-school children in a Zapotec village in Oaxaca state, southern Mexico. My visit to Kristine's field site was for the purpose of discussing the design and practicalities of her longitudinal video-observational study of Zapotec language acquisition, and to work with her on the design and piloting of some experiments. My family accompanied me on this visit, and on one occasion we went together to the village (about 60 km from the state capital) where Kristine was carrying out her study.

The basic economic activity in the village (which is typical of rural settlements in the region) is subsistence farming, but this provides only a very poor living. Family incomes are supplemented in many cases by remittances from men working as migrant labourers in the United States, and by the sale of pottery made by women. Older people are often monolingual Zapotec speakers, younger people are mostly bilingual in Spanish and Zapotec. Spanish is taught in school, but most children acquire Zapotec monolingually as their first language; this latter fact made possible Kristine's study.

On the occasion that my family and I accompanied Kristine to the village, we visited the houses of a couple of the families where a child was being studied. My 7-year old daughter Kate was invited by one of the women (whose child was a subject in Kristine's study) to sit with her and make a pot. Kate, who was enjoying looking around the house and yard, and watching the children and domestic animals, accepted with alacrity and excitement. She sat, as indicated, cross-legged next to her teacher, and attentively observed the process of pot-making, trying to emulate the techniques of (but not, it turned out, the model created by) the teacher. Neither the teacher nor the pupil were significantly hampered in their interchange by the fact that they had no language in common, and they worked together for nearly an hour while my wife and I went to watch, by invitation, the making of tortillas. When we came back to the house Kate had nearly finished making a pot. She told us she had enjoyed it very much.

The teacher at this point picked up Kate's pot and started to re-model it. Kate started to look unhappy. She told us that she liked her pot as it was and didn't want it to be changed. She took it back and started to change it again. Then the teacher picked up the pot, indicated that Kate should watch carefully, squashed it up and remade it as it should be. Kate became very distressed and started to cry. She was also embarrassed, as she realised that everyone was being kind to her, that she was lucky to have this interesting experience, and that she would be taking the pot home with her as a gift after firing the next day. She didn't want to be ungrateful, but her emotions just got the better of her. The teacher and her family members were surprised, maybe also a bit embarrassed; they could see what the problem was but they couldn't see why it was such a problem, and they laughed, more with Kate than at her, showing her the stack of pots ready for firing and indicating the one that was hers and that she would be taking away with her. As we prepared to leave, Kate was still unhappy and tearful, though she was doing her best to control her feelings.

What was going on? Here is a tentative analysis. We can assume, I think, that the setting was construed by the teacher as being, or being like, an apprenticeship situation (Lave & Wenger, 1991). The teacher, when we visited her, was already preparing to make pots, and incorporated the demonstration of her technique to the pupil into her ongoing and routine activity. For the teacher, the activity of teaching was not something which took place in a different physical and cognitive space from the activity of production: demonstrating and teaching potmaking was integrated into potmaking.

Learning by apprenticeship has several features that make it different from school learning. The most salient differences are that apprenticeship usually involves demonstration of technique as much as, or more than, instruction in knowledge on the part of the teacher; and emulation of technique more than demonstration of knowledge on the part of the learner. Apprenticeship is also frequently (though not necessarily) embedded within a process of production of material goods, that is to say within a context of non-discursive, constructive (perhaps individual, perhaps collective) praxis. Apprenticeship in pot-making (like, for example, apprenticeship in hand-loom weaving) is of this latter kind, and it involves, in the case of the Zapotec villagers, individual labour in the context of the family as the basic economic unit. Apprenticeship is always embedded within a context of some kind of productive or reproductive social practice, be it labour, or ritual, or warfare, which is not itself educational in its goals and functions.

For the Zapotec teacher, then, the demonstration of how to make pots involved the framing of her teaching in a set of interlocked and nested contexts of social practice. The demonstration is intended to induce appropriate habits of observation in the learner, which will lead to the learner emulating and eventually mastering the technique demonstrated by the teacher. The demonstration, which is communicative but largely non-verbal and gestural, is integrated into the flow of the manipulation of the clay which leads to the production of a finished pot ready to be fired. This means that the teacher's activity of teaching, while it modifies the ongoing practical activity, remains fundamentally wedded to it. In this case, the teacher, who had no language in common with the pupil, gestured from time to time to draw attention to something which she was doing, and sometimes reached over to the pupil's working space to rearrange something or to physically mould an action by the pupil. In this specifically micro-contextual sense, the teaching activity is framed by a non-discursive activity, namely the making of a pot. This is a characteristic feature of a particular modality, or style, of teaching-learning which we can call "demonstration and emulation". This style, as we have seen, is very typical of apprenticeship.

Apprenticeship, however, is more than just a particular style of learning. It is also a formal or informal institutional practice of social reproduction, whose goal is to construct, through teaching–learning, a skilled producer. Potmaking, in the Zapotec village, is a significant economic activity, and the pots produced by the teacher, and by the learner, are not just particular artefacts with use-values, but also commodities to be sold in a monetary economy. It is the economic activity which is engaged in by the family unit, as a whole, which provides the macro-contextual framing of the demonstration and emulation style of teaching which is characteristic of apprenticeship. And this framing, I suggest, frames both the learning situation and the subjectivities of the teacher and the learner.

The demonstration and emulation style of teaching is appropriate for an apprenticeship in the specific kind of production process involved in small-scale pottery making; it appropriately frames the learning process. However, the learner and teacher are also framed by a mutually comprehended goal of making pots for sale, in which the finished pot should be a commodity like other pots, which can successfully be sold. Hence, at the same time as the activity of the learner is "shaped" towards a replication of the technique of the teacher, the product of the learner's activity is also shaped and reshaped by the teacher to replicate the products that she herself makes. The learner, in a learning setting framed by the family economy, is learning not only to make pots, but also to be the kind of learner that is also and simultaneously an economic producer.

Now we can ask ourselves: how did Kate construe this setting, that is, how was it framed for her in the light of her previous learning experiences? It seemed, from her ready participation in the learning situation, that she also understood it as being framed in terms of the learning style of demonstration and emulation, and that this was a style with which she was familiar. The learner's bodily posture (sitting cross-legged, next to the teacher, so that teacher and learner could each view the other's activity in a common field of vision); her division of her attention between her own activity, the activity of the teacher, and the gaze of the teacher; and her expectation that her activity would be shaped by the teacher, were all behavioural indices (as far as I can recall without having a videotape of the interaction) of an intersubjective meshing of the teacher's and the learner's framing of the learning situation, in terms of the modality of demonstration and emulation. The learner's learning history, both informally outside school, and inside school, in activities such as cooking, gardening, drawing, sewing and so forth, had prepared her for participation. She had already learned to be a learner in the learning frame of demonstration and emulation.

It is worth making the point here that knowing how to be a learner, in this or any other frame, is not something which is already "given", let alone innate, even though there is little doubt that human infants and young children are born with a biological disposition to imitation learning (Tomasello, Kruger & Ratner, 1993); and even if (as is likely) the demonstration and emulation learning style is universal in human cultures. The learner's cognitive neurobiology, a product of human evolution, has prepared her for the specific task-management skills (distribution and sharing of attention; division of task responsibility; imitation of motor movements; execution of motor acts oriented to visible end-states; acting on cue) which are demanded by the learning frame. However, learning both to recognize the appropriateness of the learning frame, and to position herself and be positioned as learner in the complementary structure of roles (teacher and learner), is something that requires prior experience of, and induction into, such learning situations.

The micro-contextual frame of teaching–learning by demonstration and emulation is one which can be encountered not just in homes, but also in schools in Western cultures, and particularly perhaps in primary-level education in many schools, such as Kate's, in which craft and handiwork sit side-by-side in the curriculum with literacy and numeracy. However, the learning modality of demonstration and emulation is quite differently framed by the macro-context of Western educational institutions, than it is by the macro-context of apprenticeship. In the first place, and most obviously, learning (in whatever modality) in schools is not framed by economic production. The child who is learning to make a pot in a European school setting is not learning to become a producer, at least in the economic sense of the word. Rather, I would suggest, in the context of Western educational discourses, the child is learning to become a creator. That is, the child is positioned in the macro-frame that frames the micro-frame of teaching and learning, not as emulator and producer, but as emulator and creator.

My suggestion, then, is that, for the learner, the micro-context of teaching–learning by demonstration and emulation was framed by a macro-context which is quite different from that of apprenticeship: namely, the macro-context of creative self-expression. When young children, in our Western culture, engage in activities resulting in material products, in a micro-frame of teaching–learning by demonstration and emulation, whether in or out of school, they are not producing commodities for sale. In fact, they are not producing, in any significant sense, artefacts with use-values at all, even though they may be label their products as such ("look at my vase!"). Rather, learners produce artefacts whose socially and discursively constituted value is predominantly expressive. The vase or drawing produced by a child, in Western culture, is valued by the child's educators — teachers and parents — as an expression of the child's creativity, and as such of its individuality. In this discursive frame, the non-discursive process of production of an artefact is orientated towards emulating technique, not strictly to replicate a standard displayed by the teacher, but to create variation and to express individual aesthetic value, or simply individuality in itself. Thus, the discourse of individual expression positions the learner as an expressive–creative individual, in such a way that learners learn, at the same time as they learn technique, to invest the product of their activity with their own expressive individuality.

Now, we should note here that a great deal of the discourse of child-centred education not only emphasizes the fostering of creativity in the child, but simultaneously naturalizes this creativity as an "already there" attribute of the learner: children are said to be naturally creative, they are spontaneous and creative learners. My contention is that we are encountering here exactly the same kind of naturalization of the learner, in the discourse of child-centred learning, as we do in cognitive discourses which focus on the "already there" processing abilities of the learner. In positioning the learner as a creative emulator, the child-centred discourse presupposes the learning subject as being precisely that kind of learner whom the teaching practices which it recommends and prescribes "fit". And my further contention is that such a naturalizing, and naturalistic, discourse, involves exactly the same kind of fiction, or fabulation, as does the naturalistic discourse of cognitive and information-processing theories.

There is, of course, a difference. Cognitive discourses presuppose a learner who is an ensemble of "processes without a subject", whereas child-centred discourses emphasize the whole child, and the unity of the child's cognitive, social and emotional development. It is fair to say, then, that child-centred discourses position learners as having a subjectivity which is absent from cognitive discourses. Nevertheless, both these discourses are founded upon a naturalizing presupposition which posits the learner as already there. In the one case, the naturalistic stance is scientific, and in the other, it can be viewed as "Romantic", in the sense that there is an obvious historical affinity between child-centred educational discourses, and the Romantic discourses in literature and philosophy which, in reaction against "dehumanizing" scientific naturalism, foregrounded the claims of authentic selfhood and subjectivity. Both these discourses, it should be said, and has been said, express typically modern concerns about the truth of human nature: couched in terms of the valuation of, on the one hand, objectivity, and on the other hand, subjectivity.

I think we can say about the naturalness of the learner's creativity more or less the same as we can say about the naturalness of the learner's processing systems. That is, that the question is not whether attributing creativity to children is in itself right or wrong. Certainly, there is no reason to suppose that attributing creativity to children, and to their learning, necessarily contradicts the naturalistic stance of cognitive discourses (although it may contradict the "passive learner" stance of, say, behaviourism). In fact, cognitive theorists as different as Piaget and Chomsky have invoked creativity as a criterion of human learning capacities. So, I am not attempting to contradict or deny attributions of creativity to learners, any more than I am suggesting that attributions of specific cognitive processing systems and mechanisms to learners are false.

Rather, I am saying that, in assuming the "already there" creativity of the learner, the naturalizing discourse of the creative learner omits from consideration the specific way in which the learner has to be positioned and constructed, in such a way that a particular learning situation is framed as an occasion for the display or manifestation of creativity. Furthermore, it omits from consideration the specific, and socio-culturally situated, construction of a self from which creativity is (self-) understood as emanating, and whose creativity is (self-) understood as being expressed in the processes and products of creative activity. The creative learner, in other words, is no more innocent and natural than the information-processing learner.

The modern, Western idea of the self has been described by Geertz (1974/1984: 126) as involving "a bounded, unique, more or less integrated motivational and cognitive universe, a dynamic center of awareness, emotion, judgement and action organized into a distinctive whole and set contrastively against other such wholes and against its social and natural background" (see also Sampson, 1993; Taylor, 1989). Schweder and Bourne (1984: 192) say of this Western "inviolate self" that it is embedded in a world-view which "views social relationships as a derivative matter, arising out of consent and contract between autonomous individuals . . . protected by deeply enshrined moral and legal principles prescribing privacy and proscribing unwanted invasions of person, property and other extensions of the self."

Current debates about how much legal protection public persons require against the invasion of their privacy by the media underline the tension between our cultural norm of autonomous individuality, and our equally strongly held normative belief that each individual should be free to learn and profit (in ways which they should themselves be free to determine) from all and any available knowledge and information. Such notions (of privacy, autonomy, and the interiority of the self) are by no means universal; nor did they always exist in Western culture. The experiential and psychological horizon of the medieval European was organized in terms of quite different norms and values than those to which modern Europeans and North Americans adhere, and there are many extant cultures in which concepts of self and identity are quite different from those which we tend to consider as natural and self-evident. For example, in many cultures there are clear prohibitions upon the transmission of specific kinds of knowledge either to outsiders, or to non-authorized members of the culture. The violation of such prohibitions is not conceptualized by the members of the culture as a breach of the privacy of individuals, but rather as a transgression of the secrecy attached to a ritual practice which is essentially bound up with the cultural identity of the group. The transgression is against the society, and not against a particular individual.

In contrast, the framing in Western educational practices of the process of learning by observation and emulation, as involving the expression of a deeply interior creative self, is one way in which learning particular skills involves at the same time learning how to be a particular kind of learner, an autonomous individual learner in an individually orientated learning culture. This "learning self" meshes with the pre-theoretical, or practical, cultural norm of what it is, in general, to be a person, with certain rights and obligations entailed by personhood. We can now understand Kate's emotional response to the unexpected turn that her lesson took, as being a kind of protest against what she felt to be the invasion of her-self. Something as trivial as the squashing-up of a pot, which for the learner (but not for the teacher) was an extension (or expression) of her self, signified for the learner a violation of deeply-internalized norms of learning, selfhood and identity.

3 Transforming Practices and Self-creation

What I have suggested is that while, for non-Western apprentice learners, the micro-frame of teaching–learning by demonstration and emulation is canonically situated within a macro-frame of economic production, for Western school learners the same micro-frame is canonically situated within a macro-frame of creative self-expression. Hence, even though the micro-frame is common to the productive activities engaged in by apprentice learners and by creative learners, the products of these activities acquire a different significance for the learner in each of the two different macro-frames. In both cases, the products, as well as being physical artefacts, are simultaneously semiotic vehicles mediating the construction of the subjectivity of the learner in the discursive and practical macro-frames. This involves, on the one hand, a labour process within which the subject is positioned as producer of impersonal commodity value, and on the other hand, a creative process in which the subject is positioned as the originating source of personal expressive value. These two differently situated subjectivities correspond to different experiential and discursive positionings and perspectives (although, in the case of what we commonly refer to as "artisanship", aspects of both of them may be combined).

If, however, the very notion of creativity as a natural attribute of the learner is viewed as a historical and social construction; and if the learning self is viewed as an outcome or product of the practices which situate the learning process; then what, we might ask, becomes of the agency of the learner, and, ultimately, of human free will? This is a critical question that is often, and understandably, directed towards social constructionist theories, which seem to imply an implacably deterministic view, in which the individual's acts are orchestrated by already existing discourses.[2] Although I do not have a general answer to this extremely complex question, I would claim that understanding subjectivity in terms of its construction in signifying and discursive practices offers a new way of understanding the process of self-creation. In support of this contention, I offer a piece of classroom ethnographic data, which I have borrowed from Smolka, De Góes and Pino (1997: 159–161).

Bete Carrera and her cowboy hat (Smolka et al., 1997)

Situation: the house corner of a primary school classroom, with props including a cowboy hat.

Background: Beto Carrero is the eponymous cowboy hero of a popular play centre called Beto Carrero World.

Characters: Alcione, Thaís, Camila (5–6 years, girls). Alcione is in the role of Thaís' daughter.

1. Alc: You were, you were . . . Do you want to play with this hat? (puts hat on Thaís' head, who takes it off again and puts it aside)

2. Alc: Then give it to me, give it to me, Thaís! (picks up the hat)

3. Tha.   Honey, mom doesn't like hats (Alc. puts the hat on again and looks at Tha.)

4. Tha.   You look pretty! (Alc. laughs. Camila takes the hat from Alc.)

5. Tha.   Veronica (writing down the name she has given herself)

6. Tha.   What's your name? (to Alc.)

7. Alc: My name is . . . mine is Bete, Bete Carrera

8. Cam.   Mine is Bete Carrera too.

9. Tha.   Ahn . . . it can't be. Then I'm called . . . Bete.

What I want to draw attention to here is the way in which, like the pot in the previous episode, the cowboy hat in this episode functions as a semiotic mediator of both the activity of the children (their game), and the construction of their subject positions in the game. Theories of situated learning, inspired by Vygotskian cultural–historical psychology, have frequently emphasized the importance of artefactual objects —

from cups to computers — as tools for thought, and as material embodiments of cognition. Objects, in this view, are semiotic mediators of developing cognition, and of its canalization towards socially standard, or canonical, rules and representations (Sinha, 1988). Less attention, however, has been paid to the way in which artefactual objects semiotically mediate the positioning of subjects in discourse and in discursive practices.

As Smolka et al. point out, the cowboy hat, qua artefact, remains a hat, and it is never used by the children as anything other than a hat. At the same time, the cowboy hat "became"— or, rather, came to signify — more than the canonical rules of object-usage inscribed in it *qua* artefact:

> Through language, the children created Bete Carrera (Turn 7), the feminine of Beto Carrero . . . Language allows for this specific appropriation, for such a construction and transformation; it allows for a "performance" that synthesizes old and new modes and models of acting. Through language, it is possible to become another, to become homo duplex (Vygotsky, 1989: 58), or, in fact, multiplex. In this consists the dramatic character of human experience. (Smolka, de Góes & Pino, 1997: 161)

If we reflect further on the pot and the cowboy hat, we can see that the signification of both of them involves their constitution as imaginary objects. The pot, for the greater part of the interaction which was directed to producing it, and teaching/learning how to produce it, had a "virtual" or "potential" mode of existence, as a model or a goal image shaping the activity of the participants. The hat, too, was imagined by Thaís (Turns 1-4) as being an ornamentation, or fashion accessory; but then re-imagined by Alcione (Turn 7) as a particular cowboy hat, indexing a specific and imagined identity (that of the imaginary Bete Carrera). The difference between the imaginary pot and the imaginary possessor (Bete Carrera) of the cowboy hat, is that while the former is in some way independent of linguistic discourse, the second (as Smolka et al. point out) is brought into being by means of discourse. The "surplus meaning" invested in, and signified by, the cowboy hat, was constituted in and through language.

Vygotsky himself drew the analogy between signs and tools: a sign is like a tool, he said, although where the tool is directed towards the transformation of external, material reality, the sign is inner-directed, towards the transformation of psychic reality, or mental activity. The episodes of interaction which I have discussed, show us that the converse is also the case: not only can signs be seen as tools (or artefacts), but also objects can be seen as signs, or signifiers, which knit together the position and perspective of the subject with the contexts of discursive and non-discursive practice which meaningfully frame learning and other activities.

Both the being-created pot, and the successively re-imagined hat, thus functioned as signs on two simultaneous but different semiotic planes. At the first level of semiosis, they were embodiments or material representations of intentional and inter-subjectively co-ordinated activity, in familiar contexts of social practice (of learning, and of play). At this level, it is their canonical, or socially-standard, signification which is "appropriated" enactively by the participants in the activity (the child makes a pot or puts on a hat). And at this level, the construal of the object, and the positioning of the participating subjects in the activity, are intersubjectively shared and non-contested.

At the second level of semiosis, the objects were invested with a "surplus meaning", a signification which went "beyond" the canonical status of the objects to construct, confirm, or disconfirm the subjective positionings and perspectives of the individual participants within a more comprehensive, discursively constituted frame. At this second level, the signification which is "appropriated" by the participants amounts to a second-order construal of the situation from their differentially positioned perspectives. The pot meant, at the first level of semiotic mediation, the same for teacher and learner; at the second level, it meant different things for each of them. The cowboy hat, likewise, meant different things at different steps in the interaction for the girls playing with it.

The pot episode exemplifies an encounter between two different ways of construing the situation, whose confrontation is unproductive and leads nowhere in particular. The cowboy hat episode, by contrast, exemplifies a negotiated and mutual re-construal of the situation, in which the participants briefly become, through their own signifying agency, characters in a new activity and a new discourse: a little play or drama. I suggest that it is here, in the collaborative construction of novel meanings through the discursive re-imagining of the familiar, that we should seek the basis both of a general account of freedom in human agency, and of the way in which developing human beings create and re-create themselves as learners. I would also suggest (without having the space to elaborate on it), that in employing notions like imagination, embodiment, construal and perspective, the "situated" approach to learning and cognition can find a common ground with some of the most important developments in contemporary cognitive science, notably cognitive linguistics.[3]

4 Conclusion

We are by now accustomed, as I have said, to thinking of artefacts (such as computers in classrooms), as well as words, ideas, notations and representations, as tools for thought (Waddington, 1977). I have suggested, however, that artefacts may also be thought of as signs, just as signs may be thought of as tools or artefacts. Seen in this way, the meaning of an artefact is not fixed only by its socially standard use-value, privileged as this canonical meaning may be. A further analogy with language may be illuminating here. Cognitive semantics emphasizes that words in natural language are often polysemous, having many different meanings, some of which are derived from more basic or prototypic meanings by processes of metaphoric and metonymic extension (Lakoff, 1987). Think, for example, of the different meanings of the word "on", in expressions such as "on the table", "on the wall", "on the hour", "on duty", "on drugs", and so on.

In a similar way, artefacts may also acquire multiple meanings, at different levels of signification. These may also be derived by metaphoric and metonymic processes: "Bete Carrera's hat" is perhaps best seen as a metonymic sign for a complex recategorization of the hat, the play situation, and the subjectivities assumed by the play participants.

The various different meanings of artefacts correspond to different discursive patterns and practices, different subject positions within the discourses embedding the artefact, and different construals of the situation. Situatedness, from this point of view, is not given once and for all, but is continually contested, negotiated and re-established. It is this process of contestation, negotiation and discursive re-imagining which leads to novelty; and if novelty is to be valued in learning and teaching, so must the process be which brings it forth.

In a nutshell, then: artefacts, because they are semiotically complex and multi-layered signs, embody discursively situated frame-values, as well as canonically intrinsic use-values and monetary exchange-values. Even complex (and intrinsically symbolic) artefacts such as computers do not "announce" all their symbolic values independently of the "enunciation" of these values in the discourses and practices which contextualize and frame them. It would be naive to believe, for example, that learners can enactively appropriate information technology as a "tool for thought", independently of its signification in the discourses which frame the encounter between learner and the technology. A computer is not "just" a computer, any more than a pot is "just" a pot. As an artefact, a computer is a tool with a wide and flexible range of uses and applications. But as we are also quite aware, as a commodity, a computer (or mobile phone) may also be a status-symbol, an indicator of social identity. The frame-values of an artefact involve, then, not just the object itself, but also the subjective position and perspective of the user/maker/appropriator of the artefact. Subjective position and perspective, in turn, implicate the discursively-constructed self, the subject-identity of the participant.

As I tried to show in my analysis of the "pot episode", the particular "affordances" of a learning situation cannot be seen independently of how both the learning task, and the identity of the learner, are discursively constructed — both within the given situation, and in the light of the past experience of the learner. The "pot episode" also illustrates the fact that one's identity as a learner, like all aspects of one's identity, is both a cognitive and an affective construction. When we think about sites of learning, we should be aware that they not only afford "tools for thought", but also implicate what the cultural theorist Raymond Williams called "structures of feeling". Fully taking account of the proposition that learning is rooted in recurrent patterns of embodied experience (Johnson, 1987), means recognizing that structures of knowledge are inseparable (at least from the point of view of the learner) from structures of feeling: it is this inseparability which constitutes the subjectivity of the learner. "Learning how to learn", as Bateson (1973) called it, is, therefore, also a question

---

Note: In this chapter I follow convention and use the term 'Western' to refer to the cultures of, or originating in, Europe and North America. The term is a rhetorical trope derived from the European 'Orientalist' preoccupation with contrasting Greco-Judaic 'rationality' with Eastern 'mysticism' (Said,

of learning (or failing to learn) to be a learner, in a particular nexus of cultural and discursive practices.

1985). For foundational texts in cognitive linguistics, see Lakoff (1987), Langacker (1987) and Johnson (1987); and for attempts to relate cognitive linguistics to situated approaches to cognition, see Sinha (in press) and Zlatev (1997).

4

Negotiating Identities and Meanings in the Transmission of Knowledge: Analysis of Interactions in the Context of a Knowledge Exchange Network

Nathalie Muller and Anne-Nelly Perret-Clermont

1 Introduction

At the cross-roads of cognitive and social psychology, research trends on learning are now developing that are characterised by their interest in the social dimension of cognition. Inspired by the psychology of Jean Piaget, and exploring the perspectives of Russian psychologists such as Lev Vygotsky (1978) and his followers, as well as the advocates of symbolic interactionism, notably George Herbert Mead (1934), this trend suggests a definition of the learning process and of development which emphasises the role of interactions and of symbolic systems as these exist in specific cultural and social contexts.

Following the questioning of the perspective according to which cognitive processes were conceived as residing inside the individual, research inspired by socio-cultural and/or socio-cognitive approaches have paid attention to the ways in which thinking and cognitive development relate to social processes. However, initially contextual factors were de facto considered only as moderators. Today, some studies pave the way for another step by highlighting the intimate link between features of context and the acquisition of knowledge. Thus, context does not appear to be a moderating factor only, but rather a constituent of thought itself. This approach has led us to take into consideration the notions of representation and meaning. Indeed, the subject, considered as agent of his/her own development, interprets the different constituents of the context in which he/she is, and assigns them particular meanings. From this perspective, cognitive objects are constituted in the here and now of a specific context, and they should not be considered as individual faculties that can be identified independently of the situation in which they emerge. Consequently, the subject's actions must be related to the context of the activities to be performed, the purpose of the activity and the broader social context of these activities (Cole & Scribner, 1974; Donaldson, 1978; Hundeide, 1985; Lave, 1991; Light & Perret-Clermont, 1989; Resnick, Levine & Teasley, 1991; Rogoff, 1990; Säljö, 1991; Schubauer-Leoni & Grossen, 1993; Wertsch, 1991).

The interdependence and the reciprocal co-constitution of context and cognitive activities, although they are the subject of a lot of research in the social psychology of learning, are difficult to conceptualise. The very definition of context (what does it consist of? what are its boundaries?) as well as the relationship between indexicality (in the ethno-methodological sense) of the interaction and cultural meanings are notably subject to debate. Some authors emphasise the continual recreation of the social system in and through communication during the interaction: activity is its own context (Mehan, 1993). Others consider that meaning, while being defined and negotiated in the interactions, is historically constituted between persons engaged in socio-culturally constructed activity. Lave summarises this perspective by arguing that "any particular action is socially constituted, given meaning by its location in systems of activity" (Lave, 1993, p.18; see also Säljö & Wyndhamn, 1993).

Pursuing this line of research (Grossen, 1988; Grossen, Liengme Bessire & Perret-Clermont, 1997; Iannaccone & Perret-Clermont, 1993; Perret-Clermont & Nicolet, 1988; Schubauer-Leoni et al., 1989), the analyses presented in this chapter will shed light on some aspects of these issues. They follow an interactionist perspective, which presents the agents as constructing the meaning of the situation while they interact, taking into account the role played in this construction by the societal and historically generated system to which the agents refer (Dreier, 1993). Our observations of situations of transmission of knowledge are made in what is called a knowledge exchange network (cf. below), and the activities within such networks will give us the opportunity to explore the interdependencies between cognitive and social transactions in the construction of meaning.

2 Issues

Knowledge exchange networks, which now exist in several European countries, give people, irrespective of age, sex, profession, socio-economic origin, etc., the opportunity to meet in order to — as it were — give and receive various sorts of knowledge. In French, a knowledge exchange network is referred to as a *Réseau d'échanges de savoirs* and in order to keep the connotations of this expression, signalling a network and a meeting place in which knowledge is communicated, we will use the expression knowledge exchange network (KEN). The originality of this system, aside from its ambition of democratising and putting to advantage any kind of knowledge, lies in the particular idea of reciprocity on which it is based. Indeed, to be part of a network, every participant is asked to be prepared both to give and to receive knowledge, and thus to be both teacher and learner.

Beyond the ideological debate, this type of social innovation seems to be a favourable field for psychologists of learning. As it exists in parallel with the official institutions of formal schooling, it permits us, by means of somewhat of a detour, to take a fresh look at the school system as an institutional arrangement for the transmission of knowledge.

This research project has been carried out in the Strasbourg network. However, in this chapter, we will not compare the two models for knowledge transmission (school and KENs) as such. Starting from data collected in the field in Strasbourg, we will rather analyse how the KEN and its ideology are interpreted and put into action, as it were, by its members. We will see that the context is both structuring and constructed, and that the result (as in the case of school) is not always in line with the initial ideological inputs.

The Strasbourg KEN will be described at two different levels: first at the general level of the organisation of the exchanges of knowledge and skill, then at the level of the interactions.

(a) We will first examine to what extent the context of the network contributes to structuring and producing one particular type of exchange for the transmission of knowledge. The exchanges are set in a specific context, and they are linked to a specific ideology. Which part does this symbolic context (involving reciprocity and the sharing of knowledge between experts and novices) play in the organisation of the exchanges? In this perspective we will observe what kinds of knowledge become objects of exchange and how the basic principle of reciprocity is sometimes understood in a very particular way.

(b) We will then focus on the analysis of some salient forms of social interaction taking place in the network. We will explore the way in which the micro-context of the interaction is constructed by the participants in accordance with their representations of the broader social context(s). In this respect, the analysis will concentrate on three questions:

   (i) First of all, we will investigate the representations which different agents have of a situation of exchange of knowledge in the network. Does the situation have the same meaning for all participants? Which representations of the network are involved in the organisation of the didactic relationship?

   (ii) Some studies in psychology have shown that in problem solving interactions or knowledge transmission interactions, the agents (teachers, pupils, experimenters, subjects, etc.) are constantly positioning themselves in relation to the others, according to a set of explicit or implicit rules, legitimised by the institutions (such as school) to which they belong (Donaldson, 1978; Flahaut, 1978; Rommetveit, 1992; Schubauer-Leoni, 1988; Valsiner, 1994). This leads us to take a closer look at the way in which the members of a network give each other positions (*Places* [in French] is the term used by the social psychologists Flahault (1978) and Kerbrat-Orrechioni (1988) in their analyses of how participants in interaction create positions for each other) as the interaction proceeds, and how the roles of what is called provider and recipient as defined in the network are interpreted and negotiated during the interaction. In other words, how do the individuals refer to the network institution and how do they position themselves in relation to the other participants?

   (iii) Finally, we will study the status of what we will refer to as the object of knowledge, and the way in which it is itself subject to negotiation. The manner in which the object of knowledge is treated in an interaction depends not only on the agents' representations as regards the network context, but also on the identity issues at stake in the very situation.

Dealing with these questions in our analysis, we will see that the context of the network, far from being a mere container, is itself subject to interpretations and is constructed through the interactions between the agents.

3 Methods of Data Collection

In order to understand what was going on in these situations of transmission of knowledge we intended to study, it was important to grasp the general functioning, the starting point, and the aims of the organisation of the Strasbourg KEN itself as well. A participant observation approach made it possible for the researcher to take part in the events that characterise this institution and to be initiated into the habits that make up the everyday experiences of the members (Garfinkel, 1967).

The fieldwork took 4 months (in the spring and summer of 1994). This time was devoted to observation and more or less active participation in the following activities: in weekly meetings of the organising members, in six exchanges as receiver, and in the setting up of knowledge exchanges as an organiser. In general, only the presence of the tape recorder indicated the special status of the researcher, who had introduced herself as such at the beginning of the meeting (Muller, 1994).

4 The Strasbourg KEN as a Structuring as well as a Constructed Context

4.1 General Background
4.1.1 The Networks

The KENs were set up for the first time in France in the 1980s (Claire and Marc Heber-Suffrin are considered the founders of this movement; see Heber-Suffrin & Heber-Suffrin, 1992), and they are now also developing rapidly outside France. Their origin is to be found in a number of ideas and circumstances both in the socio-economic domain (acknowledgement of the rise of unemployment and problems of social exclusion as well as of the essential role played by qualifications in social integration) and in the pedagogical domain (acknowledgement of a certain inability of the official pedagogical structures to cope with the new social demands). The original organisation of this type of transmission of knowledge is also of interest in terms of what has been referred to as "the vulgarisation of science". Some scholars see these networks as possible opportunities to pass on scientific knowledge more or less directly from the expert to the citizen rather than passing via intermediaries that might distort or vulgarise the results or the scientific modes of reasoning and, in fact, increase the social distance between scholars and lay audiences (Jurdant, 1992). The concrete manner in which science and its results are made available to broader audiences is ambiguous. Ideologically it is often claimed that popularisation of scientific results provides lay audiences with an opportunity to share knowledge normally only available to a limited group of specialised professionals. But empirical studies show that this is often not the outcome of such attempts. For instance, only those who are already well educated appear to be interested in — and seem to understand — this kind of knowledge. In spite of the attempts to build bridges between professionals and other groups, such information that implies vulgarisation of science increases the social distance.

4.1.2 The Strasbourg Network

Generalities. The Strasbourg KEN was founded in 1992. In 1994, when our field work was carried out, it had about 130 members from the Strasbourg area. The members of this KEN are drawn from different socio-professional backgrounds (it seems, however, that most of them can be considered as middle to upper-middle class). A team of seven or eight people runs the network (manning the office once a week, setting up the exchanges and leading their first session, etc.). The list of the kinds of knowledge offered and requested includes some three hundred topics. Some examples that illustrate the diversity of the knowledge offered would be the following: learning how to stage a theatre play outdoors; knowledge of Strasbourg and its surroundings; Esperanto; job-seeking techniques; how to set up your own business; tango; how to repair a bike; knowledge of mushrooms; etc.

Basic principles. The principles of operation used in the running of the Strasbourg KEN are similar to those applied in other KENs: the exchanges are based on reciprocity (each teacher is invited also to be a learner, and, in turn, each learner is invited to offer his/her knowledge in some area of expertise); the exchanges are free of charge; the different sorts of knowledge offered have equal value (computer science is 'worth' the same as Mauritian cooking); the relationship between the members is one of equality (there is no hierarchy in the network); the value of the person is reasserted through his/her knowledge.

While most networks develop in disadvantaged areas of big cities with the aim of recreating social cohesion and solidarity ties, the organisers of the Strasbourg network claim a special characteristic, namely that a priority of this network is to make it possible for the participants to teach and learn scientific forms of knowledge. However, the idea of integration, held dear by Claire and Marc Heber-Suffrin, is still present in the Strasbourg network. Indeed, a pamphlet presenting the network reads: "The network is an exceptional instrument of social development providing warmth, friendliness and solidarity, and promoting new contacts."

In the Strasbourg KEN, this emphasis on the transmission of knowledge (that is to say, on technical and scientific education, as opposed, in the eyes of the informants, to an emphasis on the creation of a more general warm and friendly atmosphere) is related to the risks of so-called scientific vulgarisation that have been raised by the present president of the network. The development of the Strasbourg KEN has therefore been motivated by ideas both of a social nature (to recreate social relationships between people) and of an educational nature (to make scientific knowledge generally accessible), as well as by a desire to act and achieve concrete results in the area of knowledge reproduction.

As we have just seen, the Strasbourg network is driven by a particular philosophy or ideology. However, the question must be raised of the awareness of these principles among the participants: how are they informed about them and how are they initiated into them? The network, through its organising team, publishes documents which are handed out to the people interested, the media, the local authorities. There is also the network journal, which lists the sorts of knowledge available at the moment, provides information on the exchanges taking place and on the activities offered by the network. From these documents, as well as during the office hours, the participants can come into contact with the organisers' representations of the aims of the network and how it should function. As far as the exchanges are concerned, it must be noted, however, that great freedom is given to the members, who, after a first session in the presence of a co-ordinator, are left alone to agree on the content of the exchange, its duration and location. No teacher training course is planned for those acting as providers, nor, it must be added, for the co-ordinators. The network can therefore be seen as a framework, where outside definitions as to the purpose of the activities are not strongly imposed. How do the members interpret this context, and which meanings do they assign to it? Are the ideological values of the organising team put into practice by the participants? Two examples below will present the problematic relation between the organisers' representations and the members' concrete activities.

Scientific exchanges. One of the special characteristics of the Strasbourg KEN, according to the staff, lies in its aim to provide an opportunity for scientists and members of the general public to meet. In this respect, some people involved in the field of scientific research have been contacted and have registered as members. In what way does this scientific spirit, as it were, appear in the network?

Among the 300 types of knowledge offered, only eight seem explicitly to have to do with a scientific domain: telephony; low-voltage electricity; biology; energy and environment; electron microscopy; archaeology; the study of stars; entomology; principles of radio emission and reception. Moreover, it seems that these types of knowledge have been chosen in a very limited number of cases only. In general, the exchanges which have actually taken place were devoted to English, German, computing, simple and friendly cooking, Mauritian cooking, rose and tree pruning, transactional analysis, the conduction of press conference, jogging, car mechanics, origami, etc.

This observation is puzzling. Whereas exchanges between scientists and members of the general public usually occur through the medium of texts and images (via journals and newspapers, TV, films, and so on), the network ensures a genuine interaction in the sense of exchanges between individuals. Moreover, following the principle of reciprocity, the scientist does not appear only as the one who knows, they can also be in the position of the receiver. In return, the receiver of scientific knowledge, hence, is continuously confirmed that they know things that their partner does not know and that they will also be able to teach these things. Why, then, do scientific exchanges not take place?

In this regard, the notion of social representation associated with different types of knowledge (see also the notion of valorisation developed by de Abreu (1995), and in this volume) can help us to understand this phenomenon. Indeed, different phenomena that make up a society are subject to social representations (Moscovici, 1961), which are linked (although in a complex manner) to attitudes and behaviours. A rich number of social representations (partially associated with social class membership) are associated with scientific knowledge (Bourdieu, 1976; Jacobi, 1987; Jurdant, 1973, 1993). For example, for some people, scientific knowledge is associated with images of complexity, mystery, taboo, eccentricity, scientists locking themselves away in an ivory tower, seriousness, etc. Social representations attached to scientific knowledge (which should be studied in detail) make this knowledge symbolically inaccessible to some people. These representations cannot be changed just by the fact that this knowledge is actually accessible. We can, therefore, put forward the hypothesis that the participants' pre-existing social representations present some resistance to the plans of the network organisers.

This interpretation also leads us to raise the question of the functions that this social system can fulfil: what are the expectations of those who register in a KEN? Do they want to be confronted with expert, highly specialised, knowledge? On the contrary, it seems that they primarily wish to acquire skills (English, computing, cooking). One may also seriously consider the central role of friendliness and atmosphere in this type of organisation: members seem to look for an opportunity to meet other people and discuss in a relaxed atmosphere, rather than facing too much unknown, which might be too threatening for their own identity.

Reciprocity. Another aspect is worth examining, namely that it seems that some members specialise in one role, that of provider or that of receiver. They thus reject the idea of reciprocity, although it is a fundamental principle of the network. For some specialised providers we have met, the explicit or implicit purpose was either to prepare to become a teacher by experiencing a similar situation, or to be back in the teacher's role, which had been their profession. These specialisations outline the identity dimension of the didactic interaction: what does it mean, in terms of social identity, to transmit knowledge and to be in the position of the one who knows? What does it mean, on the other hand, to receive knowledge and to be in the position of the one who does not know? Above all, what does it mean to be in these two positions successively? This kind of observation can also question the purposes of the KEN: what is it the KEN is offering (in terms of identity) when it expects its members to occupy either of these positions? By means of a detailed analysis of some interactions, we will attempt to understand the identity issues involved in such didactic relationships.

The Strasbourg network is driven by a particular ideology essentially consisting of the notions of reciprocity (each member is provider and receiver, in a system excluding hierarchical relationships) and of scientific popularisation. Yet, as we have argued, the agents reinterpret the framework offered by the network and recreate it in accordance with their own objectives.

4.2 Analysis of Interactions

To what extent are these processes of interpretation perceptible in the interactions themselves and not only in the accounts given by the participants when reflecting on what they are doing? How, and according to what kinds of representations, do the individuals structure the situation? Which role does the context offered by the network have in these interpretations?

The verbal protocols of a few interactions which have taken place in the Strasbourg network will serve as a basis for analysis. We will use a narrative style to report the different processes observed. The general question which will lead the analysis is the following: how do the individuals interpret the situation in which the network has made it possible for them to meet? We will examine three topics in particular:

- how do the agents see the situation and its objectives?
- how do the agents see their role, and how do they negotiate their position?
- what place is given to the object of knowledge?

4.3 The Representation of the Situation and its Objectives

In the network, which representations do the different agents have of the situation of knowledge exchange? What objectives do they pursue? Let us look at an example, albeit a rather extreme one, from the field notes.

4.3.1 The German Exchange

Having made my choice from the file listing the knowledge offered and looked for, I [N.M] played the role of co-ordinator for the German exchange. I got in touch with the receiver who said she was still interested. As to the provider, he introduces himself as a retired German teacher, enthusiast and available to give courses. A first meeting is then arranged.

Mr S. seats us in the living-room, and says that "for the first contact we will be here, but then for working, we will be in the next room". Things are clear from the start: we are not here to have tea and biscuits! The conversation starts. Mr S. introduces himself (in German): he used to be a professor in the *Hautes Ecoles*, and he is very enthusiastic about his former profession. Although he has retired, he keeps in touch with the news and still gives courses. Marie, the receiver, has attended a tourism school, and needs to be able to speak German for her profession. Though she started studying it at an early age, she needs practice. Mr S. confirms: "Übung macht den Meister!".

Mr S. talks about his past profession. He insists on the fact that the students in the *Hautes Ecoles* were chosen among the best: *die Krema der Krema*. His pedagogical method is based on repetition. Then, as a first exercise, he suggests his new student reads texts aloud several times: "it permits one to remember German sentences and enables the language to take its proper place in the palate". After checking Marie's knowledge of the abbreviations for some main German organisations (SPD, CDU, etc), he asks her what she is interested in. She speaks about places of interest in Strasbourg.

Time goes by and at some point he asks Marie: "OK, do you want to come back?". To her affirmative answer he replies: "Next time it will get serious!". For the next session, he suggests she might do an oral exercise he used to give his students as homework. In the car, Marie looks a bit apprehensive. She says: "It is quite serious! That's good, it will force me to work". But she answers very evasively when I [N.M.] ask her if it was what she expected.

This particular session seems to be a caricature of the goals and ambitions of a KEN. No effort is made by the provider to adapt in the slightest way to his student's expectations. Nor does he attempt to get out of the school frame of reference he used to work in. The objectives of the partners involved seem incompatible: for one of them, it is to be back in the teacher's role, to remember a prestigious position (teaching *die Krema der Krema*, in the *Hautes Ecoles*), insisting on the role of grammar, repetition, knowledge of the political and economic history of Germany, and so on. For the other partner, on the contrary, the purpose seems to be to acquire the communicational competence (language practice) that is necessary for her job in tourism in Strasbourg. The objectives of the exchange have been proposed by the provider and the receiver has not attempted to modify them. In this exchange, it is quite certain that the provider has not at any point changed his point of view by imagining himself to be in the position of the receiver. He therefore seems to see the network solely as an opportunity to play a role he likes. Obviously he is also at risk (this might not have been the case in the *Hautes Ecoles*) of losing his student — and as a consequence — the role he enjoys. The seeker, on her side, is engaged in a process of renegotiating her expectations. Is she starting to believe that the reality of learning is necessarily serious and formal as at school and that the KEN ideology is a dream? Unfortunately she did not answer my question. But her silence was certainly telling.

Let us turn to another example. Here the topic is writing.

4.3.2 The Writing Exchange

It was in a totally different atmosphere that the writing exchange took place. This exchange seems particularly interesting in the sense that the negotiation about the framework was put into words. The participants' expectations were listened to; the provider's suggestions were worked out from the expectations expressed.

During the first meeting, the provider (Ophélia) introduces herself: she has to write a lot in her job, but this kind of writing is of no interest to her. It is creative writing she is enthusiastic about. The receiver (Dalia) does write, but mainly in English. She would like to "work on her French writing", and to learn "how to structure a story, how you do it, the way to go about it, because I'd like to get going; yes, how to structure, how to elaborate". Ophélia asks her whether she thinks of this work on an individual or collective basis. She quotes examples of surrealist games to illustrate what she means by collective writing. As to her, she has no particular wish, she enjoys writing alone, "but to write means to be in relation with other people anyway".

Ophélia suggests that Dalia writes texts alone. But the fact that somebody, "somebody who knows", is waiting to read them would be stimulating. The proposal is accepted, but "in order to get going", says Ophélia, they should meet one day for 2 hours, so that "they spend some time together", and do some imagination games. "Even if it goes in all directions, it's got to spring up; you refocus only at a later stage".

As far as the participants' objectives are concerned, it can be noted that Dalia's expectation is based on a wish to "learn how to write", to do "exercises", to have her pieces of writing "corrected" by Ophélia. This representation of writing is in contradiction with that of Ophélia, who speaks more in terms of games, imagination, enjoyment in writing. But finally, provider and receiver reach an agreement. Not an agreement for the sake of politeness, but an agreement in which the contradictions seem to be transcended in a superior conception (Trognon & Rétornaz, 1989). In this exchange, the network's principles seem to be referred to more than in the German exchange, and they are used as mediations for the activities undertaken.

The frameworks actually created by the participants in the two exchanges are not the same. This difference can be analysed in terms of closeness to the ideological core of the network, and in terms of initiation to its principles. Indeed, the participants in the writing exchange are regularly in touch with the network's organising team, whereas Mr S. has had only one opportunity to meet the network. That meeting took place when he registered. While in the German exchange it seems that the model of reference was the school setting, in the case of the writing exchange the participants tried to create a new model of interaction, based on the statements and negotiations about the expectations. This model is close to the ideals of the network organisers.

4.4 Representations of Roles and Negotiation of Positions

In the two examples above, we have seen that through the social representations operating in the situation, it is the interpretation of particular roles and identities which is also negotiated. The analyses which follow show some of the ways in which these identities are worked out during the interaction, and the importance of what is at stake.

Research in the social psychology of learning has shown that an interactive situation set up to observe cognitive activity involves social agents whose aim is not only to fulfil the task, but also to respond to the complex social situation, that is to say to understand the experimenter's expectations and the nature of the problem, to know how to interpret their role as the interaction goes along, to handle their identity images and those of their partners (Elbers, 1994; Grossen et al., 1997; Hundeide, 1992; Perret-Clermont & Schubauer-Leoni, 1981; Perret-Clermont et al., 1983; Säljö & Bergqvist, 1997; Schubauer-Leoni et al., 1989). How do they then interpret these roles in an exchange within a KEN? Let us now observe the interaction in the course of which the positions of the agents are defined and worked out around the roles provided by the network (Flahaut, 1978; Kerbrat-Orecchioni, 1988).

By means of extracts from audio-recordings of another session, the cooking exchange, we will examine how the roles of provider and receiver are jointly constructed and worked out in the course of the exchange.

Extract 1
P. OK, so we're going to start, so we said shortcrust pastry.
S1. Yeah. Have you got your own way of making it?
P. Well, shortcrust pastry, usually I make it by guesswork; otherwise, in general, it's 50 grammes of flour, 75 grammes of butter, I took out a recipe, there.
S1. You'd told me that you made it with oil . . .

It is with these words that the so-called simple and friendly cooking exchange starts. Three women are present:

P: the provider, who is about 60, lives on her own now, but whose past life has been that of a married woman with several children;
S1: a woman, somewhat younger, but also married and a mother;
S2: the researcher, who looks younger, is single and has no children.

S1 and S2 both have already done a cooking exchange with P through the network, but separately. S1 and S2 did not know each other before this first day of the new exchange. The recipes had been chosen in a meeting between P and S2. S1, who was interested in this exchange, had been informed and joined it. The exchange is held at P's flat, around 10 am.

During the whole activity, a consistent differentiation can be observed between the provider on the one hand, and the receivers on the other hand as to the communicative roles assumed by each of the agents (in terms of who asks questions, gives explanations, assesses, organises the activities), as well as to the rules of turn-taking. Indeed, as a general rule, the provider suggests explanations and structures the time devoted to the activity. When the receivers speak, on the other hand, it is often to ask for an explanation, an assessment, or a piece of information. So, the following type of exchanges can be frequently observed.

Extract 2
S1. What will the biscuits be baked in? Will this dish do?
P. No! Biscuits are baked on a baking tray.

Extract 3
S2. What's a gougère?
P. Well, basically, it's a choux pastry with cheese . . .

From the way turn-taking is handled, it is possible to observe the setting up of the roles between the agents. In this exchange, the receivers indeed seem to play the role of learners. Thus, they address the person who they think "knows" — using linguistic signs which show that "they don't know", and so they put the provider into her role of "teacher". In this manner they produce a school-like context. It therefore seems that:

(a) the official roles (marked by the terminology used in the network) of provider and receiver are respected;
(b) these roles are not only respected by the participants, but also jointly constructed by them; each one, from her position, contributes to defining the other's position.

4.5 The Place Assigned to the Object of Knowledge and its Definition

Our analysis reveals that providers and receivers spend a certain amount of energy in order to assign, define and negotiate everyone's position in the activity. It thus seems justified to raise the question of the place assigned to the object of knowledge itself in this interactive play. This question is of interest to us as psychologists of learning insofar as the following hypothesis can be put forward: when an agreement is reached between the participants to consider that the aim of the interaction is to transmit knowledge, the individuals can test the relation to reality of the concepts they use. It is thus important to allow some conversational space for the constitution of the object of knowledge, and not just negotiate one's own positions (Perret-Clermont & Nicolet, 1988).

In the first extract from the cooking exchange quoted above, P and S negotiate the definition of the knowledge which is to be the content of the exchange. Already in her first utterance, S1 interrupts P to ask her about her "own way of making shortcrust" pastry. As a response to P's refusal to reveal her personal way of cooking "Well, shortcrust pastry, usually I make it by guesswork (. . .) I took out a recipe, there" S1 insists: "You'd told me that you made it with oil". This exchange goes on for a few turns in this manner until S1 finally accepts the definition of the task presented by P: we witness a confrontation between and an adjustment of the representations of what is the very object of the exchange. While S1 expresses an expectation, that of learning the expert's know-how, P, on her part, expresses a different definition: what the exchange is about is not a personal kind of knowledge (which, besides, is difficult to transmit, insofar as it is not quantifiable — "I make it by guesswork") but rather an official and rigorous kind of knowledge, that of the recipe.

Through this confrontation between two points of view in the course of an exchange in the KEN, is it not also a representation of the network itself that is revealed? It indeed looks as if the receiver imagined the network as a place where intimate and personal knowledge could be transmitted, that is to say precisely the kind of knowledge which cannot be found in recipe books. For the provider, on the contrary, an exchange in the network might be conceived as consisting in the transmission of rigorous, specialised knowledge.

This exchange draws our attention to two points:

(a) To find out about the place assigned to knowledge transmitted in this type of interaction, it is necessary, first, to investigate the way in which this knowledge is defined by the agents themselves.
(b) The interactive processes used in order to define this knowledge are also linked to identity issues. Indeed, it looks as if in this sequence the function of S1's interruption was not only to ask for a piece of information, but also to test, as it were, the participants' positions. Through her question, she tests P's capacity to recognise her (S1's) status as novice, but as a novice who requires from P the transmission of a specific kind of knowledge. Analogously, the provider wants to be recognised as a real cook speaking on the basis of information from recipe books. She does not want to appear as a friend presenting private knowledge.

Therefore, the definition of knowledge itself is linked to identity issues: the way in which it is defined and shared allows the participants to assign non-interchangeable positions to the different agents.

5 Discussion

From observations made in a KEN, our purpose has been to examine the relations between didactic interactions and their context. Firstly, we have shown that, at a general level, the objectives of the network, as presented to the public and endorsed by the organisers, do not necessarily correspond to those realised by the members themselves in the activities. The members interpret the framework of the network, and they re-invent it in accordance with their own objectives. Thus, the network does not appear to function as a pre-structured and given order, but rather as an activity constituted through procedures chosen by the agents themselves.

The analysis of the interactions has highlighted the role of meanings and representations as organising and structuring both the procedure for the interactions and the knowledge involved. More precisely, taking some micro-stories as a basis, three observations have been made:

(a) The roles, the identities and the status of the knowledge are defined in the course of the interaction starting from representations of the situation. These de facto draw on not only the representations of the network and its objectives in general, but also on other relevant and familiar contexts such as the school.
(b) The roles and positions of the agents in this type of exchange are co-constructed, and defined in seemingly stable and complementary positions. The positions implied by the terms provider and receiver have been effectively institutionalised by the network and they are used by the members. These terms also relate to social identities assigned to people in other social settings. Moreover, this social play of assignment of roles and responsibilities seems to represent an important part of the exchange.
(c) The status of the knowledge transmitted is inextricably interlinked with this interactive assignment of roles and negotiation of identities. It therefore appears that there is no such thing, in and by itself, as a piece of knowledge like cooking or writing to be transmitted. Rather, the definition of the knowledge depends on representations and negotiations between the agents according to their present needs and in reference to the larger official context of recipe books, school, professional activities, and so on.

These observations question the understanding of the relation between what is referred to as context and socio-cognitive processes in the construction of meaning. Let us summarise some results of our analyses.

- Context is perceived by the actors via their representations and is (re-)constructed in the interaction. It is interpreted, and these interpretations are dynamic, evolving and negotiated;

- These representations and interpretations — of the context and of the situation — are not totally created (as ex nihilo) in the situation, but are taken from a stock of common knowledge, partially shared in society, as resources to give meaning to the situation in which the members operate;
- The way the actors interpret the situation can have an effect on the construction of the object of knowledge;
- The identity dimensions are highly present in a didactic interaction. In transmission of knowledge, identity issues seem to be crucial.

In learning situations, both identities of agents and the meaning they assign to the situation are defined not only during the interaction but also with reference to the representations these elements have of the wider social context in which the network operates. Thus, learning appears to be entirely embedded not only in a particular social context, namely the interaction, but also in a broader context which contributes to its meaning and form.

It seems that the context of the network does not always provide the necessary resources for the agents to lend meaning to the situation. Some studies have shown that young children, given problems in mathematics classes, implicitly use pre-definition of what it means to take part in mathematics classes, permitting problem solution in a specific way. These pre-definitions include the idea that the problems have a given solution, the elements of the problem give indications as to how to solve it, it can be solved using mathematical rules learned at school (Säljö & Wyndhamn, 1993; Schubauer-Leoni, 1988; Schubauer-Leoni & Ntamakiliro, 1994). In our case, the network has a relatively open framework. The basic principles do not tell members much about how to interpret their role, treat knowledge, interact with partners, etc.; thus they seem to hesitate as to which meanings they should assign to the situation.

We wanted to deepen our understanding of the interdependence between cognition and context, the latter element not conceived as a set of external variables affecting the mind but as a reality discovered and constructed by actors during their interactions and thinking. Learning appears as a set of phenomena mediated by representations and interpretations.

This study has paid close attention to didactical interactions with a KEN, revealing the importance that teachers and learners give to negotiations of roles and positions. This occurs in spite of the official network ideology supposedly defining these matters. In fact, the network ideology and practices are constantly reinterpreted, negotiated and reconstructed by the actors (as are those of any institution) with teaching situations.

The observations of these transactions reveal that what is at stake is not only — or sometimes not even primarily — the transmission of an object of knowledge (defined and transformed during the interactions), but also the management of social identities given by the wider society (e.g. "former teacher"), of roles in the present situation (e.g. "provider/receiver in KEN") and of positions (e.g. high or low status).

Note: The observations made in the course of this study were carried out collaboratively between Professor Baudoin Jurdant's laboratory (Université Pasteur, Strasbourg), Jacques Perriault (Centre National d'Education à Distance) and the Department of Psychology (Université de Neuchâtel, Switzerland).

5
Real-world Knowledge and Mathematical Problem-solving in Upper Primary School Children

Erik De Corte, Lieven Verschaffel, Sabien Lasure, Inge Borghart and Hajime Yoshida

1 Theoretical and Empirical Background

Analyses of expert performance in a large variety of disciplines has shown that becoming skilled in a domain — for instance, mathematics — requires the integrated acquisition of the following four categories of aptitudes: a well-organised and flexibly accessible domain-specific knowledge base, heuristic methods for problem analysis, metacognitive knowledge and skills, and affective components such as positive beliefs, attitudes, motives, and emotions relating to subject-matter fields. In addition, to overcome the well-known phenomenon of inert knowledge one should acquire a disposition toward skilled learning, thinking, and problem solving, which refers especially to the sensitivity to situations in which it is relevant and useful to apply one's knowledge and skills, on the one hand, and to the inclination to do so whenever appropriate, on the other (for a brief review of the literature with respect to mathematics, see De Corte, 1995b; De Corte, Greer & Verschaffel 1996).

Nowadays there is ample evidence that students in today's schools often do not master sufficiently the categories of knowledge and cognitive skills mentioned above to approach new mathematical learning tasks and problem situations in an efficient and successful way (see e.g., De Corte, 1995a). For instance, students often acquire only deficient, superficial, and rote knowledge of basic concepts, and in many domains they even harbour misconceptions (e.g., "multiplication makes bigger") and defective skills (e.g., buggy algorithms); they also do not master a variety of usable heuristic methods, and in their learning and thinking processes there are very few traces of metacognitive and self-regulating activities. Interestingly, those shortcomings and deficiencies can be largely attributed to current teaching practices. But what about the more affective aspects that are involved in learning, understanding, and thinking? As compared to the cognitive components, it is only recently that researchers have begun to study more thoroughly the affective aspects that are involved in expert performance in a domain such as mathematics (McLeod & Adams, 1989). This work has meanwhile already documented that in the affective domain, too, instruction can lead to negative outcomes. For instance, Schoenfeld (1988) has shown that in a number of secondary school mathematics classes, where teaching was generally considered to be of good quality, students nevertheless acquired strange and incorrect beliefs about mathematics problem solving, such as:

- being able to solve a maths problem is mainly a matter of being lucky,
- solving an assigned maths problem should not take more than 5 minutes.

As defined by Schoenfeld (1985), such beliefs about mathematics correspond to "one's mathematical world view, the perspective with which one approaches mathematics and mathematical tasks. One's beliefs about mathematics can determine how one chooses to approach a problem, which techniques will be used or avoided, how long and how hard one will work on it, and so on." (p. 45)

Over the past years, researchers in several European countries (Belgium, Sweden, Switzerland, and Northern Ireland) have studied rather intensively one aspect of the belief system of mainly primary school children, namely their (implicit) belief that real-world knowledge is irrelevant with respect to mathematical modelling of school word problems. Indeed, these investigations have so far yielded substantial empirical evidence that students at different levels of the school system — seemingly as a consequence of current teaching practices in the mathematics classroom — develop the belief that real-world knowledge and realistic considerations should be neglected when solving arithmetic word problems. For instance, when asked how much time it will take someone to run 1 km given that his best time to run 100 m is 17 seconds, the majority of the children at the end of the primary school answer without any hesitation: 170 seconds. The most spectacular illustration of children's failure to make proper use of real-world knowledge and sense-making activities during problem solving has been reported already in the 1980s by French, German, and Swiss researchers. They administered absurd, unsolvable tasks, like "There are 26 sheep and 10 goats on a ship. How old is the captain?", and found that many primary school children came up with an answer resulting from making computations with the given numbers (for a review see Selter, 1994).

In some way these observations confront us with a kind of paradox. Indeed, a major argument for the inclusion of verbal problems in the mathematics curriculum has always been to develop in pupils the skill of using and applying their mathematics knowledge to model and solve tasks and problems encountered in real life (see e.g., Burkhardt, 1994). However, it looks as if current practice of teaching word problems in schools does not at all foster a disposition to realistic modelling and problem solving. On the contrary, instead of inviting children to rely on, and employ their real-world knowledge and experience during problem solving, school word problems are perceived as artificial tasks that are unrelated to the realities of life outside the school (Freudenthal, 1991; Greer, 1993, 1997; Reusser, 1988; Treffers, 1987; Verschaffel & De Corte, 1997).

This chapter reviews a series of recent studies that aimed at a more detailed description and understanding of upper primary school pupils' belief that real-world knowledge is irrelevant when solving arithmetic word problems, but also at exploring possibilities to overcome their unrealistic mathematical modelling. First, two studies are presented that provide strong evidence for the position that as a result of traditional mathematics education, children develop a narrow but deep-rooted and resistant view of mathematical modelling. In this view, modelling is reduced to selecting the correct formal arithmetic operation with the given numbers, without seriously taking into account their common-sense world knowledge and realistic considerations about the problem context. Interestingly, in a third study the same phenomenon was observed in a group of Japanese children. In a fourth study pre-service teachers were confronted with the same problems as the children from the previous studies. Taking into account our better understanding of the phenomenon of unrealistic modelling derived from the preceding investigations, an intervention study was carried out to test the hypothesis that it is possible to change through instruction children's (mis-)belief about the role of real-world knowledge in mathematical modelling and problem solving. Most of the studies have already been published in separate articles. Therefore, in this chapter the focus will be on the integration of the findings of the different studies, keeping technical details to a minimum.

2 The Phenomenon of Unrealistic Modelling in Fifth Graders

2.1 Study 1: Ascertaining the Phenomenon

2.1.1 Method

The subjects were 75 pupils (10–11-year-old boys and girls) from three fifth-grade classes. They were taught word problem solving using traditional rather than authentic problem situations, and realistic mathematical modelling was not systematically addressed in teaching (for a more detailed description of the study see Verschaffel, De Corte, & Lasure, 1994). A paper-and-pencil test was constructed consisting of ten matched pairs of items (see Table 1). Each pair consisted of a standard item (S-item) that can be solved unambiguously by applying the most obvious arithmetic operation(s) with the given numbers (e.g., "Steve has bought five planks of 2 metres each. How many planks of 1 metre can he saw out of these planks?"), and a parallel problematic item (P-item) for which the appropriate mathematical model is less obvious and indisputable, at least if one seriously takes into account the realities of the context evoked by the problem statement (e.g., "Steve has bought four planks of 2.5 metres each. How many planks of 1 metre can he saw out of these planks?").

The 10 pairs of problems were administered by the class teacher on the same day in two series, each containing the P-variant of five problem pairs and the S-variant of the five other pairs. Pupils were asked to write down not only their answer for each problem, but also how they arrived at this answer (e.g., by mentioning the calculations), and possible other comments they might have (e.g., explaining their stumbling block when they were not able to solve the problem, supplementing their numerical answer with some comments, criticising the problem statement, etc.).

2.1.2 Analysis

Children's reactions to the problems were analysed in two ways for evidence of the activation and use of real-world knowledge and realistic considerations about the problem context:

(a) by distinguishing in their answers between realistic answers and non-realistic ones;

(b) by distinguishing in their computations and additional comments between realistic comments and non-realistic comments.

Table 1: Ten P-items involved in Study 1

P1	Carl has five friends and George has six friends. Carl and George decide to give a party together. They invite all their friends. All friends are present. How many friends are there at the party? (= the "birthday" item)
P2	Steve has bought four planks of 2.5 metre each. How many planks of 1 metre can he get out of these planks? (= the "planks" item)
P3	What will be the temperature of water in a container if you pour 1 litre of water at 80° and 1 litre of water of 40° into it? (= the "water" item)
P4	450 soldiers must be bused to their training site. Each army bus can hold 36 soldiers. How many buses are needed? (= the "buses" item)
P5	John's best time to run 100 metres is 17 seconds. How long will it take to run 1 kilometre? (= the "runner" item)
P6	Bruce and Alice go to the same school. Bruce lives at a distance of 17 kilometres from the school and Alice at 8 kilometres. How far do Bruce and Alice live from each other? (= the "school distance" item)
P7	Grandfather gives his four grandchildren a box containing 18 balloons, which they share equally. How many balloons does each grandchild get? (= the "balloons" item)
P8	Rob was born in 1978. Now it is 1993. How old is he? (= the "age" item)
P9	A man wants to have a rope long enough to stretch between two poles 12 metres apart, but he has only pieces of rope 1.5 metres long. How many of these pieces would he need to tie together to stretch between the poles? (= the "rope" item)
P10	This flask is being filled from a tap at a constant rate. If the depth of the water is 4 cm after 10 seconds, how deep will it be after 30 seconds? (This problem was accompanied by a drawing of a cone-shaped flask) (= the "flask" item)

When a child gave an answer to the problem that was scored as realistic or produced a non-realistic answer that was accompanied by a realistic comment, his (or her) overall reaction to that particular problem was scored as a "realistic reaction" (RR). Take, for example, the planks item mentioned above. A RR score was not only given to a child who produced the realistic answer "eight planks", but also to a child who responded with "10 planks" but who added the comment that "Steve will have a hard time putting together the remaining pieces of 0.5 metre." The code NR ("non-realistic reaction") was given for children who answered a problem in a non-realistic manner, and did not give any further realistic comment. (For examples of realistic and non-realistic reactions to the 10 P-items, see Verschaffel et al., 1994.)

2.1.3 Hypothesis

The overall hypothesis of the study was that due to their extensive experience with an impoverished diet of standard word problems, and to the lack of systematic attention to the mathematical modelling perspective in their mathematics lessons, pupils will demonstrate a strong tendency to exclude real-world knowledge and context-bound considerations from their endeavours to solve the problems of the P-items, and — consequently — will solve them as if they were not at all problematic.

2.1.4 Results

The observed data convincingly support the hypothesis. As predicted, the pupils demonstrated a very strong overall tendency to exclude real-world knowledge and realistic considerations when confronted with the problematic versions of the problems. In total, only 128 out of the 750 reactions to the P-items (= 17%) could be considered as realistic (RR), either because the pupil wrote a realistic answer or made an additional realistic comment. Very similar results have been reported by Greer (1993) and Reusser and Stebler (1997) for Northern-Irish and Swiss children, respectively.

Nevertheless, substantial differences in the amount of realistic reactions (RR) were found between the ten problems, varying from 0% for the rope problem (P9 in Table 1) to 59% for the balloons item (P7). Besides P7, the buses item (P4) also elicited considerably more RRs than all other items (the highest percentage for those others being 20% for the friends item (P1)). A plausible explanation for the relatively high number of RRs on P4 and P7 is that in these problems the modelling difficulty is restricted to the last phase of the solution process, where the pupil has to make sense of the result of the arithmetic operation, namely a quotient with a remainder (Silver, Shapiro & Deutsch, 1993). In all other P-items the underlying realistic modelling difficulty requires context-based adaptations at the beginning of the solution process, namely the construction of the mathematical model, rather than at the end of it (for a more detailed discussion see Verschaffel et al., 1994).

To have an indication of the inter-individual differences in the disposition toward (non-)realistic modelling, we also counted the total number of RRs on the ten P-items for each pupil separately. Fifty-nine out of the 75 pupils (= 78%) reacted in a realistic way to less than three of the 10 P-items. Almost all RRs of these pupils were found on the buses (P4) and the balloons (P7) problems.

2.1.5 Discussion

While Study 1 provides empirical evidence for the strong tendency by pupils to exclude real-world knowledge when solving word problems, its findings need to be put into perspective because of an important methodological limitation of the data-gathering technique used, namely a collective paper-and-pencil test. Indeed, one could argue that during their (private) solutions of the P-items some pupils may have activated real-world knowledge which was not reflected in their written answers, simply because they finally decided to react in a "conformist" rather than a "realist" way in line with their beliefs and conceptions about "the prevailing rules and conventions of the game of school arithmetic word problems" (De Corte & Verschaffel, 1985). For instance, Wyndhamn and Säljö (1997) have shown that when a problem of this kind (P6) is introduced as a topic of discussion in group work, rather than presented as a mathematics task in the traditional, narrow sense, pupils are likely to make realistic considerations. Thus, this procedure of using a traditional test only may have led to a significant underestimation of the number of realistic considerations in this investigation. Therefore, we set up a second study in which we assessed whether two rather simple forms of scaffolding during an individual interview are sufficient to transform children's non-realistic solutions into realistic ones.

2.2 Study 2: Do Weak Forms of Scaffolding Help?

2.2.1 Method

Study 2 consisted of two stages (for a more detailed report, see Verschaffel, De Corte, & Lasure, in press). In the first stage, seven matched pairs of word problems used in Study 1 were collectively administered to 64 fifth-graders from three classes. Based on the results on this test, the five most "realistic" and the five most "non-realistic" problem solvers from each of the three classes were selected to participate in the second stage of the investigation. During this stage, which took place one or two days later, these 15 realistic and 15 non-realistic problem solvers were individually administered the same seven problem pairs once again; with respect to each P-item solved with a NR on the paper-and-pencil test the following interview procedure was applied.

First, the pupil was asked to read aloud the problem followed by his own NR written down on the answer sheet. Then the child was confronted with the written notes of a fictitious classmate who had responded in a realistic manner. If the pupil still stuck to the initial NR (i.e., "10" with respect to the planks problem) after this weak form of scaffolding, a stronger scaffolding in the direction of realistic modelling was provided by stimulating and helping the pupil to realise the concrete problem situation (for instance, by explicitly asking the child to make a drawing of the planks).

2.2.2 Analysis

For all 30 pupils and for all seven P-items the following kinds of data were used for analysis:

(a) the initial reaction to the problem during the paper-and-pencil test (as RR or NR);
(b) the reaction to the confrontation with the RR of a fictitious classmate (= weak scaffolding);
(c) eventually, the reaction to the second and strongest form of scaffolding.

The pupils' initial number of RRs to the seven P-items of the paper-and-pencil test can be conceived as a measure of their actual level of realistic modelling, i.e., the level at which they can perform independently. On the other hand, their sensitivity to scaffolds toward realistic modelling of the P-items originally answered with a NR can be considered as reflecting their zone of proximal development (Vygotsky, 1978) or their learning potential (Hamers, Sijtsma & Ruijssenaars, 1993), i.e., those behaviors which they cannot yet perform independently but that they can manage with support from others. To measure this learning potential a score of 2 (the weak form of scaffolding is sufficient to change from a NR to a RR), 1 (the strong form of scaffold is needed), or 0 (neither form of scaffolding works) was given for each P-item not solved in a realistic manner during the paper-and-pencil test. The mean of these scores was considered as a measure of a pupil's learning potential, i.e., a score indicating how much a pupil gained from increasingly stronger forms of scaffolding toward realistic modelling.

2.2.3 Hypothesis

First, it was hypothesized that the overall number of RRs generated on the seven P-items of the paper-and-pencil test would not differ significantly from the percentage found in Study 1, which was 17.

Second, although a positive effect of the scaffolding on the number of RRs was anticipated, we expected, nevertheless, that this overall percentage of RRs at the end of the individual interviews would still be quite low, due to the fact that the tendency toward non-realistic modelling is very strong and resistant to change.

Third, the so-called realistic problem solvers of the P-items from the paper-and-pencil test would be more sensitive to the two scaffoldings during the interview, and would, therefore, benefit more from these two forms of support than the non-realistic problem solvers.

2.2.4 Results

The results on the collective test confirmed the first hypothesis: only 16% of all the reactions of the 64 pupils to the seven P-items of the collective test were classified as realistic. This percentage is almost exactly the same as the overall percentage in Study 1 (i.e., 17%). The percentage of RRs of the 15 realistic problem solvers (39) on this test was significantly higher than the percentage (8%) of the 15 non-realistic problem solvers.

In accordance with the second hypothesis, a significant effect of the two forms of scaffolding was found in the 30 pupils (15 realistic and 15 non-realistic problem solvers): the percentage of RRs increased from 23 during the paper-and-pencil test to 57 at the end of the individual interviews. An additional analysis revealed that both scaffolds contributed equally (and significantly) to this increase: from 23 to 40% RRs, and from 40 to 57% as a result of the weak and the strong scaffolding, respectively. But, notwithstanding the significant effect of the two scaffoldings, the cumulative percentage of RRs at the end of the individual interviews was still alarmingly low, as predicted. Indeed, in 43% of all cases, the pupils still reacted in an unrealistic way even after the second and strongest scaffolding. Moreover, a qualitative analysis of the pupils' reactions to the two forms of scaffolding indicated that pupils frequently stuck to their NR after the first and the second form of scaffolding, because they did not seem to grasp the line of reasoning underlying the RR (see Verschaffel et al., in press, for an illustration).

Finally, the results also confirmed the hypothesis that the realistic problem solvers would benefit more from the two scaffoldings than the non-realistic ones. The mean learning potential score (maximum = 2; minimum = 0) of the 15 realistic problem solvers (0.88) was significantly higher than the score of the 15 non-realistic problem solvers (0.52).

Thus, while the results of the second study strongly confirm elementary school pupils' disposition towards non-realistic modelling of school arithmetic word problems, they also show that this tendency is very strong and resistant to change (see also Reusser & Stebler, 1997). In addition, the findings provide evidence for the existence of inter-individual differences in this disposition, by showing that pupils of different levels of actual performance with respect to realistic modelling, differ also in terms of their learning potential, in the sense that their sensitivity differs to scaffoldings toward realistic modelling on problems they could not solve properly themselves initially.

2.3 Study 3: Comparing Belgian and Japanese Children

On the basis of the overall picture that derives from cross-cultural research on mathematics learning and teaching over the past decades (for a summary see De Corte, Greer & Verschaffel, 1996), one might suspect that Japanese pupils would perform significantly better on the P-items used in the previous studies as compared to European children. But, the findings of a few recent, more focused studies (e.g., Cai & Silver, 1994) indicate that the differences in performance on this type of problem may be rather small, or even non-existent.

Against this background, an investigation was designed to examine Japanese children's performance in realistic mathematical modelling in comparison with the Belgian pupils tested in Study 1. A second purpose was to establish the effect on Japanese pupils' test results of a similar kind of weak scaffolding as used in Study 2 with Belgian children, namely a very explicit hint at the beginning of the test in order to induce a more mindful and realistic approach to the problems. This hint warned the pupils that the test would contain some items that are difficult or impossible to solve, and invited them to comment on those items explaining why they are unsolvable. (A more detailed description of the study is given in Yoshida, Verschaffel & De Corte, 1997.)

2.3.1 Method

The subjects were 91 fifth graders from three classes in the same Japanese school located in a suburb of a middle-size city. They were randomly assigned to two conditions: the 45 children in condition 1 received neutral instructions as in Study 1; the 46 pupils in condition 2 were warned of the presence of problematic items in the test.

The paper-and-pencil test that was administered to the pupils consisted of literal translations in Japanese of the ten pairs of items used in Study 1, except that names of children were replaced by typical Japanese names. As an illustration, Figure 1 shows the Japanese translation of item P2 from Table 1.

The procedure for administering the test in condition 1 was the same as in Study 1. In condition 2 the students were given in addition the warning instruction, which was written at the top of the test booklet. Children's reactions to the problems were also scored in the same two categories as in Study 1: realistic reactions (RR) and non-realistic reactions (NR).

2.3.2 Results

No statistically significant differences were observed between the percentages of RRs over the 10 P-items in the three groups: the Japanese children in the conditions 1 (15% RRs) and 2 (20% RRs), and the Belgian pupils from Study 1 (17% RRs). With respect to the separate items, only one comparison was statistically significant, namely the difference for the planks item (P2) between the Japanese children in condition 1 (0% RRs), and the Belgian pupils (13% RRs).

Thus, the comparison of the performance of the Japanese fifth graders in condition 1 with the Belgian children tested under the same condition in Study 1 revealed that the tendency to neglect real-world knowledge while solving school word problems was equally strong in both countries. With regard to the second purpose of this investigation, the results indicated that the warning instruction used in condition 2 to induce realistic modelling was not successful. This finding is also in line with the weak effects of the scaffolds observed in Study 2, and similar results in Swiss children reported by Reusser and Stebler (1997).

2.3.3 Discussion

The results of the three studies reported above provide strong support for the hypothesis that upper elementary school pupils have developed a strong belief that real-world knowledge is irrelevant when solving school word problems. Indeed, these studies do not only show that children's answers to the P-items only rarely reflect adjustments for contextual constraints based on the activation of real-world knowl-

---

邦子さんは、それぞれ2.5 mの長さの板を4つ買いました。
これらの板から1 mの板がいくつとれるでしょう？。

Figure 1: Example of the P2 problem (Kuniko has bought four planks of 2.5 m each. How many planks of 1 m can she saw out of these planks?) as it was presented in the study.

edge. Studies 2 and 3 also yield evidence that children's belief about the irrelevance of real-world knowledge in problem solving is very strong and resistant to change. This derives from the observation made in the Belgian as well as in the Japanese sample that brief, but explicit, forms of scaffolding aimed at inducing more realistic modelling, have very poor effects.

The parallel between the findings in Belgium and Japan suggests that the tendency to neglect real-world knowledge when solving school arithmetic word problems is independent of general cultural and curricular differences. At the same time, this induces the hypothesis that there may be a kind of universal classroom mathematics culture involving a number of common rules and beliefs about approaching and solving arithmetic tasks and problems, one of them being the more or less implicit belief that real-world knowledge is irrelevant when solving word problems in the classroom.

This latter hypothesis points to the possible causes of the phenomenon of unrealistic mathematical modelling in upper elementary school pupils. Indeed, at present the development in children of a strong belief that real-world knowledge is irrelevant in solving school word problems is often attributed to the following prevailing and related characteristics of the typical instructional environment and culture in today's mathematics classroom: (1) the impoverished and stereotyped diet of standard word problems which can always be modelled and solved through the straightforward application of one or more arithmetic operations on the given numbers; (2) the nature of the instructional activities relating to word problems, especially the premature imposition of the formal arithmetic approach toward word problems by requiring that pupils must identify and execute the correct arithmetic operation to solve a word problem, and, more generally, the absence of systematic attention in teaching to the modelling perspective as one of the building blocks of a genuine mathematical disposition (De Corte & Verschaffel, 1985; Freudenthal, 1991; Greer, 1993; Kilpatrick, 1987; Reusser, 1988; Säljö, 1991; Verschaffel & De Corte, 1997). However, until now the empirical evidence to support the claim that children's unrealistic modelling is due to these features of the mathematics classroom culture is rather scarce and anecdotal, and, therefore, not very compelling. The two remaining studies attempt to contribute to filling this gap in two ways: (1) by focussing on a major component of the classroom environment, namely the teacher, and more specifically by analysing (student-) teachers' own beliefs about the role of real-world knowledge in problem solving (Study 4); (2) by testing the hypothesis that it is possible to develop in primary school children a disposition toward more realistic mathematical modelling by immersing them into a more appropriate and powerful teaching-learning environment (Study 5).

3 Student Teachers' Beliefs about the Role of Real-world Knowledge in Mathematical Modelling

In this study pre-service teachers' beliefs about the relevance of real-world knowledge in modelling school word problems was studied by analysing: (1) their own spontaneous responses to a set of word problems with problematic modelling assumptions

(P-items); (2) their evaluations of pupil answers to the same problems that do or do not take into account real-world knowledge (for a more detailed presentation see Verschaffel, De Corte & Borghart, 1997).

3.1 Method

Participants were 332 pre-service elementary school teachers from three Flemish teacher training institutions. About two-thirds ($n = 228$) had just started their first year of teacher training, while one-third ($n = 104$) were in the third, and last year of their pre-service training. The same paper-and-pencil test as in Study 2, consisting of seven standard and seven problematic items, was used. This test was administered twice to all the pre-service teachers. On Test 1, the student-teachers had to answer the 14 word problems themselves; calculations and comments could be written down in a comments box below the answer box. Immediately after they handed in this test, they were given Test 2, in which they were asked to score four different possible answers from pupils to the same 14 word problems as in Test 1 with either 1 point ("absolutely correct answer"), 0 points ("completely incorrect answer") or 1/2 point ("partly correct and partly incorrect answer"). The four response alternatives to the seven P-items in Test 2 belonged to four different categories: a non-realistic answer (NA), a realistic answer (RA), a technical error (TE) and another answer (OA) such as a wrong-operation or a given-number answer (see Figure 2). At the bottom of each problem, there was a box for writing comments.

For Test 1 student-teachers' solutions to the P-items were scored using the same two categories as in the previous studies: realistic reaction (RR) versus non-realistic reaction (NR). The analysis of the pre-service teachers' reactions in Test 2 focused on the score (1, 1/2 or 0) given to the realistic answer (RA) and the non-realistic answer (NA).

3.2 Hypotheses and Questions

3.2.1 Test 1

We hypothesized that due to their continuing experience with an impoverished diet of standard word problems, and to the lack of systematic attention to the mathematical modelling perspective in their mathematics (education) lessons, even the student-teachers would demonstrate a tendency — albeit not to the same degree as the upper primary school pupils — to exclude real-world knowledge when confronted with the problematic versions of the problems, and, consequently, would solve many of these problems as if they are not at all problematic.

We also wondered whether there would be a difference between the students who had almost completed their (third year of) training, and the students who had just started their (first year of) training. But, because of the possibility of conflicting hypotheses in this respect, no specific prediction was made about the difference between both groups (see Verschaffel et al., in press).

This flask is being filled from a tap at a constant rate. If the depth of the water is 4 cm after 10 seconds, how deep will it be after 30 seconds? (This problem was accompanied by a figure of a partly filled cone-shaped flask)

3 x 3.5 cm = 11.5 cm

After 30 seconds the depth of the water will be 11.5 cm.

A

3 x 3.5 cm = 10.5 cm

After 30 seconds the depth of the water will be 10.5 cm

B

3.5 cm + 20 cm = 23.5 cm

After 30 seconds the depth of the water will be 23.5 cm.

C

It is impossible to give a precise answer.

D

Figure 2: Presentation of problems in Test 2 of Study 3.

3.2.2 Test 2

It was hypothesised that the student-teachers' scores of the realistic (RA) and the non-realistic answer (NA) to a P-item from Test 2 would generally reflect their solutions of the same problems on Test 1. Consequently, we anticipated that the student-teachers would frequently consider the NA for the seven P-items from Test 2 as the (perfectly) correct answer and, therefore, score it with 1, while the RA would frequently be conceived as an inappropriate response and, thus, be scored with a 0. We wondered again whether there would be a difference in the scores for the NAs and the RAs between the first-year and the third-year students.

3.2.3 Relationship between Test 1 and Test 2

Generally speaking, a good match was expected between student-teachers' reactions to the seven P-items in both tests. This implies the following two predictions. First, most subjects who answered a P-item from Test 1 with a NR, will give a 1-score for the NA and a 0-score for the RA for the same item in Test 2. Second, a RR on a P-item during Test 1 will typically be followed by a 0 for the NA and a 1 on the RA for the same item during Test 2.

3.3 Results
3.3.1 Test 1

As expected, the student-teachers demonstrated a strong overall tendency to exclude real-world knowledge and realistic considerations when confronted with the problematic word problems. Only 48 % of all reactions to the seven P-items from Test 1 could be considered as realistic (RRs). This percentage is considerably higher than in previous investigations with upper elementary (see Studies 1, 2, and 3) and lower secondary school pupils (Greer, 1993), where overall percentages of RRs between 15 and 20 were observed. Nevertheless, it remains disappointingly low, as it implies that in more than half of the cases, the student-teachers solved the P-items from Test 1 in a stereotyped, uncritical way, without any consideration for the realities of the context involved in the problem.

There was a significant difference in the overall number of RRs between the first-year and the third-year student-teachers in favour of the latter group. However, the percentages of RRs remained low in both groups, namely 45 and 54% for the first-year and the third-year students, respectively. This suggests that student-teachers' disposition toward realistic modelling of arithmetic word problems is at least partially influenced by the courses on mathematics education received during their pre-service training.

3.3.2 Test 2

The student-teachers' strong disposition toward non-realistic modelling was also revealed by their evaluations of the realistic answer (RA) and the non-realistic answer (NA) on the same P-items during Test 2. Indeed, the difference between the percentage of 1-scores for the NA (56%) and for the RA (47%) was significant; the same holds for the difference between the percentage of 0-scores for the NA (18%) and for the RA (47%). Thus, the student-teachers' overall evaluation of the stereotyped, non-realistic answer to the P-items was considerably more positive than for the realistic answer based on context-based considerations.

There was again a significant difference between the first-year and the third-year student-teachers. The third-year students gave significantly more 1-scores and less 0-scores for the RAs than the first-year students. Correspondingly, the third-year students produced significantly less 1-scores and more 0-scores for the NAs than the first-year students.

3.3.3 Relationship between Test 1 and Test 2

It is also interesting to examine to what extent the student-teachers' evaluations of the NAs and RAs during Test 2 matched their own performances on Test 1. This was done by analysing the scores for the RA and the NA following the 52% non-realistic reactions (NRs) and the 48% realistic reactions (RRs) on Test 1 separately (see Table 2).

The left part of Table 2 presents the distribution of the different combinations of RA scorings (1, 1/2 or 0) and NA scorings (1, 1/2 or 0) over the seven P-items of Test 2 for the 52% non-realistic reactions (NR) on Test 1, as well as the distribution of the scorings for the RA and the NA over the three scores (1, 1/2 and 0).

As expected, we found a strong relationship between the non-realistic reactions on a P-item during Test 1, and the evaluations of the RA and the NA on that item during Test 2. In 89.3% of all cases in which a NR was given to a P-item during Test 1, the NA to that item was given a 1-score in Test 2. Correspondingly, 83.1% of the NRs during Test 1 were followed by a 0-score for the RA in Test 2. The match of a 1-score for the NA and a 0-score for the RA occurred in no less than 79.3% of all cases wherein a P-item from Test 1 was answered with a NR. Apparently, the NA was scored with 1 because this response corresponded to the stereotyped, non-realistic answer the student-teachers had given themselves on this item during Test 1, and they scored the RA with 0 because they did not take into account the context-based considerations underlying this answer. Ten percent of the NRs to a P-item during Test 1 were followed by a 1-score for the RA during Test 2. This suggests that in those cases the confrontation with the RA during Test 2 had functioned as a scaffolding toward (more) realistic modelling. However, the finding that only 10% of the scorings following a NR yielded evidence for the scaffolding effect of the confrontation with the RA provides additional evidence for the strength and the resistance of the tendency toward non-realistic mathematical modelling among student-teachers. The right part of Table 2 presents the distribution of the different combinations of RA scorings (1, 1/2 or 0) and NA scorings (1, 1/2 or 0) over the seven P-items of Test 2 following the 48% RRs on Test 1, as well as the distribution of the scorings for the RA and the NA over the three scores (1, 1/2 and 0).

Table 2: Combinations of RA scorings (1, 1/2, or 0) and NA scorings (1, 1/2 or 0) over the seven P-items of Test 2 for the total number of non-realistic reactions (52%) and of realistic reactions (48%) on Test 1

	Non-realistic reactions RA					Realistic reactions RA			
NA	1	1/2	0	Total	NA	1	1/2	0	Total
1	3.4	6.6	79.3	89.3	1	10.2	3.2	7.4	20.8
1/2	3.8	0.1	2.6	6.5	1/2	42.3	2.0	1.4	45.7
0	2.8	0.2	1.2	4.2	0	32.9	0.4	0.2	33.5
Total	10.0	6.9	83.1	100.0	Total	85.4	5.6	9.0	100.0

As shown in the table, the congruence between the RRs on Test 1 and the scorings of the RA and the NA during Test 2 was less straightforward than for the NRs. The evaluations of the RA were generally in line with the reactions on Test 1; indeed, 85.4% of the RRs on Test 1 were followed by a 1-score for the RA on Test 2. But the scorings for the NA were rather surprising: only 33.5% of the subjects who reacted in a realistic way to a P-item during Test 1 scored the NA with a 0 during Test 2 (almost always in combination with a 1 for the RA). This indicates that in many cases where student-teachers reacted themselves to a P-item in a realistic manner, they were nevertheless quite understanding and tolerant to elementary school pupils who interpreted and solved these P-items without seriously taking into account the relevant knowledge about the context called up by the problem statement. According to their written explanations in the comments box, they thought that it is unfair to punish a fifth grader for solving the P-item in a stereotyped, non-realistic manner. This is illustrated by the following comment accompanying the scoring combination "1 for RA and 1 for NA" with respect to the runner-item: "I scored alternative D (= the RA: "It is impossible to answer precisely what John's best time on 1 kilometre will be") with 1 because a pupil who gave this answer knows that it is not realistic to assume that John will be able to run at his record speed for 1 kilometre. But, I also gave 1 for alternative A (= the NA: "$17 \times 10 = 170$. John's best time to run 1 kilometre is 170 seconds") because from a purely computational point of view this is the correct answer."

3.4 Discussion

This study demonstrates convincingly that pre-service teachers are themselves strongly affected by the phenomenon of unrealistic and decontextualised modelling in word problem solving. Of course, this finding does not provide direct evidence that these teacher conceptions and beliefs are responsible for pupils' tendency to exclude real-world knowledge from their solution processes of word problems. However, taking into account recent research on mathematics teaching and learning (see e.g., De Corte et al., 1996), it is plausible to assume that these teacher cognitions and beliefs concerning the role and the importance of real-world knowledge in interpreting and solving verbal problems have a substantial impact on their actual teaching behaviour, and, consequently, on their pupils' learning activities and outcomes. Therefore, if one wants to link arithmetic word problem solving to the everyday experiential world of the children — as is strongly advocated in current reform documents relating to mathematics education — it will be necessary to develop in (future) teachers the appropriate conceptions and beliefs that are needed if they are to (re-)orient their teaching practices toward realistic and contextualised modelling of problem situations and toward realistic and situation-bound interpretations of the outcomes of arithmetic operations, as components of a genuine mathematical disposition.

4 An Intervention Study: Teaching Realistic Modelling to Fifth Graders

Starting from the preceding studies about the phenomenon of unrealistic modelling, and taking into account the literature on mathematics learning and teaching (De Corte et al., 1996), an intervention study was designed. Using a pretest — post-test design with two control groups a teaching experiment was carried out to verify the hypothesis that through appropriate instruction in a powerful learning environment it is possible to develop in upper primary school children a disposition toward realistic modelling of arithmetic word problems (for a more detailed report, see Verschaffel & De Corte, 1997).

4.1 Method

Three classes from the same school participated in the study: one experimental class (E) of 19 fifth graders, and two control classes (C1 and C2) of 18 and 17 sixth graders respectively. The pupils from the E-class were taught an experimental program on realistic modelling (during the regular hours allocated for mathematics teaching). The program consisted of five teaching–learning units (TLU) of 2½ hours each. During the experiment the pupils from the two control classes followed the regular mathematics curriculum.

The three classes were administered a pre-test consisting of 10 P-items comparable to those used in the previous studies, and five standard problems (S-items) as buffer items. With respect to each problem, pupils were not only asked to write their response, but also to mention how they arrived at their answer and/or possible difficulties or worries experienced during the solution of the problem. At the end of the experimental course in the E-class, a parallel version of the pre-test was administered as a post-test in the three groups. However, in one of the control classes (C1) this post-test was preceded by an introduction of 15 minutes in which the pupils' attention was drawn to the fact that routine solutions for word problems are sometimes inappropriate when considered in terms of realistic constraints; like the Japanese children in condition 2 of Study 3, they were warned that the test contained several items for which such routine solutions are inappropriate. One month after the post-test the pupils of the E-class were given a retention test with 10 new P-problems.

Analysis of variance was used to evaluate the effects of the experimental program on children's realistic reactions to the P-problems. As in the previous studies a reaction was considered as realistic either because a realistic answer was given, or a non-realistic answer was accompanied by a realistic comment.

4.2 Major Features of the Powerful Learning Environment

The experimental course was developed in line with our design principles for powerful teaching-learning environments (see De Corte, 1995b; De Corte et al., 1996). First, the impoverished and stereotyped diet of standard word problems offered in traditional mathematics classrooms was replaced by a set of more authentic problem situations especially designed to stimulate pupils to pay attention to context-specific elements and considerations during mathematical modelling, and to distinguishing between realistic and stereotyped solutions of mathematical applications. Each TLU focused on one prototypical problematic topic of realistic modelling. For instance, the topic of the first TLU was: making appropriate use of real-world knowledge and taking into account context-bound aspects of the situation when interpreting the outcome of a division problem involving a remainder. The opening problem involved a story about a regiment of 300 soldiers doing several military activities. Each part of the story was accompanied by a question which always asked for the same arithmetic operation (namely, 300 ÷ 8 = .), but required each time a different answer (respectively, "38", "37", "37.5" and "37 remainder 4"). For an overview of the themes of the other four TLUs, we refer to Verschaffel and De Corte (1997).

Second, not only the problems but also the teaching methods differed considerably from traditional mathematics classroom practice. A variety of teaching methods was applied aiming at stimulating small-group problem solving (in mixed-ability groups of three or four pupils), as well as whole-class discussion of solutions and solution strategies, and individual problem solving. In both, small groups and the whole class, reflection was frequently evoked by asking such questions as "What difficulties did you encounter when solving this problem?", "On which points did you disagree in your group?", and "What did you learn from solving this problem?" Thereby, children's attention was focused on the proper consideration of context-specific aspects of the problem situations, and of the assumptions of the mathematical models underlying proposed solutions.

Third, a new classroom culture was created featuring systematic attention to the realistic modelling perspective as an aspect of acquiring a mathematical disposition. This was attempted by activities such as modelling, explaining, and discussing valuable problem-solving strategies, and listening carefully to children's explanations and justifications of their own solution procedures. Another major dimension of establishing a new classroom culture consisted of explicitly negotiating with the pupils new socio-mathematical norms about the role of the teacher and the children in the classroom, and about what counts as a good mathematical problem, a good solution procedure, a good response (see also Cobb, Yackel & Wood, 1992; Gravemeijer, 1995).

4.2.1 Hypotheses

First, in line with the results of Study 1 and 2, we predicted that on the P-items of the pre-test the pupils of the three classes (E, C1 and C2) would produce a percentage of RRs that would not differ significantly from the 17% RRs and 16% RRs obtained in the first and the second studies, respectively. Second, we hypothesised that the experimental program would induce in students a disposition toward more realistic and contextualised modelling and interpreting of arithmetic word problems. Therefore, we predicted a significant increase in the number of RRs on the P-items from pre-test to post-test in the E-group only.

A third hypothesis was that the positive effect of the experimental program would be lasting. Consequently, we predicted that there would be no significant difference between the percentage of RRs of the E-group on the P-items of the retention test and the parallel problems of the post-test.

4.2.2 Results

As predicted in the first hypothesis, the pupils from all three classes demonstrated a strong overall tendency to exclude real-world knowledge from their problem solutions during the pre-test: only 15% RRs were observed. The fifth-graders from the E-class produced less RRs than the sixth-graders from the C1- and the C2-classes: the percentages were 7, 20 and 18%, respectively, but the differences were not significant.

The results also confirmed the second hypothesis. There was a significant increase in the number of RRs from pre-test to post-test for the E-group from 7 to 51% RRs. In contrast, in the two control classes the progress in the number of RRs from pre-test to post-test was non-significant, namely from 20 to 34% for C1, and from 18 to 23% for C2.

Finally, the positive results of the E-class on the retention test were in line with the third hypothesis. The percentage of RRs for the items of the retention test that were parallel with the post-test was 40%, which is almost the same as the result for those items on the immediate post-test (i.e., 41%).

5 General Discussion

In this chapter, five connected studies were reported that relate to upper elementary school children's belief about the role of real-world knowledge and context-specific task characteristics in mathematical modelling of school word problems. The findings from the first study support the view that throughout their school career upper primary pupils have gradually built up a very strong tendency to solve arithmetic word problems in a stereotyped and decontextualised way: they typically apply one (or a combination) of the basic arithmetic operations to the two given numbers in the problem, without any consideration of the possible problematic modelling assumptions underlying their proposed solution. The results of the second and the third study confirm this tendency, and show that it is equally strong in Japanese as in Belgian fifth graders. Those studies also indicate that the tendency to neglect contextual elements of problem situations is rooted in very strong beliefs in pupils about the irrelevance of real-world knowledge in mathematical modelling of school word problems, which are resistant to change. The fact that the same observations were obtained in countries that are culturally very different suggests the hypothesis that there may have developed a universal mathematics classroom culture.

The investigations revealed likewise considerable inter-individual differences in pupils' actual level of performance with respect to the use of everyday knowledge in mathematical modelling (see Study 1), as well as in their learning potential as reflected in their reactions to simple forms of scaffolding aimed at triggering the activation of that knowledge (see Study 2).

The fourth study demonstrated that upper elementary school pupils' strong tendency to solve arithmetic word problems in a stereotyped way was paralleled by a similar tendency among (future) teachers, as reflected by their own spontaneous solutions to word problems and their evaluations of different types of pupil answers. Of course, these findings do not prove that teachers' own beliefs are responsible for children's tendency toward non-realistic and decontextualised modelling. Nevertheless, they are in accordance with the argument that the current mathematics classroom culture and the prevailing teaching practices induce in pupils the belief that real-world knowledge is irrelevant when solving word problems, resulting in their tendency toward non-realistic and decontextualised mathematical modelling and interpreting of problem situations.

The fifth study assessed whether it is possible to modify children's inappropriate conceptions and beliefs about mathematical modelling and about the relevance of real-world knowledge in it. With a view to inducing such a change, a group of pupils was immersed into a new classroom culture in which word problems were effectively treated as exercises in realistic and mathematical modelling. In this innovative classroom culture, which introduced new social and socio-mathematical norms, a set of challenging, non-standard problem situations was used, and a number of instructional techniques like small-group work and whole-class discussions was systematically applied in an attempt to produce the desired revision and modification of pupils' belief about the role of everyday knowledge and contextual elements in mathematical applications. The findings provide good and promising support for the hypothesis that it is possible to change pupils' beliefs and to develop in them a disposition toward the activation of real-world knowledge in mathematical modelling of school word problems.

As a whole this set of investigations illustrates the significant role and impact of situational factors in learning, problem solving, and teaching. The first four studies provide convincing support for the hypothesis that the context and culture created by the prevailing instructional practices in today's mathematics classrooms induces in upper primary school children the belief that real-world knowledge is irrelevant when modelling and solving word problems. Furthermore, the teaching experiment provides initial indications that by modifying fundamentally this classroom situation and culture, it is possible to induce in children a more realistic and contextualised approach to mathematical problems.

6

Quantifying Time as a Discursive Practice: Arithmetics, Calendars, Fingers and Group Discussions as Structuring Resources

Jan Wyndhamn and Roger Säljö

1 Introduction: Signs as Intellectual and Practical Resources

The general background of the work to be reported here is an interest in the relationship between human sign systems and the uses these systems are put to in different social practices. Systems of signs, i.e. natural language, different artificial and/or institutional languages, mathematical expressions and so on, serve as resources for communicating about the world, and they codify objects, events and relationships in ways that are appropriate for particular activity systems. In some instances, sign systems have a very broad applicability, such as in the case of the metric system which is used in many social practices and for a variety of purposes (measuring length, weight, volume and so on). In other cases, signs are 'local' and used only in a much narrower class of activities. This would be the case with, for instance, the system of classifying plants created by Linnaeus, or the modes in which sizes of sheets of paper and envelopes are described in terms of A1, A2, A3, and C1, C2, C3, respectively.

Thus, systems of signs never merely reflect an outside reality, they place it in perspective in a manner which is suitable for a particular class of activities (Voloshinov, 1930/1973). Even though systems of signs to the experienced user may, and often do, appear as 'natural' and 'obvious', they are of course human inventions and they have a history. Hutchins (1995), for instance, describes how parts of the terminology for navigation that is used on board modern naval ships have a history that dates back several thousands of years. In celestial navigation the names of the stars that serve as reference points were used already by navigators from ancient Mesopotamia. To the experienced user sign systems are transparent and they serve as efficient structuring resources (cf. Lave, 1988) that give actors access to information of a certain kind about a situation or an event. For the less experienced user, the specific ways in which symbolic expressions of a particular system of signs refer to an outside reality may be difficult to realise. The specific modes in which signs are co-ordinated with an outside world are something that has to be learned, and, as a matter of fact, learning and the acquisition of expertise in complex societies (Hannerz, 1992) to a large extent involve becoming familiar with the rules according to which signs can be productively co-ordinated with the world and with each other. In this sense, human learning is a discursive phenomenon and even physical activities such as mending a modern car engine or building a house are increasingly dependent on conceptual constructions that are discursive in nature (Bruner, 1986, 1990;

Edwards, 1997; Harré & Gillet, 1994). Even the heavy physical labour of the blacksmith when forging the hot iron is informed by a rich number of conceptual distinctions that concern such matters as quality of the material and when it can be forged, the demands and preferences of the client and so on (Keller & Keller, 1993).

2 Quantifying Time

The concrete context in which this sociocultural perspective (cf. Wertsch, 1991, 1998) on cognition and action has been implemented in this study concerns how pupils acting in a school setting are able to solve a particular class of problems in which they have to quantify statements about time. The measurement of time is a good example of a fairly complex system of signs, or rather, a number of systems, that are sociocultural in nature and that have evolved over a long time (cf. Wyndhamn, 1993, for a brief presentation of the historical development of calendars and the measurement principles underlying these). It is also an area in which one can find varying measuring systems used in parallel in different activity systems in society. Four o'clock in the afternoon can be written and spoken about in different ways; it could be referred to simply as 4, as 4 pm, in military circles it would be 1600 hours and at sea the sailor might prefer to use bells. In previous studies (Säljö & Wyndhamn, 1988), we have analysed how pupils interpret the concept of 'week', which has a number of different meanings in different contexts; a calendar week is 7 days while in everyday language the most common referent of a week is 5 days (such as in a school or working week), and still other uses of this term are possible. When performing arithmetic calculations on tasks such as "A cow produces 18 litres of milk per day. How much milk does the cow produce during one week?" and "Lisa goes to school for seven lessons a day. How many lessons does she have a week?", the referents of the term week vary in accordance with the way in which the concept is used in different social practices.

Thus, as is the case with any linguistic representation of reality, there are conditions that determine what is an appropriate manner of referring to time and of estimating time. Competent performance requires some familiarity with these conditions and, in this sense, reference is never mechanical. The specific task which we have used as a starting point for some empirical studies of how pupils estimate time was taken from a national evaluation of mathematics knowledge in the Swedish school (Pettersson, 1990; Reuterberg, 1989). The task was formulated as follows:

> How many days are there from [*från och med*] March 24 until [*till och med*] June 18 if March and May have 31 days and April 30 days?

In all its simplicity, this task and the outcomes it results in are very interesting. In the national evaluation the frequency of correct solutions was very low. In grade 3 (when pupils are 9 years of age) only 3% of the pupils could solve the problem correctly, and in grade 6 (age 12) 7% of the answers were correct. In a follow-up study on a much smaller sample, Hagström and Sani (1991) used this task on pupils/students from grade 4 and up to the first year of upper-secondary school (when students are 17 years of age). The results obtained by these authors confirm that this task is very difficult to deal with; in grade 4 the percentage of correct estimates was 3, in grade 5 the corresponding figure was 8%, in grade 6 7%, in grade 7/9 the average was 8%, and finally in first grade at upper-secondary school, 20% of the students managed the task. Thus, there is very little evidence of improvement during comprehensive school in the ability to solve this problem, and even at the upper-secondary school level the result is remarkably low.

In addition to these observations, Reuterberg (1989), in an item analysis of this particular task in the national evaluation, found that its discrimination index was poor (i.e. an index which indicates how much a specific item contributes to the variation on the test as a whole, or in other words, the extent to which the item contributes to separating poor and good performances on the test). It furthermore showed a low correlation (point biserial correlation) with the performance on the entire test. In their interpretations, the various authors argue that this item does not measure competence in arithmetics but rather the ability of pupils to interpret the terms 'from' [från och med] and 'until' [till och med]. As one of them puts it, it is "language [rather than mathematics] which is the obstacle." (Pettersson, 1990, p. 102). From the point of view of a strict within-mathematics perspective, this particular item thus seems to lose its relevance. Still, if one assumes that systems of quantification and mathematical reasoning are modes of structuring reality in a purposeful manner, it is of some interest to scrutinise why the pupils 'make mistakes', and what it is in this item that is so problematic when closing the gap between signs — whether they be linguistic or mathematical — and statements about an outside reality. The question we will be exploring in the empirical part of this study can be formulated as: is it possible to, metaphorically speaking, position pupils differently in relationship to this kind of task?

3 What is the Problem: Reasoning as Practice

The answer given by the overwhelming majority of the pupils in the empirical studies is 86 days instead of the expected 87 days. The pupils seem to calculate $7 + 30 + 31 + 18 = 86$ rather than $8 + 30 + 31 + 18 = 87$. The decisive move which results in the miscalculation is when pupils establish how many days there are from March 24 until March 31. The interesting question is; how is this mistake produced? One answer would of course be the one given in the analysis of the national evaluation; the pupils are lacking in some kind of competence, whether linguistic or mathematical. On the other hand, this explanation in our view accounts for very little in concrete terms. Rather, it merely begs the question.

In many respects, this item is typical of tasks produced within the genre of word problems in school designed to be suitable as exercises. The pupils thus are operating in an 'as if' situation in which they are supposed to consider the relationship between an outside reality — the number of days between two points in time — and statements made about this reality. The item has a content that is expressed in words, while at the same time this content is arbitrary and generated for the purpose of serving as an exercise in mathematics learning. In order to avoid unnecessary mistakes, the knowledge of the world that is necessary is minimised by providing the number of days in March, April, and May. This 'purifies' the task and makes it into one that supposedly deals with mathematical reasoning. However, in spite of creating this specific — and in some sense decontextualised — task in which mathematical reasoning can serve as its own context, the pupil has to recontextualise (Linell, 1992) what he or she is doing and provide a reasonable set of assumptions or premises through which the text is made meaningful.

The mistake in calculating can arise in different ways. One would imply that the pupil identifies the task as one that has to do with mathematics, and he or she reads through the task without worrying too much about what the particular figures used refer to. The problem is not perceived as particularly difficult, and the pupil may not give too much attention to what the subtraction $31 - 24$ really means and what question is answered by it. The item is perceived as a mathematics task, somewhat disguised, but on the whole of the expected kind. Another way in which the incorrect answer can be produced would be if the student finds the task problematic and still decides, after consideration, that March 24 should not be included in the calculation.

Let us look a bit closer at the first alternative, which is empirically the more likely one. From a mathematical point of view, numbers differ in character depending on their function. One mode of dividing numbers is to view them as cardinal numbers, ordinal numbers and as measurements. A cardinal number answers the question "how many?" An ordinal number tells us "which number in a certain order" that a given item has. Cardinal and ordinal numbers are used when the objects we refer to are discrete. Measurements are associated with continuous variables. In this particular mathematics task we find all three kinds of numbers. The question is to be answered with a cardinal number ("how many days?"). The numbers 24 and 18 mentioned in the text are ordinal numbers. They refer to a certain day of a certain month; the 24th day of March and the 18th day of June. The numbers 30 and 31 can also be interpreted as measurements, since they indicate how long a particular month is (measured in terms of the number of days). If we enumerate the days in March, using the ordinal numbers, we get; the 24th, the 25th, the 26th, the 27th, the 28th, the 29th, the 30th, and the 31st. We have listed 8 days (8 in this case is thus a cardinal number). The subtraction $31 - 24$, however, does not produce the answer 8, since it presupposes that the numbers used for the calculation are cardinal numbers or measurements (i.e. positions on the number line). This mix of the kinds of numbers involved results in the 'mistake'. This is perhaps even more clear from the fact that there is no day 0 in the same sense as their is no year 0 in our calendar. The year 1 BC is followed by 1 AD.

This problem of the missing day in situations like this is codified in some contexts in everyday language. In Swedish, for instance, we take the statement "a week from today" (*idag om en vecka*) to be synonymous with the expression "today 8 days" (*idag åtta dagar*). For a period of 2 weeks, however, we use the expression 14 days. In French, on the other hand, the distinction is maintained in both cases: *aujourd'hui en huit* and *aujourd'hui en quinze*. In Swedish, the expressions used for referring to the beginning and the end of the time period indicate more clearly than in English that the two days — the first and the final — are to be included when estimating the length. Thus, a translation that would preserve this meaning would be "from and including March 24 until and including June 18".

4 The Study: Experimenting with Pupils' Positioning

Viewing mathematical reasoning as a resource for structuring situations — rather than as a series of abstract mental skills — in different kinds of social practices (cf. Säljö & Wyndhamn, 1990, 1993; Wistedt, 1994), we made a series of studies in which the positioning of students in relationship to problems of this kind was manipulated. The studies can perhaps best be described as some kind of naturalistic experiments. The basic idea behind the variations introduced was to provide pupils with different perspectives on the situation and with different premises for dealing with it, although the entire event takes place in the context of the school.

4.1 Method

Three different types of situations/problems were given. The pupils were told to work together in small groups with three members in each. The participating group were given one of the following three tasks that can be described as variations of the problem above.

A. You are to solve this task together:
    How many days are there from March 24 until June 18?
    (When working with the task, pupils were given a calendar)

B. Read this problem first:
    How many days are there from March 24 until June 18 if March and
    May have 31 days and April 30 days?
    It has been shown that most pupils count incorrectly
    when solving this problem.
    What mistake do you think they make?

C. Solve these two problems together in the group:
    1) LOVELY SUMMER HOLIDAY!
    Spring term ends on June 5. Autumn term starts on August 17.
    How long is the summer holiday?
    June has 30 days. July has 31 days.
    2) THE BALL IS IN THE AIR!
    The European championships in soccer will be played in Sweden between June 12–June 26.
    How long does the soccer championship last?

As can be seen, the variation in the premises that the pupils were given concern, in the first case, the introduction of a calendar as an aid in the situation. Also, the problem was slightly reformulated here by leaving out the information regarding the number of days in April and May. The calendar provided was a page with all the days, weeks and months of the year in question. In the second case, the pupils were alerted to the fact that most people tend to get the wrong answer to this problem. Their task was thus redefined as one of finding a difficulty rather than merely solving the problem (which they had to do as well, of course). In the third condition, a similar problem was constructed, but this time it was formulated using the pupils' own life world as a resource by asking for the length of the summer holiday and the length of the up-and-coming soccer championships to be played in Sweden. It should be noted that problems C1 and C2 require different forms of reasoning when establishing the length of the time periods asked for. In C1, neither of the dates mentioned (June 5 and August 17) should be included, since the first date refers to the final day of the spring term and the second to the first day of the autumn term. In C2, both days of the soccer championships should be included, since the championships started on June 12 and finished on June 26 with the final.

Each task was presented to nine different groups. The empirical material thus comprises 27 groups taken from five classes in grade 5. In all, 81 pupils (aged 12) participated. The groups were formed using previous achievement in mathematics as a criterion. The regular class teacher placed the children in one of three performance levels in mathematics; high, average, and low. This implies that for each of the tasks A, B, and C, there are data from three groups at each performance level.

All conversations in the groups were tape recorded in the regular school environment, either in the class room or in a neighbouring room specifically dedicated to group work. All interviews — which followed a common pattern — were carried out by one of the authors (JW). The conversation started with a question, asking whether the pupils used to make estimates of time in some situations (cf. excerpt below). Following this, the task was presented. It was presented in writing to every member of the groups. One pupil was told to read the problem aloud. The pupils had access to paper and pencil whilst discussing the problem. The responses given by the pupils guided the interventions, if any, made by the interviewer. When the group members had agreed upon an answer, one of the pupils was asked to write it down on a slip of paper. The conversation ended by asking the participants if they knew of any trick for remembering the number of days of each of the months. All 27 conversations/interviews were transcribed in as much detail as possible. The transcriptions have been used for the analyses below.

4.2 Results

In the presentation of the results, the different groups will be referred in the following manner. AH1, AH2, and AH3 refer to groups with high-achieving pupils working on task A. AA1, AA2, and AA3 for average achievers working on task A, and AL1, AL2, and AL3 for low-achieving students. The groups working with tasks B and C will be referred to in an analogous manner.

We will start by giving some general quantitative information on how the groups worked and the results they arrived at, and after that we will go in some detail into how the groups deal with the problem of quantifying time in this kind of situation.

4.2.1 Task A

In this condition, where the participants had access to a calendar, all nine groups arrived at the correct result — 87 days. This result contrasts heavily with the outcome in the national evaluation. In the case of one of the groups with high achieving members, AH1, the correct result was given only after the interviewer had pointed out that the text had to be read carefully. The members of one of the groups with average achievers in mathematics, AA1, hesitated for a long time before finally deciding on 87 days as their answer to the question. How the time was used during the problem solving can be seen from Table 1.

At a general level, the table reflects the strategy that most groups made use of. Each member counted on his/her own and in silence, and after this the result was discussed collectively. The fact that the groups with the low-achieving pupils (AL1, AL2, AL3) used more time to come up with a first suggestion for an answer can be explained by their using a strategy in which the days were counted one by one. Making the calculation 8 + 30 + 31 + 18 is of course much less time-consuming.

4.2.2 Task B

In this case, where the pupils were informed about the difficulty of the task as a premise for their work, two of the groups with low-achieving pupils (BL2 and

Table 1: Time used before giving first suggestion for answer and total time used until deciding on the final and written answer

Group	Answer	First suggestion for answer after seconds	Written answer after seconds
AH1	87	30	420
AH2	87	135	165
AH3	87	60	120
AA1	87	60	480
AA2	87	60	435
AA3	87	125	255
AL1	87	200	375
AL2	87	110	285
AL3	87	115	300
Median		110	300

AH, high achievers; AA, average achievers; AL, low achievers.

BL3) never managed to start a discussion. Therefore they were given a calendar by the interviewer as an extra resource. The outcome is presented in Table 2.

As can be seen, five groups (including the one operating with a calendar) gave the answer 87 when working on their own. The remaining four groups answered 86 days. Three of these four groups deliberately — and after consideration — excluded March 24. The fourth group intended to include March 24, but still ended up with 86 days. In seven groups the discussions of why people make mistakes when calculating claimed that the problem had to do with whether or not March 24 should be included. Thus, it does not seem difficult to realise that this is the critical issue; the problem seems to be one of taking it further from this realisation. In two groups (BL1 and BL3) the members did not come up with any suggestion as to why people make mistakes with this problem. In the continued discussion with the interviewer, all groups arrived at the conclusion that 87 days was the correct answer.

4.2.3 Task C1

In this case the problem of estimating the length of a time period was put in the context of figuring out the length of the summer holiday. Thus, this problem was in a sense close to the concerns of the pupils, and here both the two dates mentioned should not be included in the calculation. Only one group (CL2) could come up with the correct answer, 72 days. This group was thus one of those with low achieving students. The majority, six groups, ended up with 73 days, which implies that they included either June 5 (which was the last day of school) or August 17 (first day of autumn term) in their calculations.

After providing their initial answers, the groups were asked by the interviewer to read the text carefully and to reconsider their reply. As can be seen from the table, six of the groups (and including CL1 working with a calendar) arrived at 72 days. The

Table 2: Answer to task B

Competence	Group	Answer
High	BH1	86
	BH2	87
	BH3	86
Average	BA1	87
	BA2	87
	BA3	86
Low	BL1	87
	BL2*	87
	BL3*	86

*Indicates that the group had access to calendar while working.

Table 3: Answers to task C1 given initially and after being asked to read text carefully

Competence	Group	First answer	Second answer
High	CH1	73	72
	CH2	68	67
	CH3	73	72
Average	CA1	73	71
	CA2	73	71
	CA3	73	72
Low	CL1*	–	72
	CL2	72	–
	CL3	73	72

*Indicates that the group is using a calendar.

time used for presenting their initial reply varied from 50 seconds (CH1) to 200 seconds (CL3), and the median was 130 seconds.

4.2.4 Task C2

In this case, pupils were required to calculate how many days the forthcoming sports event would last, and here the first and the final day would both have to be included. The correct answer, 15 days, was given by four groups (CH1, CH3, CL1 and CL3) more or less immediately. The remaining five groups responded with 14 days. After being told by the interviewer to check their replies (see below), four of these groups modified their answer to 15 days. One of the groups with members that were average achievers in mathematics, CA1, did not change their initial response of 14 days. But in eight of the nine groups, the members were able to establish that the time period asked for was 15 days.

4.3 Reasoning and the Mastery of Discursive Classifications

To reason is a process of using signs in a systematic manner within discursive boundaries, and of co-ordinating signs with an outside reality and/or with other signs. In principle, the pupils were found to be using two different strategies for calculating the number of days. One strategy involved using a subtraction, as $31 - 24$ in task A and B or $26 - 12$ in task C2. A second strategy involved counting one day after the other using the calendar or one's fingers as resources, or simply enumerating the days orally. Table 4 illustrates how the two models were distributed across groups and tasks, and whether or not the groups managed to give a correct response (without

Table 4: Strategies for establishing length of time period

Competence	Group	Subt	Count	Group	Subt	Count	Group	Subt	Count
High	AH1*		X	BH1	X		CH1	**X**	**X**
	AH2*	X	X	BH2		X	CH2	X	
	AH3*	X	X	BH3	X		CH3	**X**	**X**
Average	AA1*	X	X	BA1	**X**		CA1	X	
	AA2*	X	X	BA2	**X**		CA2	X	
	AA3*		X	BA3	X	X	CA3	X	
Low	AL1*		X	BL1		**X**	CL1*		**X**
	AL2*		X	BL2*		**X**	CL2	**X**	**X**
	AL3*		X	BL3*		**X**	CL3	X	X

Subt = subtraction; Count = counting; * = the group is using a calendar; X = strategy used; **X** = correct first answer.

assistance from the interviewer). The correct answer (X) in the third column refers to task C2.

The table shows that when the calendar was offered as a resource in the problem solving situation, pupils utilised the strategy of counting the days in the calendar. This was true for all groups except AH1 (which used subtraction and where pupils did not arrive at a correct answer). In three cases (AH1, AA1 and CL1) a strategy that requires that the number of weeks in the period and the number of days in a week was used in addition to those listed in the table. It is also obvious from the table that the strategy of counting the days as a single strategy or in combination with subtraction gives the correct answer. The counting strategy, when used by the high achievers, resulted in correct answers in all cases. When subtraction was used as the only strategy, on the other hand, none of the four groups of high achievers that did so came up with the correct answer. Amongst the average achievers, five groups used subtraction as the only strategy, and three of these did not arrive at the correct answer. Among the groups of low achievers, subtraction is used by two groups only. It is worth noticing that when these groups use the counting strategy, they are as successful as the high achievers in finding the correct answer.

The cognitive activity in which the pupils are involved is discursive in nature and implies, as we have already pointed out on several occasions, the co-ordination between signs of different kinds. The discussions in the groups illustrate how the pupils are attempting to fit the problem into a frame that makes sense in the here-and-now in which they are acting. Thus, in order to be able to solve the problem it has to be construed as an instance of a problem of a particular kind, and this requires considerable interpretative activity for the participants (Wistedt, 1994; Wistedt & Martinsson, 1996). Let us scrutinise how this is done by means of one example from group AA2, which in many respects is typical as far as the type of reasoning engaged in is concerned as well as the difficulties that pupils encounter. The discussion can be described in terms of four phases.

Transcript of discussion of group AA2
Participants: Arnie, Bert, Cilla and the interviewer

Phase I

The interviewer	Arnie	Bert	Cilla
Our discussion will be about time. Do you ever measure time in any situation, for instance how long something lasts?			Sometimes in the text books in maths
		When you are going away . . . and when it's Christmas and such	
			. . . on holiday . . .
I will give you a problem here. It is about time. Please read it!			
		[Pupils are reading the text]	
What day is it today?	The 24th of March		
How many days are there from today until the 18th of June?			Are we supposed to count with this [March 24] day?
That's for you to find out yourselves		You did say June, didn't you? Not July	18th of June is a Thursday
Yes			
	A Thursday yes		Yes

Phase II

The interviewer	Arnie	Bert	Cilla
		86 days	
You are supposed to discuss and arrive at a common answer. You should talk to each other.			

## Quantifying Time as a Discursive Practice

Phase II (*continued*)

The interviewer	Arnie	Bert	Cilla
	Look here! 8 + 30 31 that is 69 . . . plus 18 is 87. Yes 87		87 or 86 it is

Phase III

The interviewer	Arnie	Bert	Cilla
		8? . . . It can't be 8 in the beginning . . . must be 7 . . . hehe	
	I counted with this here day from the beginning	Otherwise it'll be 7	86 or 87 . . . 87
Can you agree on an answer?			
		It all depends on if you count this here day [March 24]. If you take that one, it'll be 87 otherwise its 86	Mm
Why do you include this day [March 24]?			
	It has just begun		
		(inaudible comment)	
	I think we should include this day		
			Yes
		I think so too	This day before the next day
			Mm
Is there anything in the text which makes your interpretation reasonable or likely?			
		It says from . . .	
	. . . from [*från och med*]		
			Yes

Phase IV

The interviewer	Arnie	Bert	Cilla
How many days do you include from June?	18		18
	It says until		

Phase IV (*continued*)

The interviewer	Arnie	Bert	Cilla
	(inaudible)	It is 8 to begin with . . .	
			It is 86, isn't it? But 8 + 24 is 32 . . .
You are choosing between 86 and 87?			
			12 [weeks]. What is 12 times 7?
	84		
		Yes	86 it is then. There are 12 weeks between 13 and 25. And 2 days left . . . Tuesday until Thursday
Which one do you decide on then?		Let's take 86	
	87 days it is. Do you have an answer? (turns to I)		I think so too
Can that be so? Just any?		We can take any	
		Yes	
	It has to be exact, you are going away . . . if the plane leaves on a Tuesday, one can't get there on a Wednesday!		
Well then, which one is it?			Let's take 87 then
		Haa . . . 87 . . . There isn't much of a difference. But if you are playing hopscotch . . . Mm and you are standing on one . . . you sort of skip that . . . Mm	
What's it like when you are playing hopscotch?			
		You don't count the square you are standing in . . . its the	

Phase IV (*continued*)

The interviewer	Arnie	Bert	Cilla
		same thing here . . . you don't count that day . . . 87 or 86	
			87.5 haha
	It should be 87. It says from [från och med]		
Is that it?		Mm	Mm

The transcript illustrates the kinds of arguments that pupils consider relevant in this connection and, as was the case in most groups, the members work themselves towards an understanding of what the problem is all about. During the first phase, they read the text and attempt to establish what the problem is. Arnie and Cilla study the calendar and find out that "the 18th of June is a Thursday". Bert is already busy making his calculations. Bert also makes it clear at this point that he is uncertain whether March 24 is to be included when counting. The participants solve the problem on their own and without any explicit support from the interviewer.

During phase II, the pupils compare the answers they have arrived at. The answers produced by Arnie and Bert differ. Arnie has counted the days in March by using the calendar as a structuring resource. Bert made the subtraction 31 − 24. Cilla reflects on the problem and repeats the two alternatives. During phase III, the reason for this discrepancy is discovered during the first few exchanges; is March 24 supposed to be included or not? Arnie produces an argument for why it is to be included: 'It has just begun'. Bert and Cilla seem to agree about this.

During phase IV some new arguments are tested following Bert's doubts as to whether 8 days in March is a reasonable answer. Cilla argues against 8 days, since "8 + 24 is 32" — and since there are only 31 days in March, this mode of counting the days has to be incorrect. Cilla also chooses a new model for counting the number of days which takes as its point of departure the number of weeks during this period, and she again arrives at 86 days. This mode of estimating time uses the calendar as a structuring resource in a different manner. The number of weeks Cilla calculates by making the subtraction 25 − 13, where the numbers refer to the week numbers found in the calendar (referring to weeks in terms of their number, i.e. week 1, week 2, week 15 and so on, is a common, although quite recent, mode of talking in Sweden). This subtraction gives the right answer, but then she interprets her own words "Tuesday until Thursday" as 2 days instead of 3. Bert argues that "there isn't much of a difference" between 86 and 87, so his conclusion is that "we can take any".

At this stage the text seems to be drained of all possible interpretations, and the mathematics involved does not seem to produce any new openings or arguments for deciding on the correct interpretation. The pupils therefore start to consider other kinds of arguments. Arnie turns to the interviewer for a verdict, but receives no help. He then continues his analysis by taking the issue outside the classroom and the context of the school in order to make his point that there should be an exact answer. His aim is obviously also to counter-argue the position advanced by Bert that it does not matter whether they decide on 86 or 87 days. Arnie introduces flight schedules as an argument for why there has to be a way of deciding which number of days is correct. Bert makes an analogy with a familiar activity, hopscotch; "You do not count the square you are standing in . . . it's the same thing here . . . you don't count that day . . .". Thus, one should not include March 24. Cilla, again, attempts to suggest a middle way when she, while giggling, erroneously comes up with 87.5 days. Finally, and after 8 minutes, the group decides on 87 days. "It says from [från och med]" as Arnie puts it referring to the original text. And after this extensive interpretative work, this expression is seen as an argument for the conclusion that the first day of the period should be included.

The conversation in this interview, as well as in most other groups, follows the description given by Laborde (1990) of how a single student approaches a mathematical text. According to Laborde, the reader early on makes an assumption about what is an appropriate contextual specification of the meaning of the text. This interpretation is influenced by assumptions regarding what are reasonable 'readings' of texts of this kind that are held in advance, and the meaning ascribed is determined by characteristics of the text and its contents as well as by knowledge and previous experiences of the pupil. If the continued reading supports the initial hypothesis — which does not necessarily have to be explicitly formulated — the reader can go on according to plan. If the initial assumption is not confirmed by what is found in the continued reading, and if the concepts and procedures thus mobilised are not compatible with the information presented, there is a 'crisis'. The reader realises that the initial hypothesis must be scrutinised and reformulated. Laborde also claims that pupils often have difficulties in abandoning the initial assumptions and that they often seem to be willing to accept incompatible pieces of information and to exclude certain parts of the information provided in order to be able to sustain a particular interpretation.

In the transcript the group operated more or less as an individual reader. There is a 'crisis' — or a 'break-down' to use the Heideggerian term introduced by Winograd and Flores (1986) — when the participants present their answers to each others in phase II. We can also see how the individuals stick to their initial interpretations of the text. It is also of considerable interest to notice how pupils use intra-linguistic as well as extra-linguistic arguments in their reasoning. During phase IV Arnie and Bert introduce extra-linguistic arguments by referring to flight schedules and hopscotch. However, when concluding and formulating the final answer, it is an intra-linguistic argument that is accepted as decisive and this is illustrated by the emphasis put on the formulation "it says from" by Arnie.

In other groups, such 'break-downs' which forced the participants to reconsider their hypothesis were introduced by the contributions of the interviewer. When interpreted in context, contributions such as:

> Have you solved the problem correctly?
> Read the problem carefully once more.

> Are you sure that you have answered the question?
> Check the result!
> As a matter of fact you have miscalculated.

were decisive for the work, for instance, in all three groups of high achievers working on task C1 (CH1, CH2 and CH3).

5 Conclusions

As we pointed out initially, our interest in this particular type of task was triggered by the fact that it seems to be very difficult. As Reuterberg (1989) and Pettersson (1990) have shown the original task results in low performance, it has a low discrimination index and there is a low correlation between the success on this task and the results on more general tests of mathematical ability. That quantifying time under these circumstances and when being given this type of task is difficult is confirmed by the present study. If we look at task C1, for instance, only one of the groups is able to calculate correctly the length of the summer holiday when the pupils are working on their own. With some mild assistance from the interviewer, another five groups are able to do so. In the case of problem C2, which is a fairly simple case of finding the time period of 15 days, four of the nine groups answered 14 days. With some assistance from the interviewer, eight groups solved the problem correctly. In relationship to our initial hypotheses, this implies that even though the problem is formulated in a manner that is easily accessible for pupils, the performance continues to be fairly low.

The interesting problem, however, is how we are to understand the difficulties students have in finding the number of days. The explanation cannot be that "language is the obstacle" as Pettersson (1990) claimed, since the pupils manage the task when they have access to a calendar. Instead it would seem that it is the insertion of this problem into the context of mathematics learning that produces a major part of the problems that students have. It is as part of such a context that students are not allowed to use a calendar and it is as part of such an exercise that they resort to subtracting rather than counting the days.

At a general level it seems clear that the correct answer is produced when students are using the simpler mathematical strategy of counting the days. The low discrimination index that this type of item seems to have, i.e. its low contribution to the discrimination between high and low performance on tests of mathematical ability, can be explained by the fact that high and low achievers differ systematically in their choice of strategies for solving the problem. The low achievers more often choose the more successful strategy of counting the days, while high achievers use subtraction which generally produces an incorrect result. The task thus belongs to a special class of problems which pupils do not seem to become familiar with during their learning careers. They do not seem to become sensitive to the difference between ordinal and cardinal numbers, and they treat ordinal numbers as cardinal numbers when solving problems of this kind. Formulated alternatively one could argue that the low correlation of this item with mathematical ability is indicative of the fact that achievement is measured in such a way that the realisation on the part of students of the difference between cardinal and ordinal numbers plays little or no role for their competence in mathematics as defined by the school system. Furthermore, the claim that it is "language which is the obstacle" rather than mathematics testifies to an institutional definition of what counts as mathematics. According to this definition, the ability to realise — in context — the differences between how numbers are used is insignificant for mathematics learning. To qualify as a mathematics task, the item should not result in interpretative difficulties of this kind.

However, it would seem that this kind of problem lies within what we could refer to as the zone of proximal development to use Vygotskian language (Vygotsky, 1934/1986). Even though the pupils find it difficult to manage on their own, they are quite capable of discovering the difficulty and of sorting it out with some minor assistance from an adult (in our case the interviewer) or an artefact (such as the calendar). It is also interesting to note how the presence of an artefact such as the calendar in the situation transforms the task into a relatively simple one and furthermore removes the significance of the ability differences between the pupils. The calendar, metaphorically speaking, seems to exert some agency in the situation and it invites the students to count the number of days instead of using subtraction. Thus, the presence of a discursive tool restructures the situation and preserves the perspective on the problem which allows students to use a successful approach. The tool does a large part of the job and that is, we believe, how cognition and physical artefacts are integrated into social practices.

Acknowledgements

The research reported here has been financed by the Swedish Council for Research in the Humanities and Social Sciences. The final version of the manuscript was written while Roger Säljö was Belle van Zuylen professor at the University of Utrecht.

PART 2

Learning and Reasoning in Context
Introduction by Paul Light

In introducing the first part of this book, Roger Säljö highlighted the fact that much of psychology is unduly preoccupied with what takes place inside people's heads. In a variety of ways, the chapters making up that first Part illustrated the fact that understanding psychological activity often depends at least as much upon understanding social and cultural processes of negotiating meaning as it does on individual psychological processes.

This theme is continued in this, the second part of the book. Once again we shall see a recurrent tension between traditional psychological approaches which take little account of context, and other recent approaches which, by contrast, give context and situation pride of place in accounting for learning and reasoning.

These tensions are not addressed as a one-sided argument in favour of 'situated' learning and reasoning. The chapters making up this Part find weaknesses as well as strengths in situated approaches to both learning and reasoning. Nonetheless, the extent to which such approaches pervade contemporary analyses of both education and psychology is well illustrated here.

The chapters themselves range in subject matter from early conceptual development to adult reasoning, from the classroom to the company archive, and from children's play to the work of a systems analyst. Yet, despite this diversity the commonality of the underlying themes remains clear. Perhaps the central issue is how to reconcile the evidence for situational dependency in so much of what we do or say with the evidence for continuity and coherence in our experience and behaviour over time. Situated approaches offer an important critique of the excessive individualism of much of the psychology of the later 20th century, but that psychology still has much to offer which is of value, not least because understanding individuality itself is a continuing challenge.

Often, the critique of traditional approaches from a situated perspective is essentially descriptive rather than prescriptive. It offers a redescription of the way in which people come to act in the way that they do without necessarily suggesting other ways in which they should act. In some areas, though, a situated approach has been taken up in a prescriptive sense, as offering at least the basis of a recipe for what should be done. One such case concerns instruction.

Renkl, Gruber and Mandl (Chapter 7) explore the tension, which has been apparent for some years now, between established approaches to the psychology of instruction and contemporary situated approaches. The proponents of the latter criticise traditional methods of classroom instruction for delivering only 'inert' knowledge — knowledge which may be reproducible to order in school tests but which does not effectively engage with or support effective practice in the domain in question.

There are clear echoes here of those chapters in the previous section which addressed the nature of 'school knowledge'.

A number of psychologists and educationalists, especially in the United States, have come up with alternative approaches to learning in the classroom which seek to anchor instruction in authentic practice. For Renkl and colleagues, though, these approaches are seen to suffer a number of limitations.

Situated approaches to instruction, they argue, tend to be insufficiently specified, particularly in respect of how learners are to be supported in their exploration of learning resources. This issue is perhaps particularly critical in the computer-supported learning environments which Renkl and colleagues are concerned with.

Related to this is the criticism that situated approaches to instruction tend to see student motivation as unproblematic. Certainly there is more than a trace of the idealisation of 'intrinsic motivation' that marked the application of Piagetian ideas to education a generation ago. More generally, Renkl and colleagues argue, situated approaches to instruction make little attempt to address the issue of individual differences in ability or responsiveness to instruction. Moreover, a concept of 'activity' is employed which emphasises overt, observable behaviour. Renkl and colleagues put in a plea for those kinds of 'learning activity', such as listening attentively, which are largely unobservable but none the less important for that.

Whether these criticisms are properly directed at situated approaches to learning per se, as against certain attempts to apply them in classroom education, is a moot point. However, Renkl and colleagues render a useful service in reminding us of some of the enduring realities of educational experience and of the need to examine empirical evidence before plunging too enthusiastically into wholesale curriculum change.

The chapter by Rizzo, Marti, Veneziano and Bagnara (Chapter 8) explores the relationship between the psychology of reasoning, which typically focuses on the individual, and working settings in which most reasoning takes place in shared contexts. As Rizzo and colleagues point out, there is a sense in which not only individuals but also organisations can be said to learn. They may also be prone to forget. The collective aspects of knowledge production and maintenance, and the way in which collective knowledge resides in the practices of the workforce, have become much more widely recognised in recent years. Rizzo and colleagues argue that the computerisation of the workplace may prejudice some well established processes of collective learning, including the induction of new staff into existing patterns of practice.

Information and communication technologies have made possible the development within organisations of enormously rich databases of potentially useful information.

However, computer-based searching of databases typically 'leaves no tracks', so that each search tends to begin de novo. Rizzo et al. use the example of the archival resources of a major television company to illustrate the difficulty of organisational 'learning' and the potential for 'forgetting' inherent in this situation.

Moving to the level of a simple card-game to find an experimental analogue, Rizzo and colleagues show that even without co-operation, the externalisation of logical inference patterns in a shared space can influence the effectiveness even of individual problem solving. The gap between the experimental demonstration and the workplace problem is a wide one. However, the experiments nicely illustrate the 'tracks' we leave for one another as we think, learn and act in a common social space. The broader significance of this work is perhaps in its capacity to remind us that the world in which intelligence is exercised is a world which has been shaped, and continues to be shaped, by the intelligent activities of others.

The emphasis on logical inference in social context is continued in the following chapter, by Blaye, Ackermann and Light (Chapter 9). Drawing on both Piagetian cognitive developmental psychology and the psychology of adult reasoning, Blaye and colleagues critique the idea that logic can be seen as offering an abstract 'syntax' of mature thinking and reasoning. Their suggestion is that, on the contrary, pragmatic considerations of relevance-to-context actually seem to shape much of the thinking and reasoning not only of children but also of adults.

The developmental psychology of thinking and reasoning has been dominated for half a century or so by the idea that, as children develop, their thinking progresses from the concrete to the abstract. Sometimes this is characterised as a shift from embedded to disembedded thinking, or from contextualised to the decontextualised thinking; all these distinctions amount to much the same thing. In each case, the end state of this process is supposedly content- and context-free logical thought.

Research methods in this field have tended towards spare and simplified experimentation, in which variables are manipulated without consideration of how the context of the experiment itself might affect the child's reading of the significance of these variables. In fact, as the studies reported by Blaye and colleagues show, this same experimental approach can be adapted to investigate contextual sensitivity itself, to considerable effect.

The same issue of whether psychological development can be seen in terms of abstraction of one kind or another exercises Bliss, Ogborn, Cronin, Reader and Tsatsarelis in the following chapter (Chapter 10). The essential tension addressed in this chapter is between the Piagetian 'epistemic subject', developed through a process of reflective abstraction, and the 'actual subject' engaging in physical abstraction on the basis of particular lived experiences in a particular physical world. Although Piaget is often said to have pictured the child as a 'scientist', in his own work he made surprisingly little reference to the development of children's understanding of physical entities and their properties. Whereas Piaget's 'logical schemes' may be context free, 'physical schemes' cannot be. Similarly, while the process of reflective abstraction which generates logical schemes can be seen as an activity of a notional or 'epistemic' subject, the process of empirical abstraction which underpins physical schemes must be domain specific, and must rest on particular experiences.

Bliss and colleagues accept that abstraction in some form does characterise mental development. However, they see in contemporary situated approaches to cognition a variety of emphases (e.g. on knowing as enculturation, on learning through apprenticeship and on coming to 'take things as given') which are as relevant to children's understanding of the properties of the physical world as they are to their understanding of the social world.

The significance of a particular class of physical objects, namely machines, is explored by Ackermann in Chapter 1. Machines are 'objects that behave'; they are in some sense 'artificial others'. Computers are obviously of particular interest in this connection, since they have the capacity to simulate highly complex processes. However, Ackermann finds her starting points in simpler machines, and in the much simpler simulations evident in young children's pretend play. Like Bliss and colleagues, Ackermann finds many of her starting points in Piaget. Whereas Piaget saw the figurative side of the figurative/operative distinction as essentially static and descriptive, Ackermann shows through her examination of pretence and simulation that representations can be far from passive, static entities.

In the final chapter of the second part (Chapter 12), Eteläpelto and Light return to the issue of instruction with which this Part began. They explore the limitations of context-independent models of design expertise, and how such expertise is acquired. Eteläpelto and Light argue that design expertise is typically opportunistic in character, and responsive to context and situation. But 'context' has diachronic as well as synchronic aspects; it develops its significance over time through the life and experience of the individual. Learning is part and parcel of identity formation, and the development of a particular professional identity has considerable implications for the uses to which acquired expertise is put. Eteläpelto and Light seek to illustrate this thesis through a study of software developers and systems analysts as they move through the transition from instruction to early work experience.

At a time when there is so much public and political debate about the functions of higher education in relation to subsequent employment, it is perhaps surprising that there is not more research on the continuities and discontinuities marking the transition from study to work. Not only could such work be highly policy-relevant, it could also provide a valuable test-bed for ideas about situated learning and the transferability of understanding.

The issue of transfer is not extensively or frequently addressed in the chapters which make up this part. Underlying all of the tensions explored in these chapters — between cognition as abstract ('in the head') and cognition as inhering in the relationship between an individual and a particular social or physical context — there is an issue of 'what transfers' as individuals move through their lives, encountering and participating in one situation after another. Arguably, none of the chapters offers any wholly satisfactory solution to this problem (if problem it really is). However, each of them in its different way contributes valuable illumination of the broader issue of how learners' engagement with their physical and social worlds conditions their learning, their reasoning and their behaviour.

7

Situated Learning in Instructional Settings: From Euphoria to Feasibility

Alexander Renkl, Hans Gruber and Heinz Mandl

1 Introduction

It is a common observation that knowledge learned in instructional settings, such as schools, universities, or courses in vocational education, tends not to be used outside the corresponding contexts (Bransford, Goldman & Vye, 1991; Renkl, Mandl & Gruber, 1996; Resnick, 1987). The lack of knowledge transfer from in-school contexts to out-of-school contexts seriously challenges the usefulness of traditional forms of teaching. The primary purpose of any form of instruction should not be to enable the learners to answer questions in examinations, but to enhance their ability to cope successfully with problems of everyday or professional life. Knowledge whose use is largely confined to instructional contexts is labelled as *inert* (Whitehead, 1929; see also Renkl, Mandl & Gruber, 1996). In order to effectively tackle the inert knowledge problem, instructional approaches labelled as situated learning models have been developed.

In the next section, we present basic features of instructional models based on a rationale of situatedness. Then three important problems associated with these approaches are outlined. Finally, we propose questions for future research that have to be addressed in order to overcome these problems. It is important to note that our focus is not upon the strengths and limitations of attempts to describe learning and problem-solving from a situated learning point of view. Rather, we focus on prescriptive instructional models that aim to implement situated learning in instructional settings.

2 Instructional Models of Situated Learning

The most prominent instructional models of situated learning are cognitive apprenticeship, anchored instruction, and random access instruction.

Cognitive apprenticeship. Collins, Brown and Newman (1989) took the apprenticeship metaphor from craft domains such as tailors or midwives and elaborated it to an instructional approach for more cognitive domains. This approach emphasises the need to explicate or reify cognitive processes (e.g. strategies and heuristics) so that they can be as observable, and thus open to feedback, reflection, etc., as are the more manual skills trained in traditional apprenticeship. In addition, the learning process typically starts out with an expert model showing how to solve problems.

Subsequently, the learner works on authentic tasks while receiving coaching and scaffolding from a more expert person.

Anchored instruction was developed by the Cognition and Technology Group at Vanderbilt (e.g. 1992) in order to tackle the inert knowledge problem. A basic feature of this approach is to present complex and near-to-reality problems as a starting point of the learning process, providing an anchor for learning. The problems to be solved are embedded in interesting adventure stories presented in a video-based format. Several such stories have been developed.

Random access instruction stresses the importance of providing multiple perspectives in which the knowledge to be acquired is embedded (Spiro et al., 1991). It primarily deals with advanced knowledge acquisition in ill-structured domains (e.g. diagnosis of heart diseases, literary interpretation). These domains can be described by two basic characteristics: complexity of concepts and cases, and irregularity of cases with large variability of relevant features across different cases. Instruction grounded in this approach aims to induce multiple and, as a consequence, flexible representations of the knowledge to be acquired. This should enable the learners to apply their acquired knowledge in a wide variety of contexts.

As these three models illustrate, the specific focus of the different situated learning models differs. However, they have some pivotal instructional principles in common. All of them view instances of learning in real-life situations (on the job) as an ideal to which instructional settings should get as close as possible. Knowledge is viewed as fundamentally situated or context-bound and, therefore, the learning context should be similar to the contexts where the knowledge is expected to be applied.

Learning should be triggered by an interesting problem to be solved (problem-oriented learning). This problem should motivate the learners to assume active roles (e.g. information seeker, explorer, or problem-solver). Knowledge should be acquired in the context of immediate application to a problem solution and not in an abstract, decontextualized way. Furthermore, the problem should be authentic or, at least, near-to-reality. This means that the problems are usually complex and ill-defined as is typical of nontrivial problems in vocational and everyday life. While working on the problem, the learners should articulate their strategies and reflect upon them. Ideally, this should be done in social learning arrangements (i.e. through co-operative learning or tutoring).

Much enthusiasm has developed for learning environments that are designed according to these principles. However, it is not an easy and straightforward task to implement real-life-type learning in instructional settings such as schools or universities. Furthermore, in recent years, the situated learning models have been heavily criticised (e.g. Anderson, Reder & Simon 1996; Weinert, 1996). It has been argued that evidence for the context-bound nature of learning has been exaggerated, that the importance of learning by solving complex, near-to-reality problems in social settings has been over-emphasised, and that an unduly negative picture has been painted of traditional forms of instruction (e.g. Gardner, 1991).

In this chapter we shall argue for an intermediate position that tries to exploit and combine the strength of traditional as well as situated forms of learning. For this purpose, it is necessary to arrive at a more differentiated picture of the advantages and limitations of situated learning models. The potentials of the situated learning approach are of course manifold. For instance, it highlights the importance of the learner's recognition of the usefulness of the knowledge to be learned and focuses on the provision of opportunities to acquire competencies in coping with complex problems, reducing the problem of knowledge inertia. Thus, we stress the necessity of developing effective situated learning environments. For this purpose, however, some problems related to these ideas have to be solved. In this paper, we discuss three problems to which current instructional situated learning models have devoted insufficient attention: first, inadequate specification of the type of instructional support necessary for effective situated learning; second, over-confidence in the engagement-inducing potential of complex problems; and third undifferentiated use of the notion of activity.

3 Problems of Situated Learning in Instructional Settings

3.1 Lack of Specification of the Type of Instructional Support Necessary for Effective Complex Learning

Complex and open learning arrangements are a promising approach for promoting the acquisition of applicable (i.e. non-inert) knowledge. However, this is only true when the learners are not overwhelmed by the complexity of the learning environment. Without appropriate instructional support, only learners with excellent learning prerequisites, especially those with an elaborated domain-specific knowledge base, can profit from complex arrangements. Average learners, in contrast, need additional support, otherwise they will get lost.

To date, situated learning models have not been very precise in specifying the characteristics of effective instructional support in diverse types of complex learning environments (for first steps in this direction see Cognition and Technology Group at Vanderbilt, 1996; Lamon et al., 1996). The basic metaphor of effective learner support in situated learning environments is derived from the apprenticeship model. In the very beginning a teacher or an expert should model the skills, strategies, or whatever that are to be learned. Then the learners solve problems while receiving scaffolding support from a teacher or an expert. With growing learner competence, this support is gradually reduced.

Such an account leaves many relevant questions unanswered. For example, should the model ideally be a master model, showing smooth performance, or a coping model struggling with difficulties which are finally overcome? Which kind of specific help should scaffolding include? What is the specific rationale for gradually reducing support? Such issues have to be solved for the design of effective learner support.

It is not a trivial question to determine which type of support is necessary for effective learning in given learning environments. For example, help systems designed for support of learners in hypermedia environments are often not sufficiently used, or not used at all (Hofer et al., 1996).

Gräsel and Mandl (1993) analysed the acquisition of diagnostic competencies in medicine by case-based learning. Although modelling of adequate diagnostic strategies by an experienced physician led to the use of improved strategies by the learners when solving a transfer case, the cognitive model proved to be substantially more effective when learners independently tried to solve a case prior to learning from an expert model. Hence, it was not modelling alone but modelling in the context of relatively extensive experience of difficulties that proved most effective as a basis for learning.

Furthermore, rich evidence for the need of appropriate learner support in open learning environments is available from research on computer-based instruction. Many reports show that learners working in hypermedia environments easily get lost in hyperspace if no appropriate external support is provided. Another example of this phenomenon is the typical behaviour of people learning by exploring a computer-based simulation. In most cases they appear to act rather randomly, without any systematic strategy (Gruber et al., 1994; Njoo & De Jong, 1993). External guidance is necessary to ensure effective learning from simulations (Stark et al., 1995; Vollmeyer, Burns & Holyoak, 1996).

One of our own experiments (Stark et al., 1995) demonstrates clearly that the more complex the environment, the more important it is to provide appropriate learner support. We had vocational school students learn by exploring a simulation in the domain of business management (jeans' manufacturing). Our focus was on two experimental factors: (i) uniform vs. multiple learning contexts, (ii) guidance vs. no guidance.

In the less complex uniform learning context condition, the students had to deal with the same (standard) context each time. Learners in the more complex multiple learning contexts condition also began with the standard context, but they were also exposed to two other learning contexts, constructed in such a way that different market variables appeared in the foreground. Different sub-areas of the complete system were made salient in each of these divergent contexts. It is important to note that, in principle, the same learning possibilities appeared under uniform contexts as under multiple contexts (cf. Stark et al., 1995).

In the guided condition students were given a problem-solving scheme which they had to follow. The use of this scheme was explained in detail on the basis of a standardised typical situation, and then the students were required to relate the individual steps of the written scheme to an example. The steps involved: information gathering and data analysis; determining and justifying decisions; giving prognoses for competitors' prices as well as for own sales and profit; and finally evaluating the results (i.e. comparison of prognoses and actual outcomes). Unguided learners were not given this problem solving scheme.

As dependent variables, indicators of applicable knowledge were used: (i) system control (How much profit did the learners make in a test phase?); (ii) functionality of mental models (How precisely could the learners predict important market variables in new simulated situations?); (iii) far transfer (Which decisions were made in problem situations that were not directly related to the simulation? How sophisticated were the reasons provided for these decisions?).

Figure 1 shows the findings. It turned out that the multiple learning context plus guidance condition was the only one in which the learners performed well with respect to all three indicators of applicable knowledge. If complexity was increased without help (unguided-multiple), the results were disastrous. These results nicely illustrate the difficulties of dealing with complex learning environments when learners are left without appropriate support.

To sum up, learners are typically overburdened in complex learning environments; they need support. It is not, however, a trivial question to determine which type of learner support is necessary in any given type of complex learning environment.

3.2 Over-confidence in the Engagement-inducing Potential of Complex Problems

Proponents of complex learning environments assume that making complex, near-to-reality problems available as learning anchors induces high engagement, which in turn results in deep-level reasoning and learning (Cognition and Technology Group at Vanderbilt, 1993). However, using complex problems does not ensure that all learners will be highly engaged (Anderson et al., 1996; Hiebert et al., 1996; Renkl, Gruber & Mandl, 1996).

Learners are typically used to traditional forms of instruction and learning, and a new form of learning is not automatically accepted as a superior alternative. For

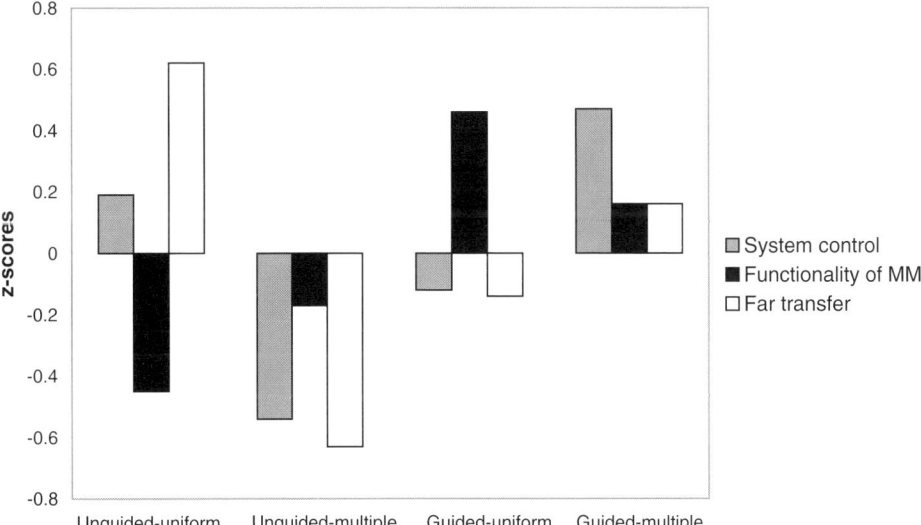

Figure 1: Learning results as a function of complexity and guidance.

example, Jacobson and Spiro (1994) found that only learners with epistemological beliefs that resemble constructivist assumptions profited from a complex hypermedia learning environment. Similarly, Huber and his colleagues have shown in several studies that only uncertainty-orientated persons benefit from complex, co-operative forms of learning (Huber, 1993). Obviously, the complex learning environments failed to induce effective engagement in all learners.

Our research group also found evidence for the relevance of inter-individual differences in learning-related attitudes (Stark et al., 1997). The data from the jeans' manufacturing simulation presented in the preceding section did not tell the whole story. The learning results in the multiple learning contexts plus guidance, which, on the whole, was the most favourable learning environment, strongly depended on the learners' tolerance of ambiguity. This finding is exemplified by the case of far transfer in Figure 2. The superiority of guided learning in a complex environment was primarily due to the success of learners with high tolerance of ambiguity.

To sum up, using complex problems does not guarantee high student engagement. Individual characteristics such as epistemological beliefs and ambiguity tolerance have to be taken into account, and procedures have to be developed for meeting the special needs of students with unfavourable motivational or attitude-related pre-requisites for complex learning.

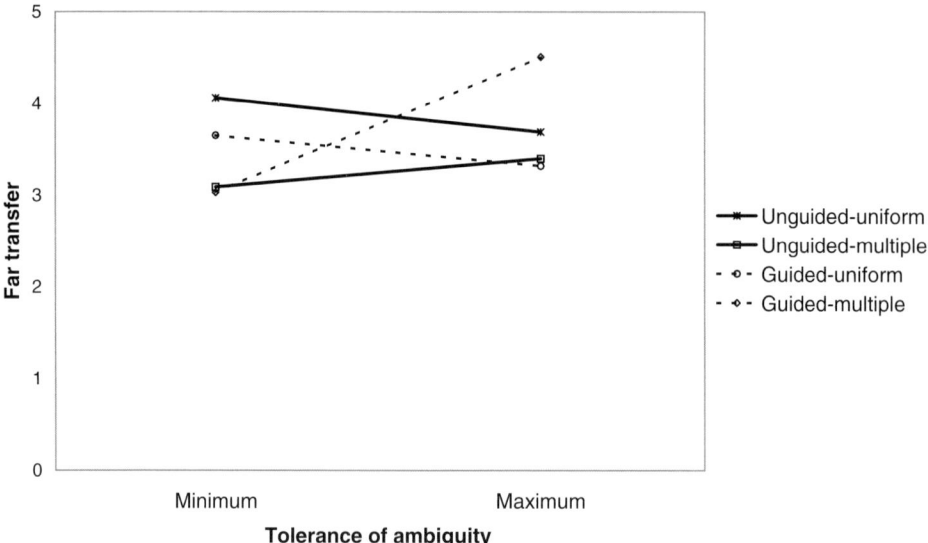

Figure 2: Regressions from far transfer performance on ambiguity tolerance in the experimental groups.

3.3 Undifferentiated Use of the Notion Activity

Situated learning models stress that a necessary condition for effective learning is that learners are active. From this standpoint, learning by listening to a lecture, for example, is regarded as a poor method of learning because learners are put into the passive role of recipients. The assumption that the learners are passive while listening to a lecture is derived from the low amount of visible and audible activities involved.

Learning arrangements in which learners actively solve problems or explain something are seen as more active, the level of activity being in effect determined by reference to visible and audible activities. Accordingly, situated learning models try to induce visible and audible activities such as problem-solving or explaining one's strategies. It is, however, important to keep in mind the possibility that non-observable mental activities related to the contents to be learned may be the prime determinants of knowledge construction (i.e. learning). By inducing observable activities, frequently some of the activities so provoked are not directed toward the contents to be learned, and may even interfere with learning (Marcus, Cooper & Sweller, 1996).

This point is illustrated by the findings of Renkl (1996). He conducted an experiment that compared learning by explaining and learning by listening. As criteria, problems of different transfer distance in relation to the learning tasks were employed: self-transfer (known problems, but with irrelevant information inserted), near transfer (isomorphic problems), far transfer (problems that were superficially similar to known ones, but where the structure was changed).

Beginning students of education learned probability calculus from worked-out examples. They were grouped together in pairs. After an individual learning phase, the two learners were brought together and assumed the role of the explainer or the listener, respectively. It turned out that the role of the listener was more favourable with respect to learning results, at least when high-level criteria of learning success were employed (Figure 3).

In the light of the prevailing assumptions in the research literature (e.g. Webb, 1991), this result is at first glance surprising. If it is, however, taken into account that the demand to explain not only induces elaborative and metacognitive activities related to the material to be learned, but also calls for management of the social (co-operative) situation, it becomes understandable why the explainers were outperformed. In accord with Wood et al. (1995) it can be argued that the explainers had two tasks: self- and other-regulation. The listeners, in contrast, could concentrate on learning the contents. Thus, in terms of the activities directed towards the contents to be learned, the listeners were obviously more engaged.

It is important to note that many studies have shown that explaining can be a rather effective way of learning (O'Donnell & Dansereau, 1992; Webb, 1991). The results of Renkl (1996) show, however, that the opposite can also be true. A more differentiated picture that takes context conditions into account has to be drawn. Moreover, the findings of Renkl (1996) illustrate that a high level of observable activity does not necessarily signify the presence of the kinds of mental activity that are relevant to learning.

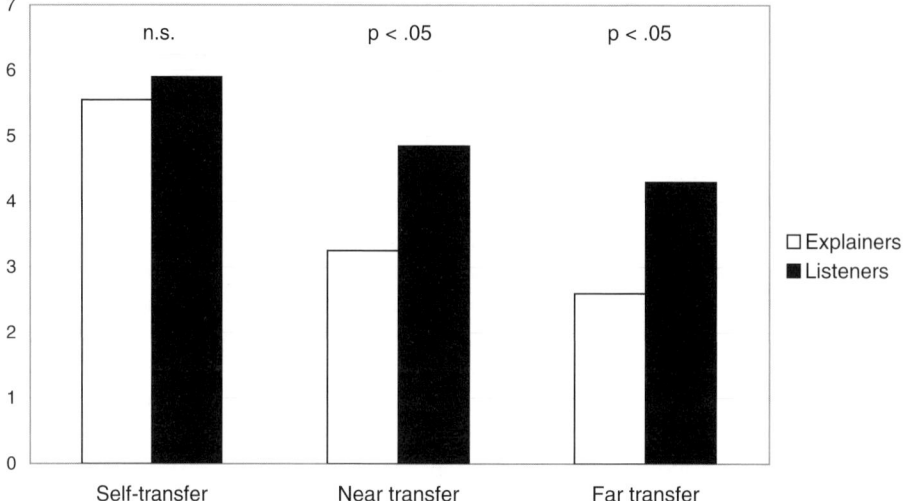

Figure 3: Learning results of listeners and explainers. ns, not significant. (t-Test for *dependent* samples were used as statistical test.)

4 Conclusions

The unresolved problems of the situated learning models discussed in this chapter highlight a number of issues that have to be drawn into better focus in future research on the design of situated learning environments, namely the need for appropriate support in complex learning, the need to prepare and motivate students for complex learning, and the need to evaluate the balance of learning-relevant and learning-irrelevant activities that are induced by complex learning environments.

In accord with proponents of situated learning models (Cognition and Technology Group at Vanderbilt, 1996), we make a plea for more attention to be given to developing appropriate means of instructional support for situated learning environments. In principle, this support can take various forms, such as scaffolding provided by a teacher or a more capable peer, or provision of a strategy for dealing with an environment. The necessary support probably depends on the specific kind of learning environment, the kind of knowledge domain and the prior knowledge of the learners. However, we lack specific knowledge on the type of support that is helpful for particular learners, goals, etc. Thus, research is necessary that aims to establish the conditions under which particular types of learner support are essential for complex open learning to be effective.

Moreover, measures have to be taken to motivate students for complex open learning. Of course, the anchor problem has to be carefully designed so that (most) students will regard it as interesting. This is not, however, sufficient. There are many more aspects that have to be considered. One major obstacle in implementing effective situated forms of learning is the lack of fit between assessment demands and complex learning arrangements (Mandl, Gruber & Renkl, 1993). If passing exams primarily requires factual knowledge, why should students be motivated to engage in complex forms of learning? Facts can best be learned by other methods.

We strongly believe that in order to motivate learners to actively engage in open learning environments it is necessary for the forms of assessment to be appropriate for the types of skills and knowledge that are fostered by situated forms of learning (e.g. application of theoretical concepts to solving problems). The main point we want to make is that although using complex anchor problems for learning may have some motivating potential there are many more important factors that determine the students' engagement in instructional settings. Therefore, it is not adequate to regard complex learning anchors as automatically engagement-inducing. The larger context, including the assessment context, has to be taken into account when learning environments are designed that aim to induce fruitful learning activities.

In complex learning environments, students are usually expected to learn complex things. In order to do so, much cognitive capacity is required and, as a consequence, learning processes are sensitive to distractions by task-irrelevant demands (Sweller, 1994). Hence, procedures for reducing learning-irrelevant requirements are necessary. One possibility is to provide instructional support, as exemplified in our study referred to earlier using the jeans' manufacturing simulation.

Another important aspect is not to regard learning by problem-solving as the ideal and only way of learning applicable knowledge. For instance, learning by example can be used to reduce cognitive load; it is not only an effective way of learning, but also a learning mode which is preferred by some learners (VanLehn, 1996) and is consistent with basic ideas of the situated learning models. There is some evidence that combining learning by problem-solving and by examples is a rather promising approach (Renkl, 1997; Van Merrienboer & De Croock, 1992). The main point we want to stress is that learning by problem-solving is not always the best method of learning, and that other modes of learning consistent with a situated approach should receive more attention.

Situated learning approaches have much to offer, but are too often associated with a dysfunctional euphoria. This generalised enthusiasm needs to be channelled into feasible programmes through research addressing the various issues we have identified here. If this can be done, we may hope that traditional forms of teaching can be effectively supplemented with (and in part even be replaced by) situated instructional units that genuinely help to overcome the inert knowledge problem.

8
Engaging with Organisational Memory

Antonio Rizzo, Patrizia Marti, Vito Veneziano and Sebastiano Bagnara

1 Organisational Learning and Memory

How do organisations use their experience to change their knowledge, routines and decision processes? Organisations routinely forget what they have done in the past and why they have done it. Often they have the information they need, but they do not know they have it, or, knowing they have it, they can't find it. A person with a memory like the average organisation would be thought to be suffering from a neurological disorder (Ackerman, 1994; Conklin, 1996).

Traditional ways to capture knowledge are: corporate forms, organisational routines (Cohen, 1991), checklists, filing systems, manuals, practices, principles, stories, etc. Most of these repositories of knowledge are the result of a design process that involves off-line representation of the activity and its manipulation in order to plan an intervention on the system. But a large amount of knowledge is in the organisational culture and in the practice of the workers (Levitt & March, 1988; Walsh & Ungson, 1991). This is especially true for modern organisations where the dominant asset is knowledge.

Recently there have been several attempts to capture the knowledge produced within organisations by their workers. The idea of an organisational memory has been addressed by some real or prototypical systems (e.g. Ackerman & Malone, 1990; Owen, 1986; Slater, 1993). These systems try to exploit the potentiality of the information technology that characterises the work environments of modern organisations so as to capture and distribute the knowledge produced during the workers' activity.

It is interesting to note that members of an organisation who work primarily at computer terminals are deprived of distributed cognition as it occurs in many other domains, where people work with tools in a shared physical space (e.g. a workshop). This makes it harder for individuals to pick up cues towards useful tools and their utilisation, and makes it difficult to share knowledge (Maltzahn & Vollmar, 1994). Indeed, co-operation and peer teaching often occur despite the existing computer systems rather than because of them. The systems are mostly designed to support an individual user, and serve to hamper traditional modalities of knowledge-sharing within a community of practice: it is almost impossible to learn a computer based activity by just looking at what the others are doing, and it is hard to find traces of other people's process activity.

2 A Specific Problem: the RAI Database

Especially in database design, most development to date has focused on how to make the user of a database feel that they are the sole user, in other words to make the other users completely invisible (Twidale, 1995). Only recently, with the rapid development of the World Wide Web and the related problem of digital libraries, there is some consistent effort to rethink the nature of databases and their use within a community of users. But the technology for information retrieval for large collections has remained basically unchanged for 30 years. A technology that runs on commercial systems designed in the 1960s and 1970s is still serving millions of Web users today (Schatz & Chen, 1996)

We are currently facing the design of a prototype multimedia database for RAI, the Italian state television company (Marti et al., 1997). RAI possesses a huge archive of archives, each one containing material produced and/or broadcast by the RAI channels. Over the years, the original archive has been split into smaller specialised archives in the attempt to ease the management, documentation and retrieval of data. Furthermore, many small personal archives have been built up by users (mainly journalists) who needed a flexible tool tailored to their specific needs. Today the management of all these different archives is not trivial, mainly because the tendency to create personal archives in order to overcome the limitation of the central database is increasing.

Documents in the existing RAI database are stored according to fixed documentation rules: (i) an analytical description containing information like title, date of production, date of transmission, credits, kind of document, e.g. comedy, drama, news, advertisement or entertainment; (ii) a synthetic description of the contents, typically a few lines reporting the plot of a story, or the names of the characters of an interview, or the description of foreground photographic sequences; (iii) keywords associated with the documents, aimed at easing and speeding up access.

The users of the RAI databases are very heterogeneous, ranging from directors and journalists to students working on their theses and people who perhaps appeared in a particular television programme. The users who have direct access to the database, typically RAI personnel, spend a lot of time searching for material that might satisfy their goals. During this process, they often screen a lot of documents that might, in some way, be useful for other purposes or that are to some extent related to what they are looking for. But the lack of tools to organise the results of a search and to categorise the data according to their preferences and needs prevents efficient exploitation of the database. Furthermore, the generality of the documentation data causes the users to waste a lot of time in screening videotapes, often only to conclude that the material does not fit their needs.

The users who do not have direct access to the database are supported by research assistants; RAI personnel who receive and interpret their requests, making searches on their behalf. The research assistants are the historical memory of the database; they remember what is in the database, and do their best to satisfy the requests of the users. If the research assistants who currently work in RAI were to be replaced by other research assistants, even if they were experts in database management, the value of the database itself would change completely.

The limitations of the current system are also evident if the users are asked to re-perform a recent search. They cannot go directly to the results. Rather, they have to re-apply the search strategies that they tried the first time, spending the same amount of time or even more, since they often do not remember the successful search strategy. Furthermore, the current system does not capture the data context. One of the problems is that the value of the data in the database changes both over time (as in the case of a character who later becomes famous), and depending on the particular context of use (e.g. a repertory sequence that acquires a different meaning according to the sequences to which it is associated over the time).

We aim to make the system readware, i.e. to design a system that can dynamically capture the way in which it is used: a system able to maintain the knowledge associated with its use. At the same time we do not want to replace some human activities or shift the allocation of tasks from the human to the computer. On the contrary, we would like to support the human activities, especially the ones related to organisational learning — the ones that produce knowledge through the accomplishment of everyday activities.

The general strategy in designing organisational memory (or any other cognitive artefact) is to try to make explicit the knowledge produced within a given working environment in a context-dependent way. This strategy is supported also by experimental evidence which shows how external representations can support human problem solving (Zhang & Norman, 1994) and reasoning (Bauer & Johnson-Laird, 1993). The results of these experimental studies are important since they offer a systematic approach to the design of external representations based on the exploitation of physical, logical and cultural affordances and constraints (see also Norman, 1993). However, these principles, even if of great value, seem applicable mostly to well-defined situations, such as problem solving, deductive reasoning, or in general to situations where it is known in advance how the information should be manipulated.

The strategy typically adopted in designing organisational memory is based on the idea of a knowledge warehouse, where the information management strategies that are appropriate for individuals break down in a shared information repository (Berlin et al., 1993). Even though a lot of effort has been put into designing organisational memory, the current attempts at implementing it fail for a variety of reasons. These include too strong a focus on work product as against process and a lack of tools which make capture and re-use of knowledge transparent (Conklin, 1996). For these reasons we are attempting to support the search process of the operators by supporting the inferential process that occurs in the course of their activity.

Most knowledge acquisition results from inferential processes that occur in the course of attempting to accomplish everyday goals, not as a deliberate exercise. The idea of keeping track of all interaction between the users and the system is not even thinkable; it will produce a meta-problem (the search of the searching activities). Of course there are intermediate solutions but they still focus on product instead of on process.

Thus to face the challenge of the two limitations identified above, we propose the design of an artefact able to capture the searching process, transparently supporting the workers at RAI. This approach should not be considered alternative to the efforts focused on products (e.g. Schatz & Chen, 1996). Rather it is complementary, since it focuses on the human process.

3 Searching a Database as Hypothesis Testing

Searching a large database is mainly a hypothesis-testing activity. The analysis of the activity performed on the users of the RAI database reveals that in order to find where given items are located (or even if they exist at all), the operators make hypotheses about how such items might have been classified. A search based on the associated keywords is the quickest strategy to limit the search space and to test hypotheses.

How can we design tools that support hypothesis testing and that capture efficient strategies? In psychology most of the research on hypothesis testing has been inspired by the work of Wason (1977), who devised simple (or apparently simple) tasks such as the selection task, the THOG task, and the 2-4-6 task (for a review see Garnham & Oakhill, 1994). All these tasks were inspired by the position of Popper, who claimed that scientific knowledge acquisition is better characterised as hypothetico-deductive rather than inductive, and that falsification plays a critical role.

Research using these tasks has produced two main results. Firstly, people have severe difficulties with falsification. Subjects tend to focus on confirming instances, which can only be consistent with their hypotheses, rather than on falsifying instances, which might show that their hypotheses are definitely false. Secondly content and context can mitigate or even eliminate such difficulties (cf. Blaye et al., this volume). Explanations put forward to explain content and context effects include pragmatic reasoning schemas (Cheng & Holyoak, 1985), mental models (Johnson-Laird & Byrne, 1991) and matching bias (Evans, 1989).

Recently Sperber, Cara and Girotto (1995) proposed an explanation of content and context effect in the framework of the relevance theory (Sperber & Wilson, 1986). Sperber et al. claim that their account can explain all the previous results obtained with the selection task. Their results do not provide evidence for or against different theories of human reasoning (e.g. mental models or pragmatic schema). But, they maintain, their results show that "people are nearly-incorrigible *cognitive-optimists*. They take for granted that their spontaneous cognitive processes are highly reliable, and that the output of these processes does not need re-checking" (Sperber et al., 1995, p. 90). If they are right, we are back to the difficulties associated with falsifying hypotheses.

Of the three tasks devised by Wason, the selection and THOG tasks are mainly related to deduction, while the 2-4-6 task involves induction. It is this last task that most adequately represents the processes of exploration and discovery. The results obtained with this task have also provided further evidence of the confirmatory bias of "the failure to eliminate hypotheses in conceptual work" (to use the title of

Wason's paper). The failure to eliminate hypotheses was mainly attributed to the difficulty that people have in understanding the logic of falsification (Wason, 1977).

In a study on a computer-simulated micro-world, Mynatt, Doherty and Tweney (1977) reported that their subjects could use falsifying data once they obtained them. But the group which was instructed to use falsification did not perform better than the control group — both groups were unable to discover the rules governing the microworld. Always adopting instructions to promote a disconfirmatory strategy, but using the original 2-4-6 task, Gorman and colleagues (1984) found that subjects instructed to use falsification performed better than subjects who were instructed to use a confirmatory strategy, but not better than the control group who did not receive any instruction.

To summarise, we have:

- a claim put forward by Popper that falsification is important in hypothesis testing,
- evidence that falsification is not easily adopted by people in hypothesis testing,
- evidence that people understand the logic of falsification,
- evidence that instructions to concentrate on falsifying do not produce a critical difference.

These results are mostly related to experimental conditions where there is a single subject facing the task. Indeed, most of the psychological literature on hypothesis testing concerns reasoning in the head of the individual. This has been reported as one of the main limitations of the psychological literature of reasoning, since it does not allow humans to make explicit one of their more particular features: the externalisation of knowledge and the manipulation of such externalisations (cf. Lave, 1988; Vygotsky, 1929; Zhang & Norman, 1994).

The cognitive activity we would like to support occurs in contexts where people very often share their space and time, and they have at least access to the results of their peers' behaviour. But so far most of the traces left by other people are not in the digital world, but in the physical world in the form of notes, lists, messages, advice, modification of layouts and so on.

To recap, all workers are constantly involved in learning and reasoning processes, and it is exactly these activities that potentially promote organisational learning. Such learning can be seen to emerge from the communication process among the workers in a community, often triggered by specific and situated problems (Hutchins, 1995). Organisational learning is critically dependent on the communication flow among the workers and it is mediated by the traces left by the worker's activity, both in synchronous and asynchronous ways.

It is just on these existing properties of a community of practice (cf. Resnick, Levine & Teasley, 1991) that we would like to act, trying to support the existing cognitive process. Or, at least, we would try to avoid hampering them, as sometimes occurs as a result of the technology-centred approach to system design. This approach tries to offer a system so user-proof that the business can cope not only with fewer workers but also with less skilled workers (Adler & Winograd, 1992).

Thus, together with the contextual inquiry and task analysis performed on the field we think it is necessary to investigate more thoroughly the cognitive processes we would like to support. To this end, we decided to investigate hypothesis testing processes adopting an experimental paradigm that would represent in an appropriate micro-world some of the main features of the working context.

We investigated the behaviour of subjects playing a card game: the New Eleusis. We have chosen the New Eleusis for two main reasons. Firstly it is a real game that is played for fun and produces engagement (there is in fact a society for the Eleusis players). Secondly, it was explicitly designed to model the scientific search for truth in a community of researchers, and thus contains all the elements of a distributed activity. Furthermore it has already been adopted as an experimental task with minor adjustment (Gorman et al., 1984). In particular Gorman et al. showed that subjects instructed to adopted a falsificatory strategy, performed better that subjects instructed to adopt a confirmatory strategy, as for the 2-4-6 task.

In the following section we report on the first of a set of experiments currently being carried out in our laboratory. In this experiment we adopted just one of the rules used in Gorman et al. (1984); and we investigated whether the confirmatory bias and the *cognitive-optimist* attitude would have as deep an impact in distributed cognition settings as it has been shown to have on isolated minds.

4 Experimental Paradigm: Eleusis

In the Eleusis version played by our subjects the experimenter acts as God or as Nature and invents a sequence rule, not revealing it to the player(s). The rule concerns the allowable sequence in which the playing cards may be laid out on the table. To start the game the experimenter places an uncovered card on the desk and gives the remaining card deck (103 cards) to the first (or only) player. The player chooses the card that he wants from the deck and tries to place it on the right of the first card. If the card respects the sequence rule, the experimenter does not say anything and the card remains on the right of the previous one. If the card does not respect the sequence rule the experimenter says No (Nature can only say No!) and the card is placed below the previous one.

The game continues until the player or players are sure that they have found the sequence rule or until the cards run out. Thus, correct cards continue in a straight line and incorrect ones go off at right angles. The cards on the table thus correspond to a sequence of tested hypotheses. Figure 1 shows a typical round of Eleusis at an early stage, drawn from one of the test sessions.

4.1 Experiment

The experiment was designed to compare single players with players who, even though playing by themselves and not being allowed to talk among themselves, nonetheless share the game spatially and temporally with two other players. To allow this comparison, two conditions were devised: the Solo game and the Triplet game. In the Solo condition the subject plays the game by himself or herself, facing only the

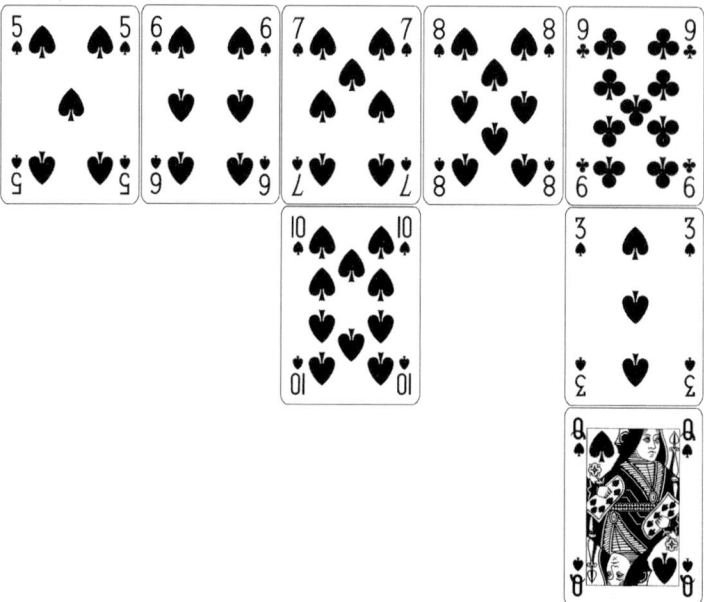

Figure 1: Eleusis at an early stage. On the main line the cards that fit the rule, on the vertical lines the cards that do not fit the rule.

experimenter who acts as Nature. In the Triplet game each one of three players in turn receives the card deck, playing a card and then passing the deck to the next player.

In the Solo condition the player should have full control of the sequence of cards that will be displayed and can explore hypotheses without interference. By contrast, in the Triplet game the player(s) have much less control of the sequence of cards since it is necessary to wait for the two cards played by the other two players before playing one's own card. At the same time, players in the Triplet condition are exposed, in a sense, to the hypotheses of the other players as reflected in their choices, and those choices may offer falsifications of the players' own hypotheses.

The study was conducted with 78 male and female students of the Communication Science course of the University of Siena and with colleagues at the Biomedical Centre of the CNR in Rome. They were randomly assigned to one of two conditions, namely Triplet ($n = 18$) and Solo ($n = 24$). In both conditions, the participants received the same written instructions in which the game was explained. They were given pen and paper with which they had to write their guesses about the rule. They could write as many guesses as they liked, but they were asked not to make any corrections to their earlier guesses, instead writing each new guess in sequential order. Participants were told that they would not receive feedback about their guesses until the end of the experiment. The experiment ended when the single participant (in the Solo game) or all the participants (in the Triplet) declared that they felt sure about their guesses, or when the card deck was finished. In the Triplet, the participants kept playing even if they felt sure about the rule until all the players declared their final guess. The rule adopted was: 'a difference of 0, 1, or 2 must separate adjacent cards'.

The major dependent measure was whether the group (the three persons in the Triplet game) or the single player (in the Solo game) solved or failed to solve the rule. Additional dependent variables were the time in minutes spent to complete the game, the number of played cards that did not fit the rule, the number of played cards that did fit the rule, and the number of cards that were left unplayed at the completion of the game.

The major dependent measure (solution of the rule) was analysed with a 2 (Condition: Solo, Triplet) × 2 (Solution: Yes, No) contingency table. The further dependent variables (Time, Fitting Cards, Not Fitting Cards, Not Played Cards) were each one analysed with a one way analysis of variance (ANOVA).

Participants in the Triplet game performed significantly better than in the Solo game ($\chi^2 = 9.84$, d.f. = 1, $p < 0.002$). All the participants in the Triplet were able to discover the rule (Triplet condition: Solved 18; Not Solved 0), whereas more than 40% of the participants in the Solo game failed to find the correct solution (Solo condition: Solved 14; Not Solved 10).

This difference is also reflected in the way in which the cards were played in the two conditions. Participants in the Triplet game played more cards that did not fit the rule than subjects in the Solo game (Triplet 34.4 vs. Solo 25.4; $F = 6.26$, d.f. = 1, $p < 0.02$) and fewer cards that did fit the rule (Triplet 38.5 vs. Solo 49.4; $F = 5.14$, d.f. = 1, $p < 0.03$). There was no difference between the two conditions for the numbers of cards not played ($F = 0.6$,) or for the time to complete the task ($F = 0.7$).

Moreover there are also differences between the conditions even if we consider only the participants who correctly identified the rule ($n = 18$ in the Triplet, and $n = 14$ in the Solo). Participants in the Triplet game again played more cards that did not fit the rule than participants in the Solo game (Triplet mean 34.4 vs. Solo 18; $F = 20.19$, d.f. = 1, p < 0.0001) but they did not play fewer cards that fit the rule ($F = 0.5$). There was also a difference between the two conditions for the numbers of cards not played (Triplet 31 vs. Solo 50; $F = 8.68$, d.f. = 1, $p < 0.006$) and a trend toward a difference for the time to complete the task (Triplet 37.4 vs. Solo 39.7; $F = 3.44$, d.f. = 1, $p < 0.073$).

On the whole, these results suggest that falsifying cards were very important to solve the rule. In the Triplet game, where all the participants identified the rule, there were more cards that did not fit the rule than in the Solo game. Correct (confirming) cards might seem to have an opposite role, since in the Solo game there were significantly more correct cards played. But this is not necessarily true; there was no conditions difference in the number of correct cards played, when we take into account only the participants who discovered the rule.

The critical difference between the two conditions seems to be the number of falsifying cards played beyond a given number of correct cards. It seems that the participants in the Solo game who did not solve the rule, even after playing more cards and taking more time, played mostly confirming cards. At the same time those participants who did identify the rule in the Solo game were able to discover the rule even though they played fewer cards overall and also fewer discomfirming cards than the participants in the Triplet condition. This suggests that their starting hypotheses were already in the right direction.

Participants in the Triplet game had the chance to observe the cards played by their co-players (both accepted and rejected cards) and this could have produced evidence that violated their own hypotheses. But this, even if helping the subjects in the Triplet game, cannot explain the higher number of falsifying cards played in this condition, since each player might still be expected to try to confirm his or her own hypothesis. One possible explanation of the higher number of incorrect cards played in the Triplet condition is that the participants, once exposed to discomfirming evidence produced without their direct involvement, were in a good position to understand and exploit the logic of falsification. Thus, adoption of a successful falsification strategy seems to be facilitated by playing the game in a distributed context. Each player places some constraints upon the others in a reciprocal way, thus moving all toward the correct solution of the rule.

5 Conclusion

Knowledge is the most useful asset of modern organisations. Knowledge assets belong mostly to people and become the organisation's asset only through their application, capture and re-use. Organisational memories are one of the attempts to provide an answer to the increasing requests for support for knowledge sharing within an organisation.

The dominant strategy in devising organisational memories is to capture the products of workers' activity, to make them explicit and to store them in a common repository. However, even if this strategy seems sensible it has two main limitations when applied to the design of organisational memories for databases. First, focusing on workers' products it leads to a meta-level of data, a shift of the problem. Second, information management strategies that are appropriate for individuals break down in a shared information repository. Faced with the specific problem of designing the organisational memory for the users of a large multimedia database, we are considering not only the products but also the processes associated with the production and use of knowledge.

The activity of the users of a database involves mainly hypothesis testing. The most relevant contributions to hypothesis testing produced in psychology are based on the three tasks proposed by Wason in the 1960s. Even though a massive amount of research has been conducted on these research paradigms, the results are not yet adequate to provide strong heuristic principles for devising tools to support hypothesis testing. The most important finding is still that people have difficulty falsifying their hypotheses (Sperber et al., 1995; Wason, 1977). However, we have here provided some preliminary evidence that this limitation may be exacerbated by the usual experimental procedure of working with a solo participant. In the more common situation of participants co-acting in shared space and time (a distributed cognition context), results indicated that players understood and exploited the logic of falsification.

The design of the organisational memory for the users of a multimedia database necessitates the study of the processes associated with the production and use of knowledge in a community of practice. The concept of vicarious learning refers to the fact that learners and reasoners are typically working in an environment populated by other learners and reasoners. In other words, the world in which intelligence is exercised is a world which has been shaped, and is continuously shaped in various way by the intelligent activities of others. In this environment new communications and new ways of facing problems evolve naturally from the synchronous and asynchronous interactions of people (Hutchins, 1995; Light et al., 1994).

Hypothesis testing in this environment of distributed cognition is almost more relevant to everyday thinking and reasoning than the hypothesis testing abilities of a single, isolated subject. Yet almost all that cognitive psychology has to offer in this field relates to the individual, isolated subject. The experiment reported here shows that results cannot simply be generalised from the individual level to the level of the working group, even where, as in our experiment, the group works without conferring. The design issue raised for the RAI database concerns the need for a more ecologically sensitive approach to studying cognitive processes, which respects the nature of distributed activity. This is of paramount importance if psychology is to contribute more directly to the evolution of cognitive artefacts.

9
The Relevance of Relevance in Children's Cognition

Agnes Blaye, Edith Ackermann and Paul Light

1 Introduction

Most developmental theories describe cognitive growth as a move away from 'intuitive' towards 'rational' thinking, or from everyday cognition towards scientific reasoning. Whether grounded in action as in Piaget's theory, or mediated through language as in Vygotsky's theory, constructivist models of human intelligence are essentially science-centred and logic-orientated. The lengthy path towards higher forms of reasoning or 'formal operational thought' is seen as proceeding from local to general, from context-bound to context-free, from externally-supported (or 'embodied') to internally-driven (or 'mentalised'). Accordingly, children's cognitive achievements are portrayed in terms of an increasing ability to emerge from here-and-now contingencies (characteristic of practical intelligence), an increasing ability to extract knowledge from its substrate (i.e. from contexts of use and personal goals), and an increasing ability to act mentally on virtual worlds, carrying out operations in the head instead of carrying them out externally (Ackermann, 1991).

Piaget and Vygotsky attributed varying roles to direct and mediated experience in the process of moving beyond the concrete. Yet, both shared the commonly held view among developmentalists that higher forms of reasoning emerge from people's ability to separate *what is known* from *how it came to be known* and *where it may best serve*. Consistent with this view, research methods themselves have become 'disembodied'. Experiments designed to study people's reasoning or logical capabilities are carefully stripped of the messy dynamics inherent in actual situations of use, and children's achievements are gauged by their comformity to logical norms or logico-mathematical canons at the expense of their pragmatic or functional relevance.

In reclaiming the deeply grounded, experience-based, and adaptive nature of human cognition, the situated approach to human learning challenges such classical views on cognitive development. Children do not think in a vacuum. Instead, they develop ever more sophisticated strategies as a means to handle complex situations. Children, like adults, progressively shape and sharpen their thinking so that it supports their purposes and augments their potentials. Current research on cognitive development addresses this idea by showing that children's reasoning and decision-making strategies are, indeed, deeply grounded in specific contexts of use, and thus cannot be gauged according to logical criteria alone.

Two broad empirical approaches can be used to study children's embodied thinking. On the one hand we have ecological studies which preserve the complexity and naturalness of the settings which people inhabit, on the other, variations on classic laboratory tasks can be used to highlight the discrepancy between logical and pragmatic constraints. Following this second approach, three aspects of children's growing reasoning capabilities will be considered in this chapter: conditional reasoning, categorisation, and the comprehension of negative sentences. In each of these fields, recent work has pointed to children's thinking being more flexible and better adapted than would be supposed on the basis of a Piagetian account of the development of operational thinking.

Studies on pragmatic development in particular strongly suggest that children's responses to many classic experiments, far from being based upon logic alone, are heavily informed by concerns of appropriateness and efficacy within specific contexts of use (Ninio & Snow, 1996; Siegal, 1991). Our purpose, then, is to revisit some of the 'mainstream topics' of cognitive development, with the idea in mind that a pragmatically grounded 'rationality of purpose' might be more characteristic of children's thinking than 'rationality of process'.

As mentioned earlier, studies on human reasoning have generally been conducted under highly controlled experimental conditions, in an attempt to eliminate the messiness inherent in practical everyday situations. Wason's Selection Task and Sentence Verification Task and Piaget's *épreuves operatoires* (and more specifically his task on class inclusion) are typical examples of such an attempt. In what follows, we discuss the psychological implications of using these paradigmatic tasks as means to study children's conditional reasoning, their understanding of class logic, and their processing of negatives. First, we shall briefly introduce the tasks themselves.

Wason's Selection Task. If a conditional rule is expressed conventionally as 'If p then q', evaluation of the truth or falsity of the rule must logically focus upon the conjunction of 'p' and 'not q', since only this conjunction can falsify the rule. The selection task invites participants to specify which of four possible cases (conjunctions of 'p' or 'not p' with 'q' or 'not q') need to be investigated in order to establish the truth or falsity of a conditional rule.

For example, suppose that four cards showing respectively a 9, a 4, a D and an E, are laid upon the table, it being known that each card has a number on one side and a letter on the other. The participant's task is to establish the truth or falsity of a conditional rule stating that "If a card has an even number on one side then it has a vowel on the other side" by turning over the minimum necessary number of cards. The correct response is to turn over cards corresponding to 'p' (i.e. the even number, 4) and 'not q' (i.e. the consonant, D), since only with these two cards could the discovery of what is on the reverse side falsify the rule.

Wason (1966) established that this task is surprisingly difficult, even for highly educated adults. However, it soon became apparent that logically irrelevant aspects of task presentation (namely the concrete contents and context of the rule) had significant effects on the difficulty of the task. For example, Johnson-Laird, Legrenzi and Legrenzi (1972) showed that if the rule was framed as a postal rule saying that if an envelope is sealed it needs a certain value stamp on it, most adults could correctly select which envelopes needed to be checked (for a review of research on 'thematic' versions of the selection task, see Evans, Newstead & Byrne 1993). Explanations for such effects range from general inference rules through heuristics and mental models to domain-sensitive rules of various kinds.

Piaget's Class Inclusion Task. According to Piaget (Inhelder & Piaget, 1964), when children are presented with, for example, a bunch of flowers (for instance eight tulips and four daisies) and asked whether there are "more tulips or more flowers", most children below 8 years old mistakenly reply "more tulips". Piaget considered this error as an indication of a lack of mastery of classificatory logic, since the superordinate class logically cannot contain fewer exemplars than one of its subordinate classes. Older children's correct responses and justifications were seen as an indicator of mastery of the relevant logical structure, namely, additive class grouping.

Current research challenges Piaget's assumption by showing that success in response to the question of quantification of inclusion is neither a sufficient nor necessary condition to an understanding of at least some aspects of class logic (Markman, 1978; Bideaud & Lautrey, 1983; Houdé, 1990). Moreover, a few studies dealing with conversational aspects of the interview during this task revealed that improving the setting and reducing violations of conversational rules can reveal success at much younger ages than expected by Piaget (Politzer, 1993; Siegal, 1991).

Wason's Sentence Verification Task. Just as in the field of conditional reasoning, Wason (1961) played an important early part in shaping the direction of empirical research on people's processing of negative propositions. He pioneered the use of the Sentence Verification Task, in which subjects are shown a series of images representing familiar scenes. Scenes typically involve an agent (man, woman, child) doing something (like painting, ironing, washing, reading) to some object/s (door, book, window). Each image is accompanied by a sentence which describes elements of the scene, using a negative form (e.g. "The woman is not painting the door"). Respondents, in this case adults, are required to make a truth value judgement based upon sentence–picture comparisons. Most theoretical debates have focused on the nature of the information processing operations carried out by subjects in solving this task (for a review see Evans, 1982).

Researchers have long ignored the fact that Wason himself had explicitly admitted the low external validity of sentence-verification studies for the study of negation. Here again, current research, mainly from linguists, has clearly established that a purely syntactic analysis reveals only a small part of what is at stake in people's understanding of negation (Ducrot, 1980; Moeschler, 1990). Many examples show a discrepancy between an interpretation of negatives based on formal logic and one based on a 'logic of use'. A sentence like "Everything that sparkles is not gold", would have to be interpreted by a logician as meaning "nothing that sparkles is gold" whereas people clearly use it to mean: "some things that sparkle are not gold". Moeschler referred to Sperber and Wilson's theory of relevance (1986) to offer a model of uses of negatives considered as specific forms of speech acts (Austin, 1962; Searle, 1969). This work remains undeveloped, however, and very few empirical studies, at least on children, have been reported.

Having introduced these three tasks, we shall now turn to a more detailed analysis of examples of recent psychological research in the three domains for which these tasks have become paradigmatic.

2 From Truth Value to Relevance Judgements: Deductive Reasoning in Children

Accounts of the development of deductive reasoning have usually taken their starting points from a Piagetian account of cognitive development (e.g. Ward & Overton, 1990). According to Piaget, reasoning development culminates in the achievement of formal operational thinking only in early adolescence (Inhelder & Piaget, 1958). Thus, if selection task performance is analysed in terms of abstract rules corresponding to formal logic we should expect a clear distinction between the performances of children and those of adults.

The main difficulty from a Piagetian point of view, of course, is the poor performance even of adults on abstract versions of the task. Ward and Overton (1990) sought to deal with this difficulty by proposing that contextual factors such as the precise nature of the conditional rule moderate the expression of the underlying competence even after it becomes available. This still suggests that the performance of pre-adolescent or 'pre-formal' children should differ markedly from that of adults. Indeed, Ward and Overton did obtain very poor performance from eleven year olds, even with thematic versions of the selection task on which older participants performed well. However, as Girotto and Light (1992) pointed out, they used a mixture of different rule types on a within-subject basis, which may have led to some complex carry-over effects.

Some of the most reliably easy thematic versions of the selection task in the adult research literature are 'deontic' versions such as the postal rule mentioned above. Here, the rule has a modal form (stating a conditional permission or obligation, for example). The truth or falsity of the rule is treated as axiomatic and the participant's task is to look for evidence of violation. For example the rule might state that "if somebody is drinking beer, then they must be over 18". Cheng and Holyoak (1985) suggested that rules with the general form "If an action is to be taken then a precondition must be satisfied" are typically solved by the application of a 'pragmatic schema' embodying rules for action or inference which are specific to this type of social regulation. It is these socially grounded rules, rather than the more abstract rules associated with formal deductive logic, that are held to support correct performance on appropriate thematic versions of the selection task.

Pragmatic schemas are envisaged as being fashioned out of children's experience of particular forms of social regulation. There is every reason to suppose that children are exposed to conditional permission and obligation rules from an early age (Dunn, 1988). Moreover, recent work by Harris and Nunez (1996) has shown that children as young as $3\frac{1}{2}$ can accurately identify transgressors of a conditional permission rule. Thus we might expect from this point of view that children, certainly by elementary school age, would show much the same responses to deontic versions of the selection task as adults.

Using a radically simplified (reduced array) version of the selection task, we have been able to show in our own research that children even as young as six can indeed do well when the rule corresponded to a conditional permission (Light et al., 1989). Even with the full selection task, we have been able to obtain competent performance from children as young as eight with such rules (Girotto, Light & Colbourn, 1988; Girotto et al., 1989; Frydman, Light & Alegria, 1999). Evans, Newstead and Byrne comment on how strikingly these findings testify to the context-dependence of reasoning both in children and in adults, showing as they do that "a task which defeats most university students in its abstract form is comparatively easy for a ten year old in a permission context" (1993, p.128).

Thus, for children as well as for adults, correct performance on deontic versions of the selection task appears to depend upon the interpretability of the rules as conditional permissions or obligations. Where such interpretation is available (whether from direct experience, explicit justification or plausible hypothesis) selections typically correspond to those suggested by a formal logical analysis, though of course, according to this account, they do not in fact result from such an analysis.

However, domain specific pragmatic schemas are not the only available option in explaining these findings. A more general interpretation of the pragmatics of selection task performance is that, in all forms of the task, participants may be relying very largely upon unreflective intuitions of relevance (Evans, 1989). Sperber, Cara & Girotto (1995) argue that the selection task is, as its name suggests, a test of selection — namely selection of potentially relevant evidence. Looked at from this point of view, Ward and Overton's distinction between the 'competence' and 'performance' aspects of the task (i.e. between the intrinsic logical demands of the task on the one hand and the 'moderating' effects of content and context on the other) dissolves. Interpreting the task is part and parcel of performing it. Whether in children or adults, Sperber and colleagues argue, rationality involves the efficient allocation of cognitive resources, and considerations of potential relevance shape this allocation.

The focus of attention thus falls not on particular types of 'pragmatic schema', but on more general pragmatic processes of comprehension which involve determining where relevance lies. Sperber, Cara and Girotto argue that the way to produce 'good' performance on a selection task is to construct the rule in such a way that it is readily interpretable as a denial of the possibility of 'p and not q' cases. It is precisely this implicit denial that makes these cases relevant, and thus supports correct selection. The deontic versions of the task achieve this by having the 'p and not q' case represent a rule violation which the participant is cued to look out for.

From a relevance theory point of view this is only as a particular instance of a more general process. By appropriate 'relevance management', easy versions of the selection task can be produced in any cognitive domain. For example, with a rule such as "If a woman has a child then she has had sex", the 'p and not q' combination (namely *women-who-have-had-children-but-have-not-had-sex*) can be highlighted by making it the object of a journalistic scoop. Here, even without any deontic element to the rule, adult performance is indeed good.

Neither explanation in terms of domain specific pragmatic schemas nor explanation in terms of a more general relevance theory offers any clear developmental story. As far as pragmatic schemas are concerned, the obvious developmental scenario would be that schemas of permission, obligation, etc. are gradually abstracted from concrete instances of experienced social regulation. A less obvious and more interesting story might posit a developmental link between the understanding of deontic necessity and the understanding of logical necessity. Harris and Nunez (1996) for example, explore the idea that 'must' in its modal sense may be a developmental antecedent of 'must' in its logical sense, and point to linguistic evidence that such terms make their first appearance in deontic contexts.

3 More Tulips or More Flowers? Categories, Contexts and Conversational Conventions.

Categorisation is generally considered as a primary way in which people make sense of their world and themselves. Research on the development of categories has long been oriented towards the formation of logical classes defined by necessary and sufficient properties (Inhelder & Piaget, 1964), or scientific concepts (Vygotsky, 1986). Bruner, Goodnow and Austin (1966) likewise characterised categories in terms of a system of hierarchical classification organised in terms of inclusion relationships.

Following Piaget's perspective, the class inclusion task has been considered for decades as the key test of access to the logic of class hierarchies. However, several studies have demonstrated that many children between 8 and 11 who succeed in this task also believe that one could, in principle, get more tulips than flowers (by adding tulips or taking away flowers; Markman, 1978). Bideaud and Lautrey (1983) proposed a distinction between empirical inclusion, as assessed by the classical class inclusion task, and logical inclusion which is acquired a few years later and assessed by the revised versions of the task proposed by Markman. Success in the Piagetian task can be achieved by considering the two classes (subordinate and super-ordinate) as disjoint collections, thus leading children to admit that one can change the extension of one of them without modifying the other. While they challenge Piaget's account, none of these studies question that logical reasoning is the endpoint of development.

These studies suggest that Piaget's test can result in false positives. Other researchers, in revisiting the task from a conversational pragmatic point of view, show that a move towards reducing referential ambiguity in the standard task can elicit correct answers in younger subjects than Piaget supposed (Politzer, 1993; Siegal, 1991). In natural language 'flowers' can both designate the super-ordinate class and any of the subclasses. Moreover, when such a word is used in context with the specific name of one of the subclasses, it is taken to refer to the complementary subclass, and not, as in Piaget's situation, to the super-ordinate class.

Politzer's analysis of children's performance in the standard task suggests that it results from the respondent's judgement of "what understanding the experimenter wants me to demonstrate", in other words, what could be a relevant answer to such a surprising question. Why then do children less than age 8 fail in this task while older ones succeed? According to Politzer, for younger subjects, demonstrating their ability to count and then compare the cardinals of the two subsets can seem appropriate. For older subjects, this is far too easy, so they judge that it cannot be the relevant answer.

In effect, they are saying: "You're asking me a trick question, but I shan't fall into the trap!".

There is, however, more to the construction of categorical knowledge than solving the class inclusion tasks or any of its revised versions. Alternative accounts exist which suggest that categorical knowledge could be organised on 'non-logical' bases, such as graded categories (Rosch et al., 1976) or schematic-based categories (Mandler, 1983; Nelson, 1986). Such research seems to demonstrate that even if people sometimes do categorise on a strictly logical basis, this is only a small part of the story. For instance, both children and adults seem to consider that members of a category are not all equivalent to one another, and that some are 'better instances' of the category than others (the so called 'gradient of prototypicality').

Research on schematically organised categories, stimulated by Mandler (1983) has been developed in particular by Nelson (1986). In Nelson's model, children's categories are envisaged as emerging from event representations corresponding to routines in their everyday life. Such schematic representations (also called scripts; Schank & Abelson, 1977) specify the range of potential elements in a category. Members of a same category, in such a sense, do not necessarily share perceptual similarity. For example, 'high chair', 'milk' and 'bib' are all linked by a common script, though they do not in any obvious sense 'resemble' one another. Rather, they are organised in spatial and/or temporal contiguity relationships.

Later on, Nelson argues, children reason on 'slot-filler' categories, grouping together items which are mutually substitutable in the same slot of a script (e.g. various types of baby food). These categories are seen as still highly contextually bound. Further development leading towards a taxonomic organisation of categories results from language development, which makes possible a progressive de-contextualisation from specific scripts. Thus, although Nelson admits the possible co-existence of these different modes of grouping, her model remains hierarchical. Early forms of categorisation remain described as experience-based and contextually bound, and the endpoint of development coincides with Piagetian logical classification.

While Rosch's and Nelson's perspectives have opened new avenues by highlighting different forms of categorisation, we would like to suggest that further steps can be taken, which could offer an image of development in which these different forms co-exist in subjects very early on, and remain co-present through adulthood.

In one of our studies (Blaye & Bernard-Peyron, 1996), 90 children at three age levels (mean ages: 5yr. 8 mo, 6yr. 11 mo, and 9yr 8 mo) were confronted with a free sorting task involving 18 drawings which could be organised either taxonomically (e.g. 'people', 'animals', 'tools') or schematically (e.g. 'the forest', 'the circus'). Both types of groupings were observed at all age levels with a non-significant *increase* of schematic groupings in the older children. No developmental hierarchy could be evidenced in terms of type of groupings. However, what appeared to develop was children's ability to switch between different modes of sorting on request.

Other recent studies confirm the idea that the type of grouping which is privileged depends largely upon factors such as instructions (Deak & Bauer, 1996), or the kinds of materials involved (Markman, Cox & Machida, 1981). In a recent study

(Lecacheur, Desprels-Fraysse & Blaye, 1996), we investigated the role of this latter factor in tests with 4 yr. 9 mo and 5 yr. 8 mo year-old children. Forty-three small plastic toys representing pieces of furniture suitable for a kitchen, an adults' bedroom and a baby's bedroom were introduced to the children. Independently of their suitability for different rooms, these objects had been divided into three subsets painted respectively in red, yellow and blue. Half of the children were asked to distribute the 'objects' in the rooms of a cardboard house and the other half in three transparent plastic bags each with a coloured sticker (red, yellow and blue). Then all the children were asked to do another sorting on a small piece of cloth (a quarter of A4 size).

All subjects produced schematic grouping in the 'rooms' condition while the majority of subjects in the 'bags' condition achieved a sorting by colour, with a significant increase in this proportion among older children. Intra-individual shifts between the first and second phase again clearly highlight contextual sensitivity of the sorting mode. Children who used a colour grouping in the first phase generally maintained their mode of sorting when confronted with the small piece of cloth, while those who had used the schematic mode produced totally disorganised collections.

The micro-context induced by particular task instructions and spatial constraints thus appears to make one mode of organisation more accessible than another. Such results have led us to the view that development of categorisation may be less a matter of the acquisition of new sorting modes and more a matter of achieving a progressively higher degree of flexibility to move from one mode to the other as a function of the task context and the experimenter's requirements.

Although not developmental in itself, the study to be presented in the next section concerning children's understanding of negatives will also offer evidence that children's processing of information does not typically conform to logical requirements but does nonetheless reflect a good sense of appropriateness and adjustment to pragmatic aspects of the situation.

4 "She Ain't no Trouble": Form and Function of Negations

From a logical point of view, the function of a negative is to reverse the truth value of a proposition: If 'p' is true then 'not p' is false. As Evans (1982, p. 25) puts it, this property of negation is fundamental to any system of formal logic, and hence of considerable relevance to the study of reasoning. However, what is at stake in any 'real life' situation that involves negations or disclaimers often seems to have little to do with truth value judgements. Double negatives like the one in our section title are unlikely to be taken as positives, though formally they should be.

More generally, if someone claims that 'A is not X', a conversational partner will not be likely to respond in terms of any kind of mental truth table. What they are likely to do, instead, is (i) gauge the pertinence of the assertion made ("what is the speaker refuting?"); (ii) consider the beliefs and intentions of the locutor ("why is he refuting X?"); and (iii) agree or disagree with the locutor with regard to a referred-to state of affairs in the world ("what do I see as being the case?").

From a pragmatic viewpoint, it is not relevant for the purpose of communication to make a judgement on the validity of a statement without also signalling agreement or disagreement with the locutor. Agreement with a disclaimer (e.g.. 'X is not blue') could be signalled by a reply such as "Yes [i.e. you are right], it's not [i.e. not blue]". More colloquially, however, one might reply: "No, it's not blue". Similarly, disagreement with the locutor could take the form: "No [you are wrong], it is blue", or, more colloquially "Yes, it is blue".

Research on the understanding of negatives by children has shown that although children as young as three have some understanding, there remain long-standing difficulties due to the interpretation of what is negated, in other words, on the locus of negation (e.g. Jakubovicz, 1971; Rumain, 1988). However, empirical research is sparse on children's development of the concept of negation (Marti, 1991). It is only recently, with the emergence of developmental pragmatics and with a growing interest in children's theories of mind, that authors like Ninio and Snow (1996) and Olson (1994) have started to pave the way for a functional–pragmatic approach to children's production and understanding of negative statements.

We recently undertook a study aimed at pragmatic contextualisation of the sentence-verification task, described earlier. Our research was based on a previous study by Bouzigue, Chamorey and Delcenserie (1994). It used a sentence verification task to test 7–8-year-olds' understanding of true and false negatives. The focus in this initial study was on the ways in which the object being negated (agent, action or recipient) affects children's understanding of negation. Children were shown a series of pictures with short subtexts, either true or false, commenting on the people, their doings, and things in the picture (e.g.: 'the man is not painting the door'). The children were asked whether the sentence, always a disclaimer of a given state of affairs, 'fits or does not fit the picture'.

Bouzigue et al.'s most striking result was that children's logically incorrect answers were generally accompanied by arguments revealing a true understanding of what was negated. This is to say, the children seemed to show an appropriate judgement on a given state of affairs. For instance, confronted with a false statement such as "It is not the lady who brushes her hair" a subject replies : "Yes, it is her — so it fits!".

In order to highlight the pragmatic underpinnings of children's judgements of negative statements, we repeated Bouzigue et al.'s experiment with two minor changes. Firstly, instead of having children read the sentences written under each picture, we introduced a puppet (of an old woman with thick glasses, manipulated by the experimenter) who spoke out the text. Secondly, instead of having the puppet speak out in a neutral voice, we endowed her with two distinct moods. In a first mood (assertive), the puppet began each sentence with: " I know . . ." (e.g. "I know; the man is painting the door"). In the second (indirect interrogative) the old woman began: "I can't see very well . . . (e.g. "I can't see very well; the man is painting the door?").

Our purpose in introducing these slight variations is to enhance the pragmatic relevance of the task. The explicit goal given to the children is now "to help the old short-sighted lady see what is really drawn on the pictures". We opted for an indirect mode of questioning as a way of keeping the actual proposition identical in both cases. These two minimal changes in the tones in which statements are produced: (i) from written only to written plus oral and (ii) assertive to interrogative, were designed to shed light upon children's sensitivity to conversational rules or maxims, and their ability to understand disclaimers within this context.

We used a within-subjects design. Three factors were combined leading to 12 types of utterances, each kind being exemplified with three different contents: Type of mood {interrogative, assertive} × truth value {true; false} × locus of negation {subject; verb; object}. Thirty-four children (mean age 7 yr. 8 mo) were then required to make 36 judgements. Our analysis focused on the relation between subjects 'yes/no' answers and their arguments. Here we consider specifically the effects of the type of mood, this being the only specifically 'pragmatic' factor.

The mood-marker "I know . . ." signals assurance on the part of the locutor. It has less a function of communication than a referential function, and hence focuses the interlocutor's attention on its propositional content. The proportion of responses congruent with the logical norm should then be more numerous in this case than in response to an interrogative mood which suggests the need for additional information or repair, and constitutes more clearly a discursive act. For these reasons, when turning to justifications of responses, we hypothesised that questions would call for correctives whereas assertives would invite the interlocutor to take a stance *vis-à-vis* the locutor (agreement or disagreement).

The results agree with our hypotheses. Thus, for example, children's typical pattern of response to a true negative changes according to whether the utterance is produced in assertive or interrogative mode. To a (veridical) sentence like "It's not the man who is painting the door", children will say "yes, it's not the man" in response to the assertive mode, and "no, it's the boy" in response to the interrogative mode. Thus, while from a logical viewpoint a single answer "yes" would be expected, the observed pattern reveals the adaptiveness of children's responses to a factor, namely mood, which is totally irrelevant from a logical viewpoint.

5 Discussion

Experimental tasks developed in laboratory settings to assess children's reasoning are typically well-defined problems in which, from the point of view of the experimenter, 'correct' responses equate to logical responses. Tasks used in the three fields discussed in this chapter are no exception. As suggested in our introduction, human rationality has also been long equated with logicality. Logic has not only been considered as the appropriate normative theory, but also as a descriptive theory of reasoning processes in general. The ultimate goal of development has been seen as being a 'rationality of process', namely "reasoning in a way which conforms to a supposedly appropriate system such as formal logic" (Evans, Over & Manktelow, 1993, p.168).

Situated approaches to cognition have served to highlight the peculiarity of laboratory tasks (in which most characteristics of the children's social and environmental 'niche' are absent), bringing about a new agenda for developmental psychology. What needs to be studied, from this point of view, is not so much the individual child's growing logic, but the growing adaptiveness of the child to his/her social and physical environment. Much work still remains to be done in this direction.

The small steps we have taken in this chapter to move away from classical laboratory situations have allowed us to create what we call a 'pragmatic contextualization' of some of the paradigmatic tasks. Keeping within the tradition of these tasks has enabled us, via comparisons of standard versus 'contextualized' versions of a task, to highlight the crucial role of pragmatic constraints in children's reasoning. A new picture of development emerges which is no longer seen simply as an age progression towards logical reasoning. Age-related changes may be better understood in terms of 'changes in the pragmatic landscape', (together, perhaps, with increasing flexibility and adaptability) than in terms of an overall shift towards more abstract or logical forms of thought.

Logical reasoning is only one amongst several different modes of adaptive reasoning available in a child's mental toolkit. Based on their everyday experience, well routinised long before formal schooling, children are skilful at using other and often pragmatically better adapted forms of reasoning. These sometimes make them fail in laboratory tasks, but these failures are often more a consequence of the presence of a pragmatically grounded reasoning capability rather than of the absence of a logically grounded form. A critical characteristic of most traditional tasks is that they require not only the activation of logical reasoning but also the inhibition of alternative and highly accessible routines (see for instance Houdé, 1995). Paradoxically, then, an important aspect of the development of reasoning skills may be a development of effective inhibition of pragmatically-based routines unadapted to most formal assessment tasks (cf. De Corte and Säljö, this volume).

The picture that emerges from the present chapter is that children, like adults, base their responses to contrived experimental measures (whether of deduction, categorisation or negation) on judgement of relevance, made within the setting of conversation and interaction framed by the experimenter. Though the responses they give may differ, it appears that the determinants of young children's responses are not fundamentally different from the determinants of adults' responses. The gradual recognition that infants and preschoolers are far more cognitively competent than psychologists had previously supposed has been accompanied by a recognition that the reasoning of adults is characterised by all manner of previously unappreciated 'biases'.

However, the extent of the mental 'toolkit' available, and the flexibility of selection of appropriate tools for particular purposes may both be subject to developmental change. Children's situations *vis-à-vis* adults also change with development, and as Politzer suggests, what changes with development more than anything else may be the child's judgement of what is most relevant to the task at hand. To this extent, the course of development would be expected to be linked in part to the child's developing mastery of the pragmatics of language and communication more generally.

The results of studies such as those described in this chapter show there is much more to rationality than logical reasoning. Children do demonstrate, even in their early years, a 'rationality of purpose', which may be defined as "reasoning in a way which helps one to achieve one's goals" (Evans, Over, & Manktelow 1993, p. 168).

What seems to be critical to children's performance (and indeed that of adults) is their interpretation of situations, and more precisely their analysis of what kind of response might be most *relevant* to the situations they encounter.

10

Empirical Abstraction and Imaginative Denial of Rules

Joan Bliss, Jon Ogborn, Orla Cronin, Will Reader and H. A. Tsatsarelis

1 Introduction

This chapter develops some of Piaget's theoretical ideas that he only named during his life-time, and develops them further. Although the starting points are Piagetian, it has been helpful to see this development in relation to the situated cognition framework.

The focus of this chapter is on children reasoning about the physical world, and the schemes that are fundamental to this type of reasoning. We all use such schemes when we try to explain why a battleship floats, or why (in England) it is warm in summer and cold in winter, or what keeps an aeroplane up. These schemes can be seen as mental tools for thinking with. When talking about knowledge of the physical world, and schemes for reasoning about it, Piaget created the valuable concept of *empirical abstraction* which he then largely neglected, a neglect that this chapter will seek to remedy.

2 Logico-mathematical Activity and Reflective Abstraction

Seeing knowledge as deriving from activity, Piaget distinguished logico-mathematical and physical activity, but concentrated on the first, in which the structure of the actions is of interest — not what the actions act upon. The notion of action or activity is a pivotal concept in Piaget's theoretical framework. At the very beginning of children's lives there are actions, sensori-motor ones. Such action schemes allow children to know, for example, that objects which disappear from the visual field do not cease to exist — objects acquire permanence in spite of their displacements.

From about 18 months onwards children can start to imagine the world in their heads, becoming less dependent on their action schemes. But they need more powerful schemes to help them think about the world. Anywhere between 4 and 6 they start to turn their actual actions on the world into mental actions. In other words, external actions become internal mental schemes, allowing children to imagine actions in the head. For Piaget these mental schemes or operations are reversible: an operation permits the imagining both of an action in one direction and in the opposite direction, and both of the result of an action and the undoing of the result.

The activity which is at the basis of the construction of these mental operations is a logico-mathematical activity. Through this activity children co-ordinate their actions. When acting on the world, they reflect on the nature of their actions rather than on the objects in the environment on which this activity operates. They learn to co-ordinate their actions and take from this co-ordination what is common to all the actions. Piaget emphasises that operations derive from the most general types of action such as ordering, uniting, classifying, rather than from just any kind of action. Such actions are crucial because, by being interiorised as mental actions or operations they are the basis of the child's cognitive structure. Underlying the logico-mathematical activity is the process of *reflective abstraction* which leads to constructive generalisation.

Piaget's formulation of how we think can be seen as describing what a person uses to think with: schemes or operations are simply mental tools. The thrust of Piaget's work is the description of how such schemes develop and allow us to adapt to the world around. But his work concentrated mainly on logical and spatial schemes and not on schemes to understand the physical world.

3 Empirical Abstraction and its Critique

Longeot (1978) was critical of Piaget's work and argued that reflective abstraction is truly for the epistemic subject — and not the natural subject — because, within this process, the objects are neutral or almost. It is only an epistemic subject who would not be concerned with the properties of the objects in the real world. Longeot's work stresses the importance of empirical abstraction .

Piaget talks about the activity of abstraction as a mental activity directing an individual's attention to specific aspects of an object or situation, and often involving a physical activty. According to Piaget, empirical abstraction extracts some properties of objects, while excluding others. However Piaget claims that the individual only gets to know these properties through the use of instruments of assimilation and accommodation (schemes, operations) constructed by the individual. In other words, at this level of the argument, Piaget does not see empirical abstraction as generating physical schemes; it generates knowledge about objects.

In spite of his work in physical causality, Piaget still did not see the necessity to postulate physical reasoning schemes, unlike his work on space where he hypothesised infra-logical spatial schemes. Basically he argued that ". . . in the construction of a causal model . . . the operations in play do not become explicative until they can be attributed to objects since it implies an understanding of how objects function" (Piaget, 1971 p. 21). In other words, the logical schemes acquired through reflective abstraction are now attributable to the physical world. Such a way of viewing the physical world is to neglect what is special about it!

However he did finally go further in work published posthumously (Piaget & Garcia, 1987). In this, Piaget began to address the development of meanings of physical entities, formulated with deceptive simplicity as understood through ". . . what can be done to them, what they can do and what parts they are made of".

4 What is Empirical Abstraction?

We propose that empirical abstraction is *not* just a means of providing the individual with knowledge about a particular object or situation by abstracting information from the objects themselves. Instead we propose that empirical abstraction allows schemes of a physical rather than logical nature to be constructed by the individual, schemes such as containment, flow, support, carry, rigidity, effort, blockage and resistance (see for earlier development of these ideas: Bliss, 1995, 1994a).

Thus, for example, a child may abstract from a physical object such as a cup that it is white, smooth and in a particular case made of china which is solid but can be broken. However more fundamental to this object is the idea that a cup is a *container*, regardless of whether the cup is white or brown, or made of pottery or china. *The aspect common to all cups is that they contain liquids.* How do children abstract this more fundamental idea, which we would term a physical reasoning scheme? What is the process underlying empirical abstraction in the construction of schemes, of 'tools for thinking with'?

Most objects, whether man-made or natural, carry social meanings. Thus, the earlier example of 'containment' could be seen as important in the use of cups in the ritual of meals. At breakfast, lunch, tea and supper, cups and glasses are constantly being filled with milk, lemonade, tea, coffee, and so on. Cracked cups get rejected just in case they break or leak and so no longer serve their original purpose. The washing up or bathing rituals, common in all households, not only bring home the idea of containment in a different context but sometimes focus more on the notion of 'flow' and water over-flowing. Since a bath overflowing will be spectacular event and cause a lot of hassle it has the potential to be a memorable event!

Similarly, the scheme of 'rigidity' becomes abstracted as children use tools, because tools such as hammers or scissors must not bend. If such objects were to bend they would not serve their purpose (Brown, 1990). Containers such as cups and glasses can also be thought about as rigid and indeed need to be if they are to be successful containers, but they are only partially rigid because of their breakability. So there are exceptions or constraints. Water can be rigid, but only while frozen. Maybe, too, chairs are seen as 'rigid' until children encounter deck chairs where the 'rigidity' is less obvious. Piaget's description of empirical abstraction proposed that it is a "repetition of the same abstractions (that) can lead by inductive generalisation to rules or laws concerning facts or reality".

But what is happening as a result of these repetitions? It must not be supposed that the process of empirical abstraction is a simple reading off from experience (Bliss, 1994b). It involves imaginative projections, imaginative substitutions, imaginative combinations, for example, the projection of the autonomous movement of living things onto inanimate things. Features such as intentionality have to be constructed (see earlier work on force and motion; Bliss & Ogborn, 1994).

Physical schemes would have to do with the creation of contexts and so would not be context free, like logical schemes. In terms of transferring across different contexts it is possible to expect a slow, rather conservative process of noticing salient features of new contexts, which would generate new and different constraints. In the above example of containers, the use of the scheme of containment in one context may not guarantee its use in another context. This movement from one context to another may be very slow. And in the example of the bath, the container scheme might be less important than the flow scheme. However, at the end of the day, children will come to recognise that baths too have a particular capacity for containment. The progression to envisaging the bus as a container may take even longer and be more difficult.

It may in fact be cognitively rather sensible to be very cautious about generalising to new contexts, contrary to some received wisdom related to the desirability of transfer. Gradually schemes abstract some of the features common to many physical situations, but they would also contain the constraints of particular situations. Empirical abstraction is domain-specific, constrained by the specific ways in which the world permits or inhibits actions.

Reasoning is and should be constrained by whether the real entities being thought about can, of their nature, be manipulated in a given way in a given context. For example, reasoning schemes could be imagined as, and carried out by, joining entities into compound entities, or substituting one imagined entity for another; each of these have evident uses in many sorts of reasoning. We would see such descriptions as vital to a more complete description of common-sense thinking and its functioning, and thus giving rise to new and different elements of concrete thinking.

5 Research Project on Empirical Abstraction

We are currently working on a research project to examine the nature of empirical abstraction and physical reasoning schemes, funded by the Economic and Social Research Council (ESRC). At the beginning of this project we stated that:

> It is easier to give examples of concrete physical reasoning schemes than it is to attempt any systematic account of them. Obvious questions have no clear answers. How many different kinds of schemes are there? How are they related? How do they develop? Why and how do people choose one scheme rather than another in a given situation?

Our study was not intended to be more than a preliminary investigation, to be used as a foundation for further research. We needed to: (i) establish a theoretical framework in order to generate a preliminary typology of common schemes and their relationships; (ii) develop a methodological framework to allow us to create generic tasks to investigate these issues; (iii) carry out experimentation with two age groups, 6–7 and 13–14 years old.

Using the following categories as a framework for task development, a group of four preliminary tasks formed the basis for the final six tasks developed during the project.

First, 'imaginative transformations' tasks asked for one entity or event to be transformed in imagination into another, seeing what is held constant at each stage of the transformation. Difficulties of transformation suggest boundaries between incompatible schemes. The emergent task here was called 'topsy turvy', and we shall return to it shortly.

Second, 'thinking of something as . . .' tasks involved trying to think of real instances as like a given scheme, for example trying to think of a gale as like the 'flow' of something. Something concrete needs to be imagined in several more general ways. The point is to see how empirical abstractions from the particular might work by moving from instance to scheme. Emergent tasks here were 'imaginative constructions', 'fitting schemes to counter-expectations' and the 'dimensions of schemes' questionnaire.

Third, 'instances of something like' tasks involved creating or identifying instances for a scheme such as 'container'. Can containment be used to think about: a train, a story or the Earth? Something general is to be thought of as being represented through several instances of different kinds, moving from scheme to instance. The emergent task was called: 'application of schemes'.

Fourth and finally there are 'instances of schemes used metaphorically' tasks . The task developed here was termed 'schemes as metaphors' and it involved deciding between good and bad examples of metaphorical use of physical schemes.

For present purposes we shall look at one task in detail: the 'imaginative denial of rules task' or what was familiarly called,'topsy turvy'.

5.1 Imaginative Denial of Rules Task (Topsy Turvy)

The task is introduced to the pupil as a game which involves making up a story about an imaginary world in which everything is 'topsy turvy'. The basic idea behind the task is that children are asked in their story to imagine carrying out a number of familiar actions, or sequences of actions, but in a world where everything is the opposite, in every imaginable way, of everything in the everyday world, so all is topsy turvy.

The interviewer gives the child a number of examples. For example, s/he could say, "In topsy turvy world I'd put my coat on inside-out, the handle of my brief case would be inside and not outside, I'd put my boots on my hands and my gloves on my feet". Thus, when the children are attempting to tell a story about an action happening in topsy turvy world they are trying to imagine carrying out an action, known to them, in its opposite form and imagining also all the consequences.

The task is a generative one, that is, it requires participants to produce a scheme or an instance of one. Generative tasks focus on possibility. The schemes are not presented explicitly, although they are partly implicit in goals set. The focus of the task is on the constraints on the changing schemes. The task has four different settings all of which are part of children's common knowledge. Thus children were asked to transform the following four familiar actions:

- making tea;
- diving into a swimming pool from a springboard;
- playing football;
- flying a kite;

In this pilot study 19 interviews were carried out with pupils between 6 and 7 years old, but not all children worked on all the above four basic actions. A substantial majority of 6 and 7 year olds carried out the first two tasks and we shall focus on these. Ten pupils between 13–14 were also interviewed, but the task was substantially adapted to suit this age group, so they will not be discussed in this chapter, except in the most general sense in the conclusions.

Our analysis does not focus on what the children actually said but on the inferences we can make (from what they said) about the schemes we consider they were operating on or with. The analysis takes the following starting points:

(i) To claim that a scheme exists it is necessary to see it used in a variety of different situations and by different people — if the claim is that the scheme is common to them all. The underlying claim of the topsy turvy task is that schemes exist if children are attempting to change them into an opposite form.
(ii) To describe how a scheme is used it is necessary to see it used by different people in a common set of situations.
(iii) As the study progressed we further speculated that schemes do not operate on their own but rather work together in clusters or packages.

For each of the four actions described above we analysed the data at two levels: firstly, inferences about the schemes with which the children were working in order to produce an opposite event or action; and secondly, the evidence, that is, that part of the child's description and or explanation that permitted us to make the inference.

Making tea in topsy turvy world. One of the most basic schemes with which nearly all nineteen children operated in their account of making tea in topsy turvy world was that of the scheme of up–down. For example, the kettle, the teapot or the cup are turned upside-down. Associated with the up–down scheme are a number of other schemes such as support, fall and containment.

Taking the upside-down scheme further, some children argued that objects which were usually on a table, and thus down, now needed to be up. As Andrew (age 7) pointed out, ". . . the floor is now the ceiling so the objects, the kettle and the cup will go up and will be upside-down as well". The idea of kettles and teapots being up rather than down was problematic for a few children because ". . . they (the objects) won't have any support". Thus they introduced the idea of partial support and suggested that these objects could float in topsy turvy world, an ingenious way of avoiding the scheme of falling. Here we see how children's reasoning connects together a package of schemes to do with upside-down, falling and support.

Most of the children interviewed thought that when objects such as the kettle or the teapot were upside-down, the water or the tea would fall out (i.e. that upside-down objects would no longer be able to contain their contents). Billie (age 6) argued that, ". . . water can't be boiled upside-down because it wouldn't stay inside". A few imaginatively thought of fixing the teapot lid so that it, too, could not fall off and so that it could keep its contents inside. Others imagined fixing a lid to the cup as a way of supporting the tea and not letting it fall out. Objects that are containers bring with them schemes relating to falling and support, so that in an upside-down world a container can no longer evidently contain.

About half a dozen children worked with the scheme 'inside–outside' which is associated with containment. Thus the handle, spout or lid of the teapot were placed inside the teapot, similarly for the handle of the cup. But for some, the tea is still inside the pot and so presented a problem. As one child concluded, ". . . you won't be able to get the tea out of the pot with the spout inside". Two others considered that the tea would be made outside the cup. Jonathan (6) argued, ". . . you could turn the tea cup upside down and put the tea on the top of the cup". Georgina, who made the same suggestion, went on to conclude, ". . . it'll go sssh sssh and then it will fall down and end up on the floor".

In summary, in making tea in the topsy turvy world pupils transfer the up–down scheme from the real world to this new fantastic world. As a result a number of other associated schemes are connected to the reversal of the up–down scheme, for example, falling, support and containment. Although for some pupils the ceiling becomes the ground, the contents of upside-down containers still fall downwards and do not fall upwards, as would be the case if up–down had been fully inverted. In this topsy turvy world the contents of upside-down containers also need ways of being supported (existing lids are fixed or lids are created and fixed). It is interesting to note that the action of pouring, which is an intended fall of a liquid in the real world, scarcely gets mentioned in the topsy turvy world.

When children are working on the inside–outside aspect of the container scheme it is noticeable that only appendages of objects move from the outside to the inside. The container (teapot or kettle) itself, in its essential structure, is not turned inside out. This may be because the container is rigid, and can not easily be imagined with its inside outside, and vice versa. Bounded spaces would have to become unbounded, and unbounded bounded. If a solution were sought through making the container in some topsy turvy way fluid rather than rigid, the scheme container would become impossible. Rigidity and containment are linked together, and so define boundaries in space which delimit some regions and leave others open.

Diving into a swimming pool from a springboard in topsy turvy world. This scenario, compared with the previous one, seems to have presented the children with more choices about what they could decide to focus on and attempt to transform in their topsy turvy world. The focus could be: the swimming pool and the water it contains; the diver and his or her action; and the nature of the water. Very rarely did children focus on all three aspects, but quite a few of them focused on two of them.

As with the making tea scenario, one of the most basic schemes with which at least half the children operated in their account was that of up–down. However, even here there are some important differences. Five children operated on the whole scenario and saw the swimming pool moving from the floor to the ceiling. But it is not clear in their accounts whether the swimming pool is still facing upwards or has also been turned upside down.

For example, Anna says, ". . . the water will be floating . . . the big, huge bathroom will be floating up as well (the swimming pool)". Similarly, Alex talks simultaneously about the diver and the swimming pool, ". . . the man would go on the roof — because the water (the pool) is in the air. He'll plop, because the water will be there as well". So for both these children the swimming pool had changed location — down becomes up and vice versa — but the pool itself has not apparently been turned upside down.

Anna introduced the idea of partial support and suggested that objects could float in topsy turvy world — the same ingenious way of avoiding the scheme falling as with the making tea scenario. It is possible that for Alex and Anna not only the scheme up–down is being operated on but also the scheme to do with containers. Swimming pools or baths are large containers without lids so if they are turned upside-down the contents will fall out.

For two children, inferences can be made from their comments on the action of falling that not only has the location of the swimming pool changed but also the pool itself has been turned upside down. But here too the fact that swimming pools are containers but without lids has to be taken into account. For Joseph ". . . the water's gonna be up, the water it would be up. No it can't keep it up (the pool can't). All the water would be falling down, upside down". Also Joseph adds ". . . that's the water and I'm drawing the people (swimmers) falling down". Later he goes on ". . . the sky would be down there (at the bottom) and the ground's upside down". Georgina also points out ". . . if the swimming pool was upside down, no-one can get swimming, they have to fall down, or they have to hold the bar, and if they all do, it'll go crack and they'll all go whee and fall down (too many people holding bar to keep them up)".

Lastly Kimberley operates on both the up–down scheme and the container scheme in an interesting way, both operating on the pool and its contents and also on the world itself. She points out that ". . . if you turn it (the pool) upside down the water would fall on the ceiling". In other words, if the pool and its water are turned upside down then it follows that because the pool is a container which has no lid its contents will fall out. But it also follows that in this topsy turvy world the water will no longer fall to floor but to the ceiling. It is a very hard thing to imagine but the logic does flow.

Thus, with these five children, we see how their reasoning connects together as a package of schemes to do with upside-down, falling and support and partial support, and containment.

For another three children the up–down scheme appeared to operate in a local manner, moving the parts of the swimming pool and its contents around. Again, as discussed above, swimming pools are containers and bring with them schemes to do with containing and to do with the nature of their contents. Water is a fluid and fills the space that it is given, or if not enclosed, it has no boundaries and so spreads out as far as it can. Children are giving the water boundaries and moving it around like an object, thus denying its properties of fluidity and making it behave differently, more like a solid or something rigid. However this transformation allows these two children to avoid the scheme of falling, with the ground still providing support.

Two other children operate on the falling scheme, changing its direction, which was a little unusual. For example, Jonathan pointed out that (when diving) "...you'd fall into the sky . . . you'd go down the sky and then you'd probably land in the water". He does not think about the water.

Turning now to children's focus on water, and its properties, some of the children make fairly simple changes not affecting the water's fluidity. For example, some state that there could be a change of colour, and one removes the chlorine. Another two children substitute milk for water, while another says, ". . . it's warm instead of cold". Finally one suggested that it would be ". . . light as a feather", while a second said, ". . . the water would still be wet but we wouldn't get wet".

Others change the water's fluidity a little by changing it into sand. For example, Carl says that in topsy turvy world there would be no water in the pool but when asked "what would you swim in?" he replied, ". . . sand, erm, you'd sink in it, it's wet sand". Later he added that in topsy turvy world, ". . . in the land when it rains, it rains with sand [from?] the sand clouds, that was the way that it was made".

In the making tea scenario we looked to see if solutions were sought by making containers fluid rather than rigid, but this proved not to be the case. However in the present context some children attempt to imagine such a transformation. For example, Iannis said that, ". . . after 5 m going deep it (the water) would be cement . . . it could feel hard when you try to hit it". When asked how the water would become hard, he said, ". . . by putting flour in it and cement, that would make it go hard. (Would it still be water?) No, loads and loads of blocks because there'd be cement."

In this scenario we can see that some children use the same kind of package as those in the making tea scenario, using packages of schemes to do with upside-down, falling and support and partial support, and containment. Others who work with local upside-down changes bring with them a package of schemes to do with the contents of the container and an attempt to give fluid objects boundaries so as to move them more easily. Others decide to overtly operate on the fluidity scheme, changing the water to a solid. In the making tea scenario, the appendages of containers could go inside. Alex reflecting on the pool and its contents, says about the water; "No it couldn't be inside out, could it?" He could not develop this idea and it can be seen that fluids, while needing to be contained, do not themselves have an inside and outside in the same way as rigid objects.

5.2 Task: General Results

In our analysis a scheme is a basic re-usable component in everyday reasoning. These basic reusable components combine with one another to form packages but there are a variety of combinations of packages. We claim that the context determines the way in which the individual puts schemes together into packages, as can be seen from the two examples discussed. Schemes appear to be constrained in terms of the possible inferences that they permit and prohibit, e.g. seeing something as a container enables getting full or becoming empty as potential states of affairs. A scheme also has a small set of potentials which provide entailments and link it to other schemes. For example, the potentials of support include the schemes: something underneath, prevented from falling, etc. Physical schemes are often used to understand phenomena in the absence of detailed mechanistic explanations.

However everyday reasoning schemes veer towards conservatism; they do not allow too much change; they do not allow reasoning about the impossible. For example, in these scenarios there are no instances, in any of the pupils' accounts, of objects being transformed into spaces, or vice versa; there is always sequence — a before and an after, and there is always causality, some cause always has an effect. In other words, the basic Kantian schemes or dimensions remain unchanged.

6 Situated Cognition

There is some resemblance between some of the ideas put forward in the acquisition of concrete physical reasoning schemes and the concepts within situated cognition. The Laboratory of Comparative Human Cognition (LCHC, 1984), when describing the cultural-practice theory of culture and cognition, said,

> ... cultural practices by which we mean activities for which the culture has normative expectations of the form, manner and order of conducting repeated or customary actions requiring specified skills and knowledge (see Scribner & Cole 1982). Cultural practices have to be learned as systems of activity. ... "Culture", "cognition" then refer jointly to behaviour assembled by people in concert with each other. It is for this reason that a cultural practice theory takes cultural contexts, that is socially assembled situations, not individual personal or abstract cultural dimensions, as the unit of analysis for the study of culture/cognition (pp. 333 & 334).

While LCHC proposes that cognitive development can be characterised by mastery of context specific knowledge, it also states that culture must arrange for 'selection' of contexts for children. They are very open in stating that they "have not said much about how within-context interactions result in within context mastery of essential cultural practices". It is here that they turn to the socio-historical school of Soviet Psychology, and Vygotsky in particular, since he explicitly connected ideas of interaction with concept development.

Rogoff (1990) develops these ideas, proposing an apprenticeship model of development, where ". . . the rapid development of young children into skilled participants in society is accomplished through children's routines, and often tacit, guided participation in ongoing cultural activities as they observe and participate with others in culturally organised practices". While her work develops this idea, with numerous rich and exciting examples, it remains at a very descriptive level and does not explain the mechanisms underlying development,

Social practices are important in that they contain ideas and knowledge which in the end must become taken for granted. That is, there is a level at which human beings

— in order to function well in life — must no longer question all things around but must be able to take these as givens (Berger & Luckman, 1966; Giddens, 1997). Piaget, in his work on memory, said that memorising is not a problem once memory is seen in relation to an individual's knowledge and intelligence, and that when we remember things, we structure them and assimilate them to what we know already. As he said, ". . . conserving (a scheme) goes without saying because the essence of generalisable assimilation is to function without ever stopping, thus to conserve itself by its reproductive and generalisation activity itself which guarantees the conservation of the scheme to which this activity leads" (Piaget, 1964, pp. 29–30).

In summary, learning about the physical world is intimately tied in with learning about the social world, its customs and practices. As we said earlier in this article we would expect transfer of learning across different contexts to be a slow, rather conservative process. The practices of the social and cultural world present numerous occasions for repeated or customary actions to occur and so for children to build up a stock of knowledge and schemes through these routine performances.

7 Conclusion

We propose that in empirical abstraction the participants all play important roles. The adult or competent peer (in the Vygotsky, then Rogoff and Lave sense) is an active part of culture, inducting children into many simple and repetitive cultural practices: cleaning teeth, washing, tea parties, going shopping, etc. By being part of the culture, they legitimate the activity that occurs in these practices.

For example, at a tea party it may not acceptable to drink from a bottle — there being a normative expectation of having some sort of appropriate container, like a cup or glass. Children join in and learn; not only do they learn about the ritual of, for example, 'who pours the tea', but they also learn about pouring tea into things that can contain. It is both a legitimate activity and one which will in the end become so taken for granted that it is obvious — we drink from cups and glasses at tea parties because they are the prototype of the best container for that sort of situation. This, in turn, means that the knowledge and the scheme become tacit.

Also in this situation children are taking on all sorts of 'social ideas' of appropriateness, etc., as well learning about the social objects themselves which are important actors on the scene. Later, when helping parents after shopping to unload and pack goods, they may be asked to fill tins with biscuits, bottles with sugar, or the fridge with fresh goods. They may even get scolded when trying to put too much of something into a container by being told, ". . . be careful, the jar is too small for all the sugar, you need to take another jar". Or when going out with parents in the rush hour they can not get on the bus because the conductor says, ". . . sorry, the bus is full".

In all these instances there are actions on objects, sometimes directly suggested by adults, but in all events in socially legitimated situations because arranged by adults and backed up by verbal interactions between participants. As the children gradually see that, for example, containment not only has to do with cups, but also with tins, fridges, later with baths, and even later still with buses and trains, they are empirically abstracting an understanding of *what can be done* to such objects, *what they can do* and *what parts they are made of*. Thus, they begin slowly to acquire powerful instruments for thinking about the physical world.

11

Enactive Representations in Learning: Pretence, Models and Machines

Edith K. Ackermann

1 Introduction

People are natural designers of tools and mediations to support their purposes and augment their potentials. They surround themselves with human-made artefacts, and as soon as they know how to walk and talk, they add their own creations to the existing cultural wealth. People also know how to project meaning into artefacts produced by others, and they use existing forms as mediations to communicate among themselves (Ackermann, 1995). As they reach the age of 2, most children actively engage in keeping track of things and leaving traces behind them (Stamback & Sinclair, 1990). They set the stages and build the props which enable them, alone and with others, to play out their feelings and ideas. They write the scenarios and invent the characters to enact their parts. Children's ability to recast events through symbolic play allows them to objectify their experience, and opens the way to a dialog between what is and what could be, a dialog through which they gain deeper understanding and broaden their experiential field.

At a time when computational objects have become an integral part of our culture, and have made it easy to run programs, model dynamic interactions, and simulate behaviors, people's ideas of what modeling is all about are radically changing, as are their ways of relating to existing modeling tools. Enactive forms of expression and communication are gaining popularity. More than in the past, psychologists and educators emphasize the specific role of enactive representations in learning. Performance and simulation are granted a new place beside language (Quéau, 1986). There can be no doubt that modeling plays a key role in learning. Models provide make-believe versions of a phenomenon which highlight, in simplified and principled ways, its characteristic features. More importantly, models are external representations which embody these features, thus making them tangible and shareable (Ackermann, 1991). From a psychological point of view, models offer the transitional spaces (Winnicott, 1989) in which people can explore, express, and exchange ideas, and run experiments, under safe conditions. Models can be static or dynamic. They can freeze or unravel processes, state or solve operations.

The purpose of this chapter is to discuss the significance of enactive representations and dynamic modeling techniques in the light of children's abilities to engage in make-believe activities, i.e. to use their creative imagination as a means to project themselves into situations and to change their stance in the world, and to endow 'objects that behave' with a life of their own. Three aspects of children's construction of — and interactions with — enactive forms of representation will be considered to highlight their specific role in learning. A first aspect concerns the emergence and growth of children's pretence and symbolic play. A second aspect deals with children's conception of — and interactions with — existing miniature models or replicas of larger objects, which are inert, yet interesting as objects in their own right. A third aspect focuses on children's conceptions of — and interactions with — machines and programs (mechanical and cybernetic devices). Unlike static models, machines and programs accomplish something on their own, thus adding value to their inputs by transforming them. The focus of the discussion is on how children build and make sense of different modeling techniques and/or devices, and how they use them to their advantage. I shall argue that putting oneself in other people's shoes (taking on their views), moving in and out of situations (changing one's stance in the world), and animating things (asserting agency) are three valuable heuristics by which children and adults make use of empathy and imagination as levers for cognitive growth (Bachelard, 1971). Exploring situations on a make-believe basis allows them to turn knowledge back into experience on safe ground, or else to situate what they know by anchoring it into past or possible contexts of use.

To conclude, I shall relate the discussion of children's views and uses of enactive representations back to the framework of existing developmental models, in particular Piaget's distinction between figurative and operative aspects of human intelligence (Piaget, 1962). As an alternative to the classical view of representations-as-descriptions and actions-as-transformations, I propose a functional divide between morphogenetic and morphostatic, i.e. generative and stabilising, functions in both action and representation. The advantage of this approach is that it re-establishes the significance of hybrid forms of enactive representations such as pretence and simulations, often left in the background by existing developmental theories.

2 Enactive Representation and Dynamic Modeling

Unlike descriptions or inscriptions, enactments are symbolic manifestations that behave. They do not merely signify, but they carry out entire sequences of actions. Enactments can take two different forms depending on who orchestrates them, and which media they use for expression. While plays and rituals are live performances, and use performers' bodies and associated props as a medium (as in theatre), programs and simulations can be thought of as automated or mechanised performance embodied in human-made artefacts (machines, automata). Not unlike a human actor who (re)casts an event on a stage, a machine — computational or mechanical — mimicks aspects of a system's behavior by electronically or mechanically executing a prescribed sequence of steps (Laurel, 1991). Thus, in spite of their obvious differences, both performers and their automated counterparts embody dynamic processes at work, and carry out actions and operations which are only stated in descriptive and axiomatic systems (Ogborn & Miller, 1994). Both can be used and, indeed, are extensively used — by children and adults — as generative modeling devices to run multiple scenarios of past or possible events. This, in turn, opens the way for safe exploration of otherwise hidden cognitive processes, usually carried in people's heads with the help of external supports, such as descriptions or figurations. Techniques of re-enactment, as expressed through pretence or symbolic play, are called *simulâcre* in French; whereas enactive representations as embodied in a mechanical or electronic device, are referred to as simulations (Quéau, 1986).

The dual nature of *simulâcres* and simulations, both representation and transformation, raises serious questions as to the symbolic nature of 'live' as well as mechanized performance. They challenge the classical divide between figurative and operative aspects of human cognition, as defined by Piaget (1962), and question the generally agreed upon distinction between representations (internal and external), seen as descriptions and inscriptions, and actions or operations seen as transformations. In effect, if representations are meant to describe, then their function is to provide a repertoire of tokens, or signifiers, which can be composed and articulated according to syntactic rules, and used to evoke objects and events in their absence. If representations, on the other hand, start generating things on their own, manipulating signs and symbols, and transforming rules, then the boundary between the two becomes unclear: from symbolic objects, or notational systems, they become actions/operations embodied in a machine.

3 Emergence and Significance of Pretence or Symbolic Play

Research in cognitive development shows that young children naturally engage in pretence or symbolic play as a way to recast events of significance to them (Leslie, 1984; Piaget, 1962; Stamback & Sinclair, 1990). As soon as they reach the age of 2, toddlers generally start setting the stages and building the props which enable them to play out, alone or with others, some intriguing or captivating events. They do so by interacting with different kinds of others, symbolic or real, co-actors or audience, who may be more or less supportive of their personal quests. As they engage in pretence, children become both playwrights and performers. They create their own characters and imagine the scenarios in which characters play their part. They distribute roles by turning people into things, or objects into living creatures. They may take on a character's role by crawling under its skin and mimicking its behaviors. Children also vary their stance in the world by becoming absorbed into situations or removing themselves, adopting a god's eye view (Ackermann, 1996). They change the scale and attributes of things, becoming *tour à tour* Lilliputians or Gullivers. Who hasn't seen a bunch of 3-year-olds drink out of empty glasses, ride on boxes, become a baby or a cat, or fly off as an airplane? Children also know how to project meaning into stories told by others. They identify with heroes in puppet shows, fairy tales, and video games. They live through their hero's joys and sorrows as if they were theirs, again under safe conditions (Turkle, 1995).

Children's ability to pretend, or play out their feelings and ideas in a make-believe setting, serves them well in many respects. It allows them to revisit past experience by modifying or subverting its outcome. It enables them to generate variations by running different versions of a scenario, gauging their consequences (Bettelheim, 1977). Most important, doing as-if and playing what-if allows them to engage in a dialog between fact and fancy, between the actual and the possible, between what is currently happening and what could happen. One couldn't think of better preparation for later forms of conditional or hypothetical reasoning (Ackermann, 1991).

People's ability to pretend and imagine does not diminish as they grow older. Nor does their tendency to dwell upon situations, to put themselves in other people's shoes, to animate things. Quite to the contrary, these abilities become more sophisticated as people evolve. Their role, as mentioned earlier, is to put empathy at the service of intelligence. In Sayeki's words, people, young and old, get in touch with situations, fictional or real, by literally dispatching pieces of self, or throwing out *kobitos* (*kobito* means little person in Japanese) which, in turn, enables them to act out and take in aspects of a situation *as if* they were in it, while remaining physically removed (Sayeki, 1989). Note that this particular form of imaginary projection is different from how it actually feels to be immersed in a 'real' situation. It is more like remembering a sunny day on the beach and seeing oneself running on the shore. While the lived experience is more like scuba-diving, its reconstruction through memory can be mentally evoked as a bird's eye view, with a projection of a 'copy-of-self' into a miniaturized scene. The view is 'removed', in that the thinker, unleashed from the actual sensation of diving, can change the scale of things and move in and out of the site at will (Bachelard, 1971). Both of these projections differ from becoming one with a situation through fusion, as illustrated in Woody Allen's movie *Zelig*. Here, the protagonist literally takes on the traits of another person. His self has dissolved into other, indiscriminately, uncontrollably (Masciotra, 1996).

The importance of creative imagination, through miniaturisation and projection of self (creation of a double-in-the-world), has been too long ignored by developmental psychologists and educators. Researchers' emphases on growth as separation and cognitive mastery as the ability to handle complex situations in detached analytical ways has made them oblivious to the idea, so clearly expressed by Sayeki, that understanding requires connectedness and that connectedness entails intrusions of self-in-world as well as intrusions of world-in-self (Kegan, 1982).

In recent years, the uses of empathy as a means for gaining understanding have been addressed by a number of scholars interested in gender differences (Jordan et al., 1991; Keller, 1985), as well as by researchers engaged in the study of children's theories of mind (Astington, Harris & Olson, 1988). The heated debates which oppose theory-theories and simulation-theories of children's 'theories' of mind (Gordon, 1996) constitute a true breakthrough in this respect. Challenging developmentalists' views on human cognitive growth as a progressive move inward (from action to thought) and upward (from concrete to abstract), researchers recognize children's ability to empathize or tune into situations as central to their ability to understand other people's mental states. Interestingly, this debate remains limited to the domain of folk psychology, and, with a few exceptions (Perner, 1996), it remains strongly polarized. Its 'either-or' tone calls for further exploration of the mutually enriching roles of simulation and representational redescription in theory building and conceptual change (Karmiloff-Smith, 1995).

4 Children and Miniature Models: Revisiting De Loache's Experiment

In a rich corpus of experiments, De Loache and her colleagues show that young children have great difficulties in understanding the representational nature of objects that are interesting in themselves (De Loache, Uttal & Pierroutsakos, 1998). A scale model of a room, for example, is salient and appealing in its own right, and treated as such by 3-year-olds.

In the well-known teddy bear experiment, a scale model of a room gives children information about a full-sized room that the model represents. In a preliminary phase of the experiment, De Loache explains to the children that the scale model is an exact mini-replica of the full-sized room, and insists that whatever happens in the model simultaneously happens in the room: "A big bear lives in the big room and a little bear in the little room. Whatever the baby-bear does, the daddy-bear does too". De Loache then proceeds to hide the baby bear in various places in the scale model and asks the child to find the big bear "who is hiding at the same place in the big room". If children recognize the model-room correspondence, they should be able to match locations which coincide between the model and the room.

To understand that the miniature replica of a room (in which things can be hidden) stands for the larger room (with its hiding place) requires, according to De Loache, a dual representation: The child needs to simultaneously represent the model and its referent. In other words, s/he has to disentangle two different functions embedded in the scale model: while it is an object itself, it also serves as a representation of the larger space. Such a dual representation is not constructed before age 3.

Variations on the teddy-bear experiment further suggest that dual representation is easier to achieve when the representational object is a picture instead of a scale model, or when the symbolic relation between model and room is altogether removed, as in the ingenious 'shrinking room' experiment. The 'shrinking room' variation shows that if 2–3-year olds can be made to believe that the scale model actually *is* the room (which has been shrunk by the incredible shrinking machine), then they successfully retrieve the teddy-bear. The make-believe shrinking operation, once made tangible, forces the model-room correspondence, thus facilitating the child's understanding of the match between the two.

According to De Loache, people generally admit that arbitrary symbol systems, such as numbers and letters, are difficult for young children because they bear no resemblance to their referents. What is less easily accepted, in her view, is that the same holds for 'realistic' symbols, such as models and photographs, which look very much like their referent. The commonly held argument is that if children recognize pictured information of a thing, then they also know that pictures represent that thing. De Loache's work clearly demonstrates that this is not the case. Much to the contrary, the more realistic the representational objects, the more interesting it is for its own sake, and the harder it becomes to treat it as something which stands for something else.

5 Models, Simulations, Microworlds

In discussing the implications of their work for education, De Loache and colleagues express their doubts as to what children may take away from watching edutainment programs, such as *Sesame Street*, in which letters and numbers are brought to life or personified, and in which they talk, sing, and participate in beauty pageants. If children have difficulties using symbols that are highly attractive, then treating letters and numbers as cartoon characters may well be counter-productive. More generally, the authors claim that turning abstract symbols into concrete objects probably makes their meaning less, rather than more, clear to young children. Does this imply, as the authors suggest, that symbol systems need to be more abstract, less lively?

Alternatively we might see the evidence of children's resistance to conceiving of an object as an ersatz which stands for something else as offering an opportunity to rethink the pertinence of correspondence theories of representation altogether (Lakoff & Johnson, 1981). Of course, children have to learn how symbols relate to the concepts they instantiate. On the other hand, no manipulatives or external representations will ever capture a phenomenon in its entirety without bringing about distortions and misunderstandings. Representations are not copies but translations, and like any translation, they transform the original. The function of manipulatives, in this sense, is not to represent things, but to provide occasions to carry out mental operations. If such is the case, a good suggestion would be to let children remain 'naive correspondence theorists'. Hence, learners should be encouraged to treat manipulatives (both static or dynamic) not as simulators but as stimulators (Resnick, 1990), or microworlds (Papert, 1980). The difference between the two lies precisely in their expected truthfulness to reality. While simulations are generally meant to be true to reality, microworlds as Papert's defines them, claim their autonomy by providing tangible and shareable externalizations of hidden aspects of a 'reality'. Their purpose is to bring to the fore for exploration, and combine in novel ways, some important underlying mechanisms otherwise unnoticed. Instead of proposing more abstract or less attractive representations, a more radical solution, indeed, would be to offer an array of manipulable and dynamic externalisations of a set of hidden principles. It is the generative power of a modeling device and not its resemblance to reality that enables learners to shed new light upon specific aspects of a complex phenomenon.

6 Machines in a Child's Eyes: Looking Like Things Yet Acting Like People

Young children's fascination with machines, from mechanical wind-up toys to music boxes, from robots to sound-sensitive flowers, may well be yet another facet of their natural tendency to rely upon enactive representations as a favourite way to explore, express, and exchange experience and ideas. Machines, as mentioned earlier, can be thought of as automated or mechanized processes embodied in artefacts that behave.

Their singularity, as children put it, is that they look like things yet they act like people (Turkle, 1984).

Made out of wood, metal, or plastic, machines 'know' how to add value to their inputs, sometimes by executing or augmenting motor activities (simple machines), sometimes in quasi-intelligent ways (control devices). Unlike miniature models, machines, both mechanical and cybernetic, act upon the world by transforming some features into something else. And they do so in consistent and reliable fashion. It is precisely their degree of autonomy that brings children to think of them as agents, and to treat them as such. From objects-to-think-with, they become in a child's eyes 'artificial others' with whom the child can share tasks, or delegate parts of a job to be accomplished. Early on, children develop their own ideas about what it takes for an object to be a machine, and what it takes for a person to put machines to their advantage and rely upon them.

7 Children and 'Intelligent' Artefacts

Research on children's conception and use of simple-minded 'intelligent' artefacts (computers, Logo microworlds, Lego/Logo creatures) shows that, early on, children tend to endow objects that behave with a life of their own, and relate to them as if they were animated. This is not to say that 5-year-olds actually believe that a computer or a robot is alive (Carey, 1985; Turkle, 1984). They know very well that it is not. Yet, in interacting with them, children nonetheless treat them as social agents capable of initiating, sustaining, and controlling behaviours.

Our own research at the Media Laboratory further suggests that young children's apprehension of computational devices is both instrumental and anthropomorphic (Ackermann, 1991). What is true for children is also true, to a lesser degree, of adolescents and adults (Martin, 1996; Resnick, 1990). Observations of adult teachers show that, to many, the question of importance is not so much how does an artificial creature work but what does it achieve? It wouldn't even come to their mind to take apart a creature to see what is inside. Instead, they take their creature as-is and explore its ways of evolving in its surrounds (Grannott, 1991). They like to understand not 'how a robot works' but 'what is the logic' behind its responsiveness — or lack of responsiveness – to the world in which it 'lives'. Instead of deconstructing them for the sake of explaining, they spend a great deal of their time optimizing their 'dance' with a creature, to gain insight into its ways of being..

Further research on children's animism by Hatano and Inagaki (1987), Carey (1985) and Steward (1982) suggests that children's tendency to attribute agency applies beyond computational artefacts to include transactions between objects in general. The most striking characteristic of children's understanding of causal transactions is that they describe the moves between interacting entities (alive or not, agent or recipient) in terms of how each controls or is controlled by another's behaviour, either through direct or mediated action. Note that in the case of direct action, an agent A does something to a recipient B, or impacts it physically, whereas in the case of mediated action, agent A *signals* something to B, and B acts or signals back accordingly. In both cases, agents at play tend to be animated, at least while currently active, and recipients tend to be objectified. In a chain of transactions, any particular object is by turns seen as an agent or a recipient, depending on whether it is perceived as generating an action from within (agent), or responding (recipient).

For a review on children's growing ability to distinguish between animate and inanimate objects and to disentangle purpose and causation — psychological and physical descriptions — when explaining the behaviours of people, machines and simple-minded intelligent artefacts, refer to Ackermann (1991).

8 Children and Simple Machines

A more recent study by Ackermann and Brandes (Brandes, 1992) on children's conceptions of simple machines brings additional support to the idea that the criteria used to determine 'machineness' are relative to a tool's ability to give back something different from what has been put in in the first place. The purpose of this study was to explore elementary-school children's sense of mechanism. To do so, we wanted to bring children's own categories and understandings to the fore. In a first session, we simply asked small groups of children what, in their eyes, makes something a machine, and how machines work. We then presented individual children with a collection of images showing instances of devices with similar functionality yet different as far as their source of power, level of complexity, and control mechanisms are concerned. We asked children which of the objects were machines, and why. Two examples of collections were: skateboard, bicycle, and car (all used for transportation); and scissors, power lawn mower, and push lawn mower (all used for cutting). Items were presented one by one, in an order chosen to foster conflicting definitions (choices of groupings and arguments).

Although children were far from unanimous as to which objects were machines, a number of regularities emerged. In the first session, all the groups produced definitions by use to characterize what makes a machine. ("A machine is something that helps you do your homework" or "go places" or "defend you against enemies", etc.). Groups' ideas on how machines work were varied yet generally vague. "Machines work because they have a motor" or "power" or "electricity" or "a mechanism". In the second session, children's categorisations of objects showed that almost everyone drew a line between machines and non-machines in terms of an object's autonomy, that is, its ability to transform an input in significant ways.

An object, in children's eyes, is a machine if it can modify what you do with it in ways that make a difference. Thus for one child scissors are not a machine because "it's you who cut". A push lawn mower is a machine because "you push and it cuts". To the question: 'what are scissors then? the child answers "a tool". For another child a car is a machine because "it has a motor". A bike is not a machine because "its you who pedal". An aircraft with a bicycle mechanism (as exhibited at Boston Science Museum) is a machine because "if you pedal and it flies . . . then it's gotta be a machine". Note that in the case of the bike, the transformation of a rotation to a translation (moving on the ground) is not perceived as significant, whereas for the airplane, the transformation from rotation to taking off the ground is indeed significant. Being able to produce this difference requires an entity capable of generating it from within. This entity is generally referred to by the children themselves as a mechanism. Yet, if the mechanism is the identifying feature of a machine it is often treated as a black box. It is only upon request that children refer to the mechanism as the 'brain', the 'motor', or the 'power'. One important finding of this experiment is that the ideal of transparency, so important to many adults, seems to leave our children cold. Their spontaneous focus is much less on how the machine works than on what it achieves (transforming inputs) and how it can be used by people (usually themselves) to do things for them.

Elementary school children's criteria for 'machineness' generally move from a focus on switches or buttons which allow a user to produce a desired result (controls, utility), to a consideration of multiple factors including its dynamic interaction with the user (autonomy, division of labour). It is only on occasion that the physical components of the machine are evoked (power, sources, motors, complexity). Thus, if children can be said to show an increasing sensitivity to underlying mechanisms, this sense of mechanism remains to a large extent psychological/functional, since its defining criteria are autonomy and agency. A mechanism is evoked whenever a tool 'knows' how to transform inputs in consistent and reliable ways. It gains intelligence as it 'knows' how to execute a series of transformations (like a washing machine), as it responds to and executes orders at a distance (like a remote control), and as it controls or regulates its own behaviour (like a wind-up robot who 'knows how to walk') or that of some features in the environment (like a watch).

9 Discussion

Developmental theories generally assert the superiority of abstract versus concrete, and internal versus externally-based forms of reasoning (See Chapter 9, Blaye et al., this volume). In defining higher forms of reasoning as being disembodied, decontextualized, and 'in the head', developmentalists de facto relegate external mediations to the role of support or substrate of inner operations. External mediations are seen as important, especially by socio-constructivists, in helping learners to shape and share their ideas, by objectifying them. However, as minds grow, external supports and mediations should somewhat disappear, much in the way that a scaffolding should be taken off once a construction is achieved, to reveal the stand-alone nature of the edifice.

In recent years, many new trends and questions are challenging such disembodied views of cognitive development (Ackermann, 1996). Research on situated cognition (Brown, Collins & Duguid, 1989; Rogoff & Lave, 1984; Suchman 1987), and advances in the fields of pragmatics (see Chapter 9 by Blaye et al., this volume) bring evidence for the idea that knowledge is deeply grounded in its contexts of use, and thus should not be detached from the substrate and substance that make it possible. Lakoff and Johnson's experiential models of human reasoning further suggest that thoughts are not simply located 'in the head' but that they live in our bodies (and its prosthetic extensions), thus materializing our modes of physical functioning in the world, and find their roots in the world (Lakoff & Johnson, 1981).

To claim that knowledge is situated does not entail, as many cognitive scientists have it, that intelligence is pre-symbolic. Semiotic activities are the benchmark of human learning (Karmiloff-Smith, 1995). People can evoke objects in their absence, and use symbols to mediate their experience. This enables them to be proactive instead of reactive, to draw lessons from the past as a way to anticipate the future, and to hold on to their goals while engaging in ever wider detours. Their ability to run scenarios in their heads allows them to foresee disasters and to avoid them, to defer gratification, which actions-in-the-world won't do.

This being said, to acknowledge the symbolic nature of human intelligence does not mean that semiotic activities need to grow inward more than outward, nor that internalization should precede or outlive externalization. Inner speech is no substitute for outer speech; operations in the head cannot replace actions-in-the-world. Both go hand in hand and build upon each other to exist. The symbolic function, in other words, is not just a transition from action to thought. It simultaneously opens the way to new forms of externalization (spoken and written language, mappings and drawings, enactments and simulations), without which no sustained thinking could ever see the light of day. People's abilities to control if and when it's best to act, or refrain from acting, to speak out or to remain silent, and to delay gratification, are no doubt major breakthroughs in human development, and all require internalization. Yet mental operations, I argue, would fade away very soon, or remain at the level of free-floating reveries, without an ongoing anchoring of what has been internalized as a means to direct one's train of thought.

In order to think in any kind of principled and sustained ways, people need to hold onto their thoughts, i.e. to freeze certain ideas in order to vary others. Making thoughts tangible by building external supports is a condition sine qua non for building self-orienting devices, or cognitive invariants. These needs do not decrease with age, nor do scientists manage without them. Scientists, too invent abacuses and create external representations as a way to systematically carry out operations on the concrete.

People's abilities to project out their experience and to mediate it (with others and through things) enables them both to immerse themselves in situations and to 'objectify' their feelings and ideas. In Kegan's words, the abilities allow them to have an experience in addition to just being one (Kegan, 1982). Indeed, once projected and encapsulated in a description or captured through enactment, a personal experience can be addressed as if it were 'other' (not-me) which, in turn, sets the stage for a novel conversation and brings about deeper understanding and connectedness.

Along with Kegan, I believe that engaging in a conversation with people or things requires that one alternates between being embedded and emerging from embeddedness; that one switches roles from being a speaker to being a listener, from holding the stage (soloists) to disappearing in the background (chorus). Learners put themselves in other people's shoes in order to see themselves through others' eyes. They vary their stance in the world by projecting themselves into situations, or tune into things, and, then, removing themselves, to see what they have done from a distance. In addition, while they converse with their own descriptions or externalisations, learners usually start a dialog with a whole range of interlocutors, imaginary or real, to whom they address their quest. It is not an exaggeration to say that while interacting with the world, a learner's mind moves both in and out of its own descriptions (first, second, third person description), and back-and-forth from itself (to include viewpoints of others).

In conclusion, then, interacting with artefacts that 'behave' is different from shaping one's thoughts through language. While language remains a vehicle of thought, no matter how articulate, machines (or programs) are mechanical (or electronic) incarnations of processes at work. Beginning as object-to-think-with, useful in a process of constructive thinking, machines become operations-embodied. From substrate or media for expressing and shaping ideas they become agents that behave. Machines have an identity of their own. While responding to our solicitations in reliable ways, they add their own contribution, and they do so with consistency. What they achieve is not just to record or freeze processes, but to carry out operations and execute calculations. Learners perceive machines as efficient artificial partners, and use them to assist in the work of learning. Machines, in other words, are agents to which learners delegate parts of a task at hand, and with which learners distribute their knowledge. No description or inscription has ever played such a role in learning

The distinction between figurative and operative aspects of cognition, as introduced by Piaget, remains limited in that it does not account for the symbolic function of actions and operations, as expressed through re-enactments and embodied in simulations. In Piaget's model, actions transform (operative aspect) and representations describe or depict (figurative aspect). Hybrids like enactments or simulations seem to find no place in either side of the dichotomy. Piaget's distinction, on the other hand, remains useful in that it allows one to differentiate between tools (which transform) and models (which represent), as well as between programs and simulations (which replicate). Tools and programs belong to the realm of the operative in that they add value to their inputs by transforming a given state of affairs. Models and simulators, by contrast, belong to the realm of the figurative because they stand for something else, which they imitate.

A useful way forward, I suggest, is to replace the divide between figurative and operative by a more functional distinction between objects or actions used to transform a situation, and objects or actions used to consolidate a situation. This distinction makes it possible for the same activity — say playing a tune on the piano — to be both representation and operation, depending on whether the purpose is to create variations or to consolidate a sequence. Needless to say, any action, over time, becomes bi-faceted, and can be used both ways. People are able to switch modes from automated (blackboxed) to manual (unravelled or glassboxed). The capacity to use one's mind to move from black-boxing and glass-boxing — from running a procedure to freezing it — is at the core of the human's ability to learn.

12

Contextual Knowledge in the Development of Design Expertise

Anneli Eteläpelto and Paul Light

1 Introduction

In this chapter we address the question of contextual knowledge and its role in relation to professional expertise in design. Design activity has in the main been analysed in terms of context-free and non-situated strategic knowledge. What we attempt here is a re-definition of design expertise from the perspective of contextual and situational knowledge.

After a brief overview of the components of professional knowledge in general, we shall consider the main features and limitations of recent cognitive approaches to specifying the nature of design expertise. We shall then consider whether a 'situated learning' approach provides a more adequate way of conceptualising design expertise. Our conclusion will be that, despite its potential to contribute to our understanding of the nature of professional expertise as a socially located phenomenon, this approach unduly neglects developmental continuities at the level of the individual. We outline an integrative approach which attempts to combine the analysis of contextual factors (arising from professionals' work settings and functional roles) with an understanding of components of design expertise arising from the individual's experiential history.

Empirical findings will then be presented bearing upon the context- and experience-based nature of design expertise in the domain of information systems design, and upon the way in which such expertise transfers from one situation to another and from the context of university-based studies to the contexts of working life. Our argument is that an adequate understanding of design expertise, its development and its transferability must encompass (i) the nature of design knowledge and its different components, (ii) the organizational and social context as a determinant of design expertise, and (iii) the longitudinal patterns of acquisition in relation to different aspects of design knowledge.

2 The Components of Professional Knowledge

In recent discussion of professional learning and competence, professional knowledge is typically envisaged in terms of three complementary components. These involve practical, formal and metacognitive knowledge (Bereiter & Scardamalia, 1993; Eraut, 1994; Achtenhagen, 1995; Bromme & Tillema, 1995). According to a rough division, practical knowledge has been associated with subject's working life experience whereas formal knowledge is understood as a textbook-based knowledge which can be acquired in schooling contexts. The third component of metacognitive knowledge has been thought to play an important role in professionals' fusing of formal and practical knowledge.

Attributes of practical knowledge include being experience based (produced through the activity of practical problem-solving), mainly procedural (manifesting itself as the know-how of problem-solving and task accomplishment) and personal (consisting of subjectively meaningful constructs). Practical knowledge is also contextually dependent (emerging from the specific contexts where experts operate), tacit and often unconscious (making expert knowledge difficult to elicit) (Sarvimaki, 1988; Bereiter & Scardamalia, 1993).

The second component of expert knowledge, formal knowledge, can be readily articulated, and this gives it particular value in communication, teaching and systematic learning. It is also essential for dealing with issues of truth and justification (Boshuizen et al., 1995). The third component, self-regulatory or metacognitive knowledge, has an important role in controlling the application of practical and formal knowledge. Self-regulatory knowledge includes the processes of attention regulation, monitoring and evaluation of ongoing activity (Bereiter & Scardamalia, 1993). Metacognitive knowledge also covers people's awareness of their strategies as well as their ability to deploy them appropriately.

Understanding the general nature of expert knowledge and its different components is an important precondition for understanding how such knowledge is acquired. For example, little is understood about the ways in which formal knowledge presented in an educational context provides a starting point for the construction of informal practical knowledge emerging from experience gained in working life contexts. In ill-defined and complex domains of professional expertise such as design, a considerable period of work experience is generally considered a necessary condition for becoming an expert. But what exactly is achieved through such experience? More generally we can ask: what exactly is involved in 'being an expert' in the field of design? In the next section we take a look at what has become the most firmly established approach to specifying the nature of design expertise.

3 Design Activity in the Cognitive Science Framework

The information processing approach in cognitive psychology has long aspired to lay the foundations of design science by specifying the general features and components of design activity (Simon, 1981). A key assumption of this approach is that there exists a uniform structure of solution for design problems which cuts across the boundaries of specific domains and disciplines.

Goel and Pirolli (1992) provided some evidence for the existence of a uniform structure of design problem spaces in three domains: architecture, mechanical engineering, and instructional design. They analysed subjects' thinking-aloud protocols during a process of simulated problem-solving in these three domains and compared the time spent on different aspects of the task. This led them to differentiate between four general phases of design problem solving: (i) structuring the problem, (ii) preliminary design, (iii) refining the solution and (iv) detailed design.

How are design problems distinct from other problems? In the psychology of problem-solving, design problems have been characterised as ill-defined and open problems (Lawson, 1986; Hoc, 1988; Akin, 1990; Guindon, 1990; Christiaans, 1992). They are seen as having an incomplete and ambiguous specification of goals, no predetermined solution path, and a requirement for integration of multiple knowledge domains. These features make design problems as a whole particularly difficult. Because of the openness of problem goals, one of the main functions of the designer is the redefinition of the task, taking into account the economic, temporal and other constraints upon the realisation of the design.

Design problems are problems with a large number of open constraints (i.e. parameters whose values are left unspecified in the problem statement), and solving a design problem is partly a process of resolving these constraints (Simon, 1973). In information systems design, in particular, important constraints arise from the perspective of prospective users. Such user constraints are emphasised by the participative paradigm of software development (Lyytinen, 1987).

Early studies of computer science and software design did not, however, give much consideration to customer constraints. Rather, they aimed to construct prescriptive models according to which the designers were supposed to work. One of the most popular models was the structural approach to program design. In this approach, the overall system is designed first, and then the system is progressively decomposed into increasingly detailed subsystems.

Empirical support for the employment of such a top-down strategy was found, for example, by Jeffries et al. (1981) who analysed the think-aloud protocols of two novices and four experts designing a book-indexing program. However, this and other similar studies are open to the criticism that the tasks selected were actually atypical, in being relatively well-structured, having well-defined goals, presenting little novelty to the designers and requiring the integration of only a few sources of knowledge (Guindon, 1990).

In later studies, more problems with top-down strategies were discovered. Parnas and Clements (1986) argue that the top-down approach is unworkable because the requirements given in task assignment are typically incomplete and vague; the projects are of such complexity that designers cannot comprehend and keep track of all the details; and, to complicate matters, designers are often biased by preconceived design ideas of varying relevance.

The opportunistic nature of the authentic design process is apparent from much of the more recent research (Visser & Hoc, 1990; Davies, 1991; Ball & Ormerod, 1995). This is especially true of the early stages of software design, in which a designer must transform informal and incomplete requirements into high-level design requirements expressed in a formal or semi-formal notation. Guindon (1990) maintains that opportunistic decomposition is better suited to handling the ill-structured problems of high-level design, where the given system's operational requirements are refined. By analysing the verbal protocols of three professionals designing a software system of realistic complexity, Guindon was able to illustrate the frequency and causes of opportunistic decomposition. These included the sudden discovery of new requirements, the development of solutions for newly discovered requirements, and 'drifting' through partial solutions.

Top-down decomposition thus appears to be a special case, a method used with well-structured problems when the designer already knows the correct decomposition. Guindon (1990) argues that the operations of a designer depend on how incomplete and ambiguous the problem specification is, how much the designer knows about the different domains that need to be integrated, how familiar he or she is with a given problem, how many structuring operations have already been performed, and of course on the interaction of all these variables.

In the wider literature on learning and cognition increasing emphasis is being placed upon contextual and situational factors. The rich and multifaceted role of context has also been clearly explicated in a recent discussion on human computer interaction (Moran, 1994; Nardi, 1996; Newman,1994), where it has been argued that all good design takes shape at the boundaries between the artefact and the surrounding context. Context, it is suggested, serves to imbue artefacts with material, economic, and social meanings (Brown & Duguid 1994a, b; King, 1994).

Such an emphasis on context is not proposed only on theoretical grounds. Studies focusing on the activities of authentic software design teams (e.g. Walz, Elam & Curtis, 1993) have also demonstrated the importance of contextual knowledge of the application domain for a successful design outcome. Techniques have been introduced with a view to improving contextual knowledge of customers and end-users in information systems development. As yet, however, there is little in the way of empirical research on the different manifestations of contextual knowledge in solutions to design problems, or on how such knowledge is related to the development of professional expertise.

4 Contextual and Situational Approaches to Learning and Expertise

'Context' and 'situation' have quite different meanings within different approaches to research on learning. In the classic cognitive view, context typically referred to environmental conditions or other specific circumstances that needed to be controlled in experimental research. Concerns about the ecological validity of cognitive psychology gradually helped to give contextual factors greater prominence. Authentic research task, presented in the relevant physical context, came to be considered as important preconditions for the generalisability of results in real-life conditions. Subsequent research increasingly emphasised the domain-specific nature of professional expertise. It became very clear that expertise in one domain does not imply expertise in another (Chi, Glaser & Farr, 1988).

Contextual constraints came to be understood as intrinsic and defining elements of the domain in question. Contextual constraints are rather differently understood within the 'situated learning' approach. Here, context is understood in the wide sense of the historically and culturally constituted conditions of learning communities (Lave, 1988; Lave & Wenger, 1991). Expertise arises through participation in the authentic activities of such communities. Individual learning processes are embedded in ongoing social practices, and should be interpreted as an aspect of the individual's identity formation as a member of a learning community.

Thus, the proponents of a radical situated cognition approach see human cognition as having a fundamentally socio-cultural and historical nature. The focus of this approach is on the spontaneous affordances of human activity which emerge from shared participation in practical situations. Lave and Wenger (1991) treat the process of apprenticeship as a prototype model of learning. Here, learning takes place in the interaction between an expert and a novice, an interaction which involves observation, coaching, and participation in real practice. In phases of practice, a novice takes responsibility for specific components of the job. In phases of guidance, the learner integrates feedback from the master.

Ideally the master, more skilled than the novice, should provide explanations for the particular strategies of working and should allow the novice to participate in decision making. Empirical studies (Radziszewska & Rogoff, 1991; Rogoff, 1990) suggest that without guidance and participation, the novice does not benefit from the interaction. The situated learning approach has achieved a shift of focus from individual cognitive processes to the social relations and discourses realised in practical communities. It has properly highlighted the social nature and location of professional learning and expertise. Since professional expertise is embedded in the socially and culturally constructed conditions of work organisations, it cannot be understood without reference to these contextual determinants. The emphasis thus falls on the need for studies to be carried out in the authentic setting of working and learning environments.

Other recent research has approached the social nature of expertise from a slightly different standpoint (Bereiter & Scardamalia, 1993; Sternberg & Horvath, 1995). An expert is person we turn to when we are in trouble; the normative character of professional expertise means that the performance of experts should match the expectations of the social context where it is realised. This approach also suggests that the contextual attributes of work organisations and their functional goals constitute a natural starting point for any attempt to understand the nature of professional expertise.

Radical situationism is open to the criticism that it neglects developmental continuities that exist at the level of individual subjects. When applied to questions of the nature and development of professional expertise, the situated cognition approach may thus lead to the reduction of professional expertise to mere workplace discourse, with no roots in the participants' experiential learning histories (Gruber et al., 1995; Nardi, 1996). Paradoxically, this can lead to a form of theorising in which individuals are treated as if they themselves had no history or context.

What is neglected in this perspective is the continuity of human experience, a continuity that includes developmental achievements which have a bearing on behaviour in a whole range of situations. By contrast, some contemporary developmental approaches regard the individual's experiential learning history as the most important determinant of their level of expertise at any given time. For example, in the sphere of developmental cognitive psychology there have been attempts to analyse contextual constraints in terms of the quality of subjects' experiential learning background and their prior knowledge, schemes and mental models (e.g. Hatano & Inagaki, 1992; Light & Butterworth, 1992). However, most such studies are concerned with child development, and there has been very little empirical elaboration of the relationships between adults' professional development and their experiential learning backgrounds.

5 Towards an Integration of Individual and Situational Factors

Recognition of the limitations of the cognitive approach on the one hand and of the situated approach on the other has produced a more balanced view of their complementary roles and a desire to achieve an integration (Guberman & Greenfield, 1991; Norman, 1993; Sternberg & Wagner, 1994). In relation to research on professional expertise, an interactive approach might lead us to ask how different kinds of contextual factor contribute to, and interact with, different kinds of individual competence. It might also lead us to ask whether contextual and situated factors have a similar role at different stages of expertise development. This is an issue which has thus far only been discussed by researchers working within the framework of skill acquisition.

Dreyfus and Dreyfus (1986; see also Benner, 1984; Berliner, 1992) have proposed a stage model of expertise development. At the beginning, activity is primarily controlled by inflexible and context-free rules. When learners progress from the 'novice' level to the level of 'advanced beginners', context-free rules will be partly replaced by situational factors. As learners acquire still more practical experience, they begin to be able to control and monitor the target activity through a flexible combination of situational factors and context-free rules. Dreyfus and Dreyfus maintain that a serious commitment to other people's perspectives cannot be achieved until the highest level of skill development (see also Eteläpelto, 1994a).

The idea that different kinds of contextual knowledge are acquired at different stages of acquisition of professional expertise has not been empirically analysed in relation to design. In other fields expert-novice comparisons have been used to illuminate changes in knowledge structures and cognitive strategies as a function of increasing practical experience. Such studies have demonstrated that experts perceive large meaningful patterns within their own domain, that they focus on the relevant cues of the task, that they represent the problem on a deeper level and have better self-monitoring skills than novices. Experts' knowledge structures are hierarchically organised and have more depth in terms of conceptual levels than those of novices. They categorise problems within their domain of expertise according to abstract high level principles, and their knowledge structures are more coherent than those of novices (Batra & Davis, 1992; Chi et al., 1988; Ericsson & Smith, 1991; Saariluoma, 1995). Parallel differences between experts and novices have been shown in our empirical analysis of computer programmers and systems analysts (Eteläpelto, 1993, 1994b).

The limitations of such expert-novice comparisons have been widely discussed in the recent literature on expertise (e.g. Bereiter & Scardamalia, 1993; Rambow &

Bromme, 1995). Often the length of practical experience is the only criterion used for defining expertise. Recent studies have, however, shown that after a certain minimum length of practical work experience, the quality of experience is more important than the length alone.

Sonnentag (1995) showed that excellent software professionals had broader but not necessarily longer professional experience than did their colleagues. They did not differ from their more average colleagues in terms of the amount of time they had spent in typical software development activities such as design, coding or testing. However, they had typically been more heavily engaged with such demanding activities as evaluations and consultations. The continuous challenges associated with such tasks are considered by Sonnentag to be powerful formative influences upon expertise. This conception of an expert career, characterised by growing challenges, is also discussed by Bereiter and Scardamalia (1993), who emphasise the sense in which good professionals are always working at the 'growing edge' of their competence.

6 Design Expertise as an Integration of Contextual and Strategic Knowledge

In our analyses of professional systems analysts' knowledge structures and problem-solving strategies, we have tried to capture the qualitative variation of professional design expertise and analyse the main sources of this variation. We have used a combination of expert–novice comparisons and longitudinal methods, focusing upon changes occurring across a 7-month period during which students were making the transition to working life. The aim here was to investigate developmental continuities in the acquisition of the strategic and contextual knowledge necessary for professional design expertise.

The expert–novice comparisons revealed that indeed experts tended to perceive the task of design and development of information systems on a much more comprehensive and abstract level than novices. The great majority of professional systems analysts, who had abundant work experience, perceived the development of an information system at the level of an organisation, whereas the novices, who had very limited work experience, mainly perceived it at the level of an individual user (Eteläpelto, 1994b).

The longitudinal dimension of the study showed that during their first practical project-based course, novices acquired a good deal of strategic competence in using domain-specific tools and methods. Indeed, the course strengthened their strategic knowledge to the point of over-emphasis. Identification with the specifics of the customer's situation tended to decrease. This resulted not so much from any reduction in consideration of the customer as from a shift to more procedural ways of taking the customer context into account. During the practical course, novices tended to move from inflexible and professional-centred strategies toward more individualised and interactive strategies such as prototyping. The adoption of such strategies actually provided for a more adequate and professional way of taking the customer context into account.

At the initial level of acquiring design expertise, the strategic component of professional knowledge formed the central focus of learning. Later, a more balanced combination of contextual and strategic knowledge was evident. In our study, the professional systems analysts who had a great deal of work experience suggested more versatile strategies and ways of taking customer perspectives into account than did the novices. Experts also identified more with customers, and considered the customer's situation in a more comprehensive way. At the highest level of expertise, contextual constraints were sometimes considered from a meta-level perspective, in questioning the meaningfulness of the whole task assignment (Eteläpelto, 1998).

Some trends in the acquisition of strategic and contextual knowledge can, thus, be distinguished. Our results indicate that systems analysts who have been exposed to formal knowledge in the university context will not be able to take adequate account of the perspectives and organisational contexts of their customers until they have acquired strategic competence in the domain in question, and become involved in the practical communities of working life. Consideration of contextual factors as regards both the professionals' own organisational context and that of their customers represents a major learning target at the initial stage of entering the working life. In the later stages of a professional's career, contextual knowledge of organisations plays a more minor role, whereas questions of professional goals and values are at the core of their continuing learning.

In all of the groups that we analysed, there was a small but significant minority that deviated from the patterns we have described. Typically they manifested an extreme identification with the customer's perspective, to the point where they made almost no use of their strategic knowledge of information systems development, adopting instead a 'soft' or anti-professional style.

When interviewed about the origins of their solution strategies, some 90% of both experts and novices responded that their solutions originated from recent experience in their work or practical training setting. This was confirmed by systematically comparing the approaches they adopted in problem solution and the roles that they had occupied in their prior work or practical training contexts. Thus, all the systems analysts, whose regular work involved them in getting acquainted with new software products and then purchasing and tailoring them to fit some need, also approached problems in information systems development in terms of purchasing and tailoring products. By contrast, those whose work (or recent practical project work) involved the production of software de novo would tend to suggest this approach to information systems problems (Eteläpelto, 1998).

What, then, do these results suggest in terms of transfer from learning experiences to solutions of design problems (cf. Mayer & Wittrock, 1996)? Project-based learning experiences resulted in transfer mainly at the level of domain-specific strategies and tools of information systems development. In some respects this is encouraging, since in the wider research literature the use of such strong domain-specific strategies has been found to characterise the problem-solving of experts in a given domain, whereas novices tend to use weak methods representing such general heuristics as means–ends analysis, applicable to a wide range of problems (Salomon & Perkins, 1989). In this respect, at least, it would seem that the practical project learning experience has successfully introduced our novices to a more expert-like way of problem solving. On the other hand, what our professional subjects transferred from their working life went beyond simply domain-specific methods and tools comprising strategic knowledge. In addition our professional subjects also tended to 'transfer' those functional roles and tasks that they had adopted in their work organisations.

Accordingly, our results suggest the need for a more comprehensive framework for analysing the question of learning and transfer from practical work experience. Especially in open-ended and ill-defined problems, such as design and development, such a framework should cover not only strategic knowledge or specific skills but also the social context of subjects' work organisation.

7 Conclusions

We have observed that, in the framework of cognitive science, design activity has been mainly analysed in terms of general cognitive processes and strategies. The goal has been to establish the general structure of design problem solving. The associated research has tended to employ highly structured and artificial 'textbook' problems which are far removed from the tasks that professional designers meet in their day to day work environments. When more authentic tasks are used and the activities of authentic working teams are analysed, the uniform structure of design activity seems much less apparent. Instead of using highly structured topdown strategies, designers employ much more 'opportunistic' ways of constructing plans. Moreover, when real professionals are studied, a great deal of inter-individual variability is apparent.

For these reasons, we have argued that the nature and development of design expertise needs to be redefined so as to take fuller account of the contexts and situations in which the professionals are working. Contextual approaches have considerable relevance to promoting our understanding of the social nature of professional expertise. Radical 'situated learning' approaches, influenced as they are by anthropology, tend to focus on how continuity works at the level of a given community or institution. They characteristically have rather less to say about how continuity works at the level of the individual, moving between communities and institutions. From a psychological standpoint the question of 'what develops' at the level of the individual is a central one (cf. Blaye, Ackermann & Light, this volume). The suggestion that professional activity has a context-sensitive nature is not inconsistent with taking an interest in the experiential histories of designers, or in the patterns of developmental change which mark the acquisition of design expertise.

On the basis of our discussion of integrative approaches, we conclude that a re-conceptualisation of professional expertise is needed, in which account can be taken both of the functional demands of organisational contexts and of designers' experiential learning histories. The transitions involved in the acquisition and integration of formal and practical knowledge have been the subject of our own research on the processes of becoming expert in systems analysis and software design. Future studies of professional expertise require this more comprehensive conceptual framework, and will need to focus on both continuities and discontinuities in the experience of professionals in their training and subsequent work environments.

PART 3

Learning with and by Machines
Introduction by Joan Bliss

The focus of this book is on human learning, reasoning and thinking in different social and technological contexts, and the manner in which these practices and artefacts together help to determine the way in which people approach situations. In the first Part there was a specific focus on the manner in which meaning is created and negotiated by people operating with cultural tools in concrete activities. While Part Two continues this theme, there is a more specific focus on how contexts constrain or facilitate human thinking and actions. More particularly, it looks at the tension between traditional psychological approaches, which take little account of context and situation and the situated learning perspective, which gives context and situation pride of place in accounting for learning and reasoning. Part Three chooses a more specific focus in that it examines how new technologies, or new forms of artefacts, are becoming part of the contexts in which we learn, reason and think. It sets out to understand the extent to which people can appropriate these new technologies as tools for their thinking, and whether or not these new approaches can enhance their learning, reasoning and thinking.

As mentioned in the Introduction to this book the term 'new technologies' is used in an attempt to distinguish between existing technologies which have been critical in the development of humanity and new forms of technology which have emerged in the second half of the 20th century. We mentioned earlier Ogborn's (1996) comment that commercialised technologies make the workings of artefacts invisible (see p. 2, this volume). For example, many Western people use aeroplanes as a fairly standard form of transport, some of us believe we know why aeroplanes fly, but few would be confident, of giving a really clear explanation of flight. Even fewer know how to fly or how to maintain them. But overall, air travel is here to stay and we no longer question it. It changes for good or for ill our patterns of interaction that we as individuals incorporate into our lives.

One of the features of the new technology is that much information and knowledge can now be contained in very small spaces and transmitted, communicated or simply carried around without a problem. The space required for its storage is extraordinary small.

The concept of storage runs through the development of technology and both Giddens and Bourdieu employ it in their analyses of the development of society. In the hunter and gatherer societies storage of food and perishable goods was of fundamental importance. In a discussion on the rites of institutions, Bourdieu (1991) argued, ". . . all groups entrust the body, treated like a kind of memory, with their most precious possessions " (p. 123). As societies and tools developed other issues of storage arose. Giddens (1997) put forward the idea:

> Storage is a medium of 'binding' time-space involving on the level of action, the knowledgeable management of a projected future and the recall of an elapsed past (p. 261) . . . Information storage, I wish to claim, is a fundamental phenomenon permitting time-space distanciation, a principle in the constitution of societies . . . (p. 262).

New information and communication technologies make it possible to retain huge quantities of valued and powerful information and knowledge without a problem; space requirements are now of an altogether different kind. Also ICT transforms the way in which knowledge can be distributed in and between learning sites. Such knowledge and information, stored in personal computers or tracked through electronic mail or internet connections, is now accessible outside the standard recognised institutions. Thus ICT challenges traditional boundaries between knowledge communities and the demarcation of knowledge domains. So we can ask: "What will count as epistemic authority in a society of distributed knowledge and how will this authority be exercised?"

Also I suspect that such IT-led visions of an educational future take too much for granted for although these new technologies offer new and powerful resources they too have their limitations. Our historical experience of earlier information technologies, like writing and print, tells us that they not only opened up new avenues for social participation but also played a significant role in the social exclusion of particular groups.

What are the implications of new technology on how we learn, take decisions, and make relationships? In particular, in what way and to what extent will computers influence our cultural and social context? For example, it is often believed in education that computers will redefine the locus of a learning institution by freeing education from the constraints of time and space.

These earlier questions make assumptions about who or what is in 'control'. While we see individuals as active, constructive and able to shape to some extent, or at least participate in, the direction of their own futures, it is also important to ask questions about the role of society and its practices in shaping futures. For example, what are the influences of human practices and customs on changing and shaping new technologies? Indeed this is the question that underlies the 'situated learning' approach of the first two chapters of this Part. Perhaps more constructively we should ask: how can we adapt new technologies to better suit individual and collective goals? As Muller and Perret-Clermont, in this volume, point out, ". . . the context does not appear to be only a developmental factor, but rather to be constituent of thought itself . . . the subject, considered as agent of his/her own development, interprets the different constituents of the context in which he/she is, and lends them particular meanings."

Ackerman, in this volume, talks about machines but rather than computers she finds her starting points in simpler machines, and in the much simpler simulations evident in young children's pretend play. She goes on to claim: "Learners perceive machines as efficient artificial partners, and use them to assist in the work of learning. Machines, in other words, are "agents" to which learners delegate parts of a task at hand, and with which learners distribute their knowledge". She goes on to argue,

". . . they (machines) become agents that behave. Machines have an identity of their own".

Computers are yet a different kind of machine from those just discussed. Computers invite interaction. They offer not only the possibility of doing things better and faster than before but also they offer the chance of doing things that we could not do before. They promise help, they assist us in solving problems, they provide us with the means to generate and create all sorts of new products — from a poem to a balance sheet. It is not so much the machine that is doing the work but the person in co-operation with the machine that produces something new; computers are a new type of tool. A computer's function will depend on the context of its use: industry, commerce, business, communications, the leisure industry, educational settings, the home, and so on. It could be argued that, unlike the machines referred to by Ackermann, the computer's identity as a machine is secondary to the identity that usually resides in the product the user creates: the academic paper, the novel, the software program, or the solution to the problem it has helped to solve.

The first four chapters in this Part involve learning with computers. Thus we examine the importance and the nature of educational settings or contexts in which computers are used: first in schools, and then in universities. Then we look at software tools: in a first instance at how ideas from situated learning have influenced their design, and in the second case, how the user's situated knowledge both enhances and hinders the use of IT tools. In the last two chapters we turn to learning by machines, and the situated agents movement in artificial intelligence which will be familiar to those in the field. While the situated agents approach may have had its roots in cognitive psychology, quite often its later developments are so different from its origins that it is not always possible to see the links. These two chapters about situated agents attempt to build these links.

Recent research would indicate that learning with computers — whether at a fundamental school level or in the discipline of, say, computer science at the tertiary level — is becoming a genderised issue in much the same way as the learning of physics or chemistry, where boys and later young men would seem to dominate the scene, at least in many European countries. Littleton and Bannert's survey of the area shows, however, that the issue is subtle at two different levels. First, Littleton and Bannert refer to the work of Huff and Cooper (1987) which points out, ". . . that it is not the computer or even the software that is at the root of sex bias in software, but the expectations and stereotypes of the designers of the software" (p. 519). Second, they analyse a number of studies, including their own, which indicate that even when software has no obvious gender connotation, a sex bias is introduced by the "particular contexts and situations which constitute girls' and boys' experience of computing activities". In other words, through the analysis of the social contexts of young people's computing activities we can begin to understand the importance of either the social situation or even the activity itself. For example, we learn that the framing of an activity as a 'game' or as a particular type of task can deter girls from or attract girls to working with computers and, of course, similarly for boys.

Similarly through Crook and Light's examination of issues related to the greater deployment of computers in order to allow teaching and learning to be more flexible and more affordable, we become aware of the importance of the analysis of the learning contexts. Crook and Light show that learning with computers in universities does not take place in a type of neutral zone; existing educational practices have a long historical development and, whether we like it or not, these circumscribe the ways in which today's students learn.

Crook and Light's studies are diverse — from self-report surveys of study practices during examination revision to students working together on academic problems. They also paid particular attention to the informal, out-of-class experience of study, for example, the possibilities for informal collaboration with peers and informal contact with tutors that arise from participating in a full time community of study. Their results give food for thought when they state, ". . . our results confirm us in the view that students are enculturated into particular communities of learning, and the resulting practices will offer resistance to the 'bolting on' of new educational technologies".

Both these chapters serve as a warning since through their analyses the authors are able to show that: "Far from revolutionising society, the computer has conformed to society, becoming another element of the status quo" (Reisman, 1990, p. 45). Yet, perhaps again they reflect the nature of the transition from an industrial to a technological society. Perhaps at the moment those who work with new technologies are, as Crook and Light point out, highly self-selecting and may still be small in number. Perhaps such studies can provide information about what is both sufficient and necessary to make the new technologies more powerful resources through being more properly integrated into a culture we understand more fully.

Huff and Cooper point out that, ". . . gender stereotypes in educational software often arise as a direct result of the designers' expectations about potential users" (Huff and Cooper, 1987). Not only are the designer's expectations critical but also the theoretical perspectives that inform his or her implementation of the software design. Luckin's chapter takes inspiration from the situated learning perspective to design a new learning environment, VIS, which is part of a larger system called TRIVAR (Luckin, 1996). VIS allows the user (children between 10–11 years) to create a mini ecosystem through the selection of animals and plants which share a common habitat; these elements are then, through a range of activities, built into a food web.

One of the key situated learning concepts which underlies the design of VIS is what Vygotsky called the 'zone of proximal development' or the ZPD. Crucial to the ZPD is the distinction Vygotsky draws between a child's actual level of development and his or her potential level, that which can be reached with the assistance of the able peer, parent, teacher or in this case, the computer. Wood, Bruner and Ross (1976), referring to the ZPD, coined, at that time, a term to describe this process of outside help.

> . . . (it) involves a kind of 'scaffolding' process that enables a child or novice to solve a problem, carry out a task or achieve a goal which would be beyond his unassisted efforts (p. 90).

Thus, understanding and implementing the strategy of an able teacher is essential to the VIS environment. Luckin has used Wood's later work (Wood et al., 1993) to implement the scaffolding approach in her software. This involves five levels of hints or scaffolding as well as ways of adjusting or differentiating the activities provided to the child, in other words, ". . . 18 possible combinations of help and activity adjustment". This is only part of the story in the design of this software — for example, the student model and the knowledge base are equally as complex.

Teaching is an activity that is often taken for granted. Most of us believe that, given we know some area reasonably well, we could teach it. Reading Luckin's chapter brings home very sharply the complexity of the activity of teaching and makes us rethink again the talents needed to do such a job. Let us hope that Luckin's software achieves at least some of the success of an ordinary school teacher!

Corruble and Bliss in their chapter present two research directions which use computational methods to aid learning about the world. In both cases, computational modelling and simulation are helpful in externalising some key assumptions embedded in individuals' internal representations of the world. The approaches are however different in the specific tasks they assist. A dynamic modelling tool IQON is used in the classroom to help students to model phenomena they experience in their daily life (traffic congestion, fitness and diet, shop management). CHARADE and PASTEUR are used to aid researchers to develop causal hypotheses from a number of observations (leprosy research, cancer research). Yet, both systems address the situatedness of the task they assist through a constructive approach to modelling, by taking advantage of the trial-and-error capability of computational simulations to resolve the discrepancy between model's behaviour and modeller's expectations. Our examples illustrate how computational tools offer the possibility of being used simultaneously as a medium and as a mirror. As medium, users can externalise their thinking through modelling, as a mirror through the simulation of the model which reflects back to the user the true face of what has been externalised.

Turning to the other theme of this Part — learning by machines, we have two chapters that address artificial intelligence's (AI) situated cognition approach. In the first of the two, Cañamero and Corruble set out the classic ideas within this perspective. Thus they describe how the situated learning ideas are developed through situated agents and come in several flavours. One emerged within the symbolic AI community in the field of knowledge acquisition and has its roots in Newell's knowledge-level hypothesis and Simon's bounded rationality principle. The other has developed within the behaviour-oriented AI community and takes ethology, evolution and biology as its inspiration sources. Cañamero and Corruble examine landmark work illustrative of both research lines, through an analysis of the underlying assumptions according to a number of criteria that allow a better definition of the similarities and differences in these areas.

Cañamero and Corruble highlight Agre's point of view (1993) that the two approaches belong to two distinct 'world views' which cannot directly interact in a scientific fashion: "each has its own coherence and internal validity and is not touched by external arguments".

In Chapter 18 de Boer and Cañamero examine more in depth situated learning in autonomous agents. They define an agent as adaptive when it is able to improve its behaviour to make it more appropriate for its environment and for the task to be performed. Thus in this context learning is seen as a form of adaptation. De Boer and Cañamero describe two types of situated agents: synthetic and robotic. They have chosen this division because there are a number of problems and possibilities particular to these two different embodiments.

With synthetic agents, or learning in software implementations, they examine three important architectures: classifier system-based Animats, behaviour networks (Maes) and a schema-based simulation of Piagetian development theory (Drescher). Some of the elusive problems which they address in relation to these agents are: explore/exploit problem or the drive for survival, that is when to stop exploring in order to exploit what an agent already knows; assignment of credit problem — or which behaviours or 'performed actions' should be reinforced; and incomplete state information, that is, as Kaelbling points out, ". . . we must make our agents adaptable when the specification of the environment and the agent's sensori motor capabilities are incomplete, incorrect, or simply at an inadequate level of abstraction" (1995).

Learning problems in robotic agents are magnified because of the physical embodiment. De Boer and Cañamero show that in addition to the problems discussed for synthetic agents it is important to understand the problems associated with processing complex sensor data, planning complex actuator actions and actually operating in real time. They illustrate these problems through an example of a simple robot learning experiment carried out at the AI Laboratory of Vrije University in Brussels. Of particular interest is the problem of over-training, and whether or not robots can 'forget'. In this instance they show that 'situatedness' allows for efficient learning but that, because of this, transfer becomes a difficult problem.

At the end of these two chapters on learning by machine, we are left with the feeling of having explored 'autistic learning'. When talking about situated cognition and attempting to give a meaning to the context, we are referring not only to the physical but also the symbolic and social environment in which an individual is embedded. Those working in AI are well aware of the long path they have yet to tread in order to be able to give an account of the 'situation' either through synthetic, robotic or other approaches. Yet their struggles to understand human learning and simulate it often help us to understand better our own learning situation.

13

Gender and IT: Contextualising Differences

Karen Littleton and Maria Bannert

1 Introduction

In recent years there has been a wealth of research focusing on the design of technology-supported learning environments (see e.g. Duffy & Jonassen, 1992; Issing & Klimsa, 1995; Sewell, 1990; Vosniadou et al., 1996). Yet despite the characterisation of classroom computing as a male domain (e.g. Beynon & Mackay, 1993; Culley, 1993; Pelgrum & Plomp, 1991; Schofield, 1995), the impact of gender issues on the design of such environments remains largely unexplored (Bamossy & Jansen, 1994). Here we present a brief review of the gender and educational computing literature, and go on to argue that our understanding of gender-related differences in response to computer technology can be enhanced by taking a situated perspective on human cognition and learning — a perspective which sees human cognition as being fundamentally rooted in physical and social action. We draw on examples from our own research with school-aged children in order to demonstrate that contextual cues and social situatedness are constitutive elements of childrens' responses to computer-based learning environments.

2 Gender Differences in the Use of Educational Technology

During the early and mid 1980s, as computers were becoming increasingly prevalent in the workplace and in schools, gender-related differences in response to computer technology became a key focus for research. Issues such as the impact of new technology on female labour and women's jobs, computer education for girls and women, and gender-related differences in the exposure to and use of computers were all investigated (e.g. Chen, 1986; Fetler, 1985; Hoffmann, 1987; Kay, 1989; Linn, 1985; Sutton, 1991). It is the work focusing on the educational use of computer technology that is of particular relevance here.

Whilst there is a huge volume of research addressing the use of computers in educational settings some consistent trends do emerge. First, it seems that during the early years there is very little difference between girls and boys in terms of how they see computers, their liking for computer technology, or their involvement in computer-based activities (Bergin, Ford & Hess, 1993; Williams & Ogletree, 1992). As they get older, however, girls' engagement with IT begins to decline (e.g. Lage, 1991) and data spanning the full age range of compulsory schooling indicate that overall computers are being used more by boys and male teachers than by girls and female teachers (e.g. Bannert & Arbinger, 1996; Kay, 1992; Littleton, 1996; Podmore, 1991; Straker, 1989.). In mixed schools boys often dominate computer activities

(Beynon, 1993) and where small groups of girls and boys are expected to work together on computer-based activities, the boys invariably dominate the discussion and joint-activity (e.g. Barbieri & Light, 1992; Culley, 1993). Boys also make more use of computers in the home than girls (Fife-Schaw et al., 1986; Lockheed, Nielsen & Stone, 1985; Robertson et al., 1995) and whilst this is primarily for games, gender differences in use are reported for all home computer activities (Linnakylä, 1996; Martin, 1991). It is perhaps not surprising, then, that significant gender-related differences in knowledge of computers, so-called computer literacy, appear throughout studies of secondary school pupils (e.g. Arbinger & Bannert, 1993; Fetler, 1985), college students (e.g. Lockheed et al., 1985) as well as postgraduates (e.g. Kay, 1989).

Data drawn from across the entire age range of compulsory schooling also reveal that girls are often less positive about computer-use than boys (e.g. Martin, 1991; Todman & Dick, 1993). Whilst generally males and females are regarded as equally competent, girls often report more negative feelings concerning their own personal involvement with computers (Chen, 1986). We should not, however, simply conclude that girls are not interested in using computer technology. Rather, it would appear that girls' responses to computers are more polarised than those of boys (Culley, 1993). For example, a survey of attitudes to computers conducted in Australian schools revealed that whilst as many girls as boys enjoyed using computers, many more girls than boys ardently disliked them (Hattie & Fitzgerald, 1988). Even where girls and boys express equally positive attitudes, both believe that boys like and use computers more than girls do (Hughes, Brackenridge & MacLeod, 1987). This belief may well be justified since there is some evidence to suggest that more boys than girls actually report liking computers (Wilder, Mackie & Cooper, 1985) and perceive computers as having a larger part to play in their future lives and career aspirations (Hattie & Fitzgerald, 1988; Culley, 1993).

Given this trend in compulsory education, it is perhaps not surprising that in higher education too computing is increasingly becoming a male preserve. Recent years have seen a dramatic decline in the number of female undergraduates studying informatics. For example, in Germany 10 years ago 19.1% of informatics undergraduates were female, now women make up just 8.6% of the student population (Ebach, 1994). This trend has been mirrored in the UK where applications by girls to study computer science at university dropped by 50% between 1978 and 1988, as did acceptances (Hoyles, 1988; Newton & Beck, 1993). Currently only 7% of UK computer science undergraduates are female (Coolican, 1997). Thus, in addition to national and international research findings regarding gendered access to the fields of science and technology, we see another potential case of 'gender-specific discrimination' (Schiersmann, 1992) arising from the introduction of computer technology into the home and classroom.

3 Explaining Difference

It is undoubtedly important that we document any gender-related differences in terms of attitudes towards, access to and participation in computer-related activities. It is, however, vital that we also derive a clearer understanding of how any such differences arise so that we can devise methods and strategies to support and encourage girls' and women's participation in computer-related activities.

There have been a variety of explanations put forward to account for gender-related differences in computer use. The construction of computing as a gendered activity and the image that the computer has within our society, where the computer seems to be defined as a machine for men, offers a recurrent explanatory framework.

Other explanations see gender differences as being inextricably bound up with individual versus collaborative modes of working. For example, Hoyles (1988) maintains that when the use of the computer is coupled with an individual or perhaps even competitive mode of working, girls tend to find the experience alienating. By contrast, when computers are associated with collaborative modes of working girls are typically just as enthusiastic as boys in their response to computer technology (Hoyles, Sutherland & Healy, 1991).

Hoyles' work also suggests differences in programming style between boys and girls using Logo, with boys adopting a more formal and closed approach compared to the open-ended and exploratory style favoured by girls (Sutherland & Hoyles, 1988). This issue has been treated in more depth by Turkle (1984) and Turkle & Papert (1990) who suggests that there may be fundamental differences in the cognitive styles of males and females which impinge upon the ways in which they relate to computers. Turkle suggests that many women find the experience of computer-use aversive, primarily because the prevailing computer culture imposes on them a masculine, top-down, formal–analytical method of working. From this standpoint the argument is that if women and girls are to become full users of computer technology then they must reject the emerging conventions concerning how computing is done, and negotiate ways of working with the computer that they find convivial (Kirkup, 1992).

Classroom observations highlight the issue of interactions between the genders in relation to computers. Girls tend to get less time on machines and (when they do get on the machines) get less help from the teachers because the boys tend to dominate both teacher and machine resources (Culley, 1988). It is from this kind of observation (together with the reportedly positive engagement with computers of pupils in girls-only schools) that the current enthusiasm in some quarters for segregating boys from girls in the context of computer-use arises.

4 Understanding Gender-related Responses to Technology: the Need for a Situated Approach

The literature detailed in the previous section clearly documents indicators of the existence of quite marked gender differences in response to computers, and a variety of hypotheses about their origin. Many of these hypotheses highlight long-term issues such as differences in respect of cognitive styles and preferred ways of working or patterns of classroom interaction. However, we argue that there is a need for a complementary approach which attempts to understand gendered responses to computer technology through a careful consideration of the particular contexts and situations which constitute girls' and boys' experience of computing activities. Only by studying the nature of these contexts will we come to understand the ways in which particular situations and contexts may either constrain or offer the affordances necessary for initiating and supporting the learning processes. Such a focus on the importance of context is inspired by a situated perspective on human cognition and learning.

As other chapters in this volume make clear, a situated approach to the study of human cognition embraces a diverse range of theoretical perspectives (see also Artman, 1995; Gruber et al., 1995; Law & Wong, 1996; Säljö, 1995). Yet, underpinning these perspectives is a recognition that human cognition and learning are embedded in specific contexts (e.g. Brown, Collins & Duguid, 1989) and are constituted through processes of inter-dependence, interaction and transaction between the person and their environment (e.g. Sternberg & Wagner, 1994). Seen in these terms, then, context comprises the physical, symbolic and social environment in which an individual is embedded.

In the section which follows we focus on children's performance on a variety of computer-based tasks, and results taken from four of our own pieces of work are presented. Each of these pieces of work serves as a striking example and powerful testament to the critical role of context in determining the level of children's performance on specific computer tasks.

5 Empirical Evidence for Contextual Determinants

5.1 Example 1: Metaphors and Images

Despite the large discrepancies recorded in girls' and boys' use of, and attitudes towards computer technology, very few gender differences in performance on computer-based learning and programming tasks have been observed. Both girls and boys work effectively on programming tasks (e.g. Finlayson, 1984; Light & Colbourn, 1987; Linn, 1985; Webb, 1984), electronic database search tasks (e.g. Eastman & Krendl, 1987), on simulation exercises (e.g. Cummings, 1985; Johnson, Johnson & Stanne, 1985) and on science tasks (Issroff, 1994). A notable exception to this trend, however, was reported by Barbieri and Light (1992) who studied children's computer-based problem solving using a route planning task known as King and Crown and found marked differences in performance (i.e. how far they got towards the correct solution within a fixed period of time) favouring boys. Similar differences were not evident, however, when we examined the relative performance of boys and girls using a structurally identical, but apparently less gender-stereotyped, version of the same task known as 'Honeybears' (Littleton et al., 1992).

Since the structure and the cognitive demands of the two tasks were identical, the reduced gender differentiation shown with Honeybears suggested that the setting of the task and the gender of the characters may have had a crucial bearing on the relative performance of boys and girls. There were, however, crucial design differences between the two studies which meant that we were unable to draw this conclusion with confidence, so we went on to make a direct comparison of the performance of girls and boys working with the two versions.

The study involved 11–12-year-olds working individually either on the King and Crown or the Honeybears problem for half an hour. The results revealed that whilst the boys' performance was little affected by the version of the software they were using, the girls' performance was substantially affected by the version of the software in use, it being far superior for Honeybears. The study was repeated with a new sample of children, and with an even closer isomorphism of the two versions (Pirates versus Honeybears) and the results showed a close replication of the previous study's findings (Littleton et al., 1993). A not dissimilar pattern of responses also emerges from a recently completed study (Littleton & Light, 1996) in which the performance of girl–girl and boy–boy pairs working for one half hour session on either the Pirates or the Honeybears software was considered. Here again the girl–girl pairs showed markedly superior performance on the Honeybears while the boy–boy pairs responded similarly to the two software types.

We are currently undertaking further analyses of the data from these studies in an attempt to understand the psychological processes mediating the gender differences in response to the Pirates problem-solving software. However, the results as they stand provide an immediate testament to the crucial roles of metaphors and images in the presentation of a computer task. Here, despite the constraints and demands of the task being identical with both versions of the problem-solving software we can see that the images and metaphors used to carry the task have a crucial bearing on girls' cognitive performance. The metaphors used to present the task to children are far from neutral; they clearly exert a powerful effect on the children's engagement with and conceptualisation of the problem, which in turn impacts on their performance.

5.2 Example 2: Contextualising the Task

In this second series of studies we illustrate the effects of varying the context in which a computer-based task is presented, without making any changes at all to the software itself. For these studies we used a specially produced piece of software modelled on a physical device which involves manoeuvring a wire ring around a bent wire frame in such a way as not to touch it. If the ring touches the wire, it closes a circuit and makes a buzz, or flashes a light or whatever. There is some history of using this task in different guises to investigate task presentation effects (e.g. Hargreaves, Bates & Foot, 1985). The computer screen version of this task involved a mouse-controlled cursor which one can think of as a section through a ring and a deformed, angular line which moves across the screen at a regular speed. The objective is to avoid the cursor ring coming into contact with the line. If it does it emits a warning tone and the line stops moving until the cursor is moved so as to free it up.

Using this software we ran a preliminary study which seemed to show that whereas girls did better when the software was presented as a test of stereotypically gender-appropriate skills (needlework and textiles), the boys did better when the test context was not emphasised. Performance was scored simply as number of collisions. In the course of working with the software, a number of the boys spontaneously referred to the software as a game, and it seemed to us likely that the superior performance of the boys in the ill defined condition resulted from their treating the software as a game. So we set up another, larger study, which included conditions in which the software was represented as a game (called Electric Eel) or as a skills test linked to strongly gendered roles (beautician and technician). This attribution was given either low salience (just a name in the title bar on the screen) or high salience (i.e. referred to explicitly and repeatedly in these terms in the introduction). These conditions were used, on a between-subjects basis, with 120 11-year-olds. The children were seen individually in a school context, by a female experimenter, and were given first a brief demonstration and practice trial and then a single attempt for real. Piloting had allowed us to set the parameters of the task in such a way that, with 11-year-olds, all would suffer at least a few collisions, despite their best efforts.

An analysis of variance performed on the frequency of collisions revealed that the boys performed significantly better than girls, but there was also a significant interaction between gender, salience, and the game versus task distinction. When the software was introduced to the children as a test (regardless of any implicit gender reference) there was no gender difference in performance. By contrast, where exactly the same task was introduced as a game, the boys' and girls' performance diverged to a point where there was a significant gender difference in performance, favouring the boys.

So, these experimental studies show that even with a single, standard piece of software (tapping, on the face of it, rather basic sensorimotor skills) children's performance can be significantly affected by the way in which that software is presented. Where the software is presented as a measure of work-related skills there is no difference at all between the performance of the girls and the boys. When the software is presented as a game, on the other hand, a gender difference favouring boys is very apparent.

5.3 Example 3: Achievement Test

Our third example comes from a comprehensive study, the purpose of which was to investigate educational software in classroom settings (Jäger et al., 1993). In the context of this study, a qualitative investigation regarding the use of a geometric software package 'Cabri Geometre' was undertaken. This study involved the detailed observation of 24 eighth graders (12 girls and 12 boys) who had used the software as part of their regular classroom activities. These observations were conducted in the pupils' classroom outside of the regular lessons. Each pupil worked individually for three sessions each lasting approximately 2 hours. After the third and final session, a test designed to assess the pupils' understanding of geometry was administered to each child individually. The test comprised of two parts: one of these required the child to undertake geometry tasks using the computer, for example, constructing a triangle. The second part comprised a standardised paper-and-pencil test, derived from a comprehensive school test battery measuring pupils' understanding of geometry.

Whilst there were interesting findings regarding the nature of the pupil's interaction with the software (in detail see Wosnitza, 1993) here we will focus on the results of the geometry test, which revealed that boys' performance on the computer task was better than that of the girls. Boys solved 60% of the computer tasks unaided, whereas the girls completed just 46% of the computer tasks correctly. This finding is interesting given that the boys and girls had the same amount of prior knowledge, as measured by a standardised geometry test prior to the first session, and that they obtained equivalent results in the standardised paper-and-pencil performance test conducted after the computerised tasks.

So, whilst the girls and boys demonstrated equal levels of achievement in the paper-and-pencil-tests at both the beginning and at the end of the study, the boys out-performed the girls in the computerised learning test. The gender differences observed can be explained by the fact that the boys experienced fewer problems negotiating the software interface. It is likely that the boys' higher levels of computer experience, achieved mainly through the playing of computer games, enabled such negotiation. Assuming that the standardised paper-and-pencil-test represents a valid measure of the pupils' understanding of geometry, we should consider whether the differences observed on the computing test arise as a direct consequence of using computer-based testing methods.

5.4 Example 4: Working with Others

As outlined earlier in this section, the term context also embraces the social environment in which an individual is embedded. Thus, our final illustration focuses on the dynamic established when boys and girls are required to work with or alongside one another on a computer-based problem solving task.

Light, Littleton and Bale (1994) studied 11-year-old children working on the Honeybears computer problem-solving task discussed earlier. The children were required to work either in co-active or interactive pairs, in either girl–girl, girl–boy or boy–boy pairings. In the case of the co-active pairs, whilst the children entered the room together, each child worked individually at one of two machines. There was no verbal interaction, and the children were unable to see one another's screens. In the case of the interactive pairs, the children entered the room together and then proceeded to work together on the problem at one computer. The children were matched on the basis of an individual on-task pre-test, both within the particular pairs and across the conditions.

The data revealed that while they were actually working together the children in the interaction condition performed slightly better than the children in the co-action condition. However, this advantage disappeared when the children were post-tested individually on a slight variant of the task just 1 week later. More interestingly, in the interaction condition the boys and girls performed equally well, and it made no apparent difference to their performance whether they were working in same or mixed gender pairs. In the co-action condition, however, where the children had no opportunity to interact with one another, the mixed gender pairs produced a marked and statistically significant polarisation of performance, with the girls in the mixed pairs performing worse at post-test than the girls in the other conditions and the boys in the mixed pairs performing much better than boys in other conditions. The absence of gender polarisation of performance in mixed pairs in the interaction condition was consistent with previous findings using this software (e.g. Littleton et al., 1992). The discovery of such polarisation in the co-action condition was all the more unexpected.

As a check on the result, we ran a further study (Light, Littleton & Bale, 1994). Using exactly the same software, we examined the performance of pairs of 11-year-old children working on the task during a single session under co-action conditions. One third of the pairings were girl–girl, one third boy–girl and one-third boy–boy. Once more, as seen in the previous study, a gender by pair-type interaction emerged, with boys performing markedly better than girls only in the mixed gender pairings. Thus, the effect which a number of educationalists have been concerned about, namely the fostering of gender differences in performance in mixed gender groupings, does occur, quite markedly, but only when the children work alongside one another without interaction. When they are actually collaborating the same polarisation does not seem to occur.

Explaining these results is far from easy. We know from their responses to attitude questionnaires and interviews that the children do subscribe to the view that boys are generally more interested in and capable with computers than girls. However, we also know that with this software the girls are actually at least as good as the boys, and in the first of the two studies the children had been paired so that (whilst they did not know it) the partners always had the same pre-test scores. We might speculate, then, that the opportunity for interaction provides some feedback to the children on how well they are performing relative to their partner; and that this feedback has a tendency to undermine their gender stereotypes. In the case of the co-active pairs, however, these stereotypes may be evoked by the situation of working alongside a partner of the opposite sex and cannot be ameliorated by feedback. In the first of these studies the two children came from different school classes so that their particular knowledge of one another's abilities was limited. The situation was such that they were well aware of one another's presence, and there was perhaps a tacit expectation that their relative achievements on the task would be made visible to one another at the end (though actually they were not). Thus the partners constitute both an audience for one another and potential competitors for one another. The absence of feedback during the session serves to maintain these relations throughout the session. In the mixed gender situation, therefore, boys are particularly challenged to achieve well, while girls may be inhibited by low expectations of their own likely relative performance.

Such an interpretation remains speculative and these results obviously need exploring further. The finding, however, serves to highlight the centrality of processes of social comparison in these kinds of learning situations. The contemporary work of Monteil and colleagues in Clermont-Ferrand (e.g. Huguet, Chambres & Blaye, 1994; Monteil, 1993), which stems from social psychology, illustrates the extent to which children's expectations of their own success relative to their classmates can impact upon their learning. Huguet and colleagues (1994) have established recently that an artificially raised expectation of success on the previously mentioned Honeybears task can indeed lead to enhancement of children's performance.

As we come to appreciate the all pervading nature of contextual effects on cognition (e.g. Forman, Minick & Stone, 1993) we need to recognise that the social context of a cognitive task embraces not only direct interpersonal interactions but also the social norms, expectations, representations and comparisons that condition such interactions. In the end, taking this wider sense of social context into account in our research may turn out to demand richer and more diverse research methods than those used in the research reported here.

6 Discussion, Implications and Future Research

The four examples discussed above clearly demonstrate the need to pay close attention to the particular contextual and situational location of computing activities which may result in gender differences in response to computer technology.

The need for such an approach is also evident when one considers the results of other research studies which report the crucial role of context in determining performance on computer-based tasks. For example, results from the Minnesota Computer Literacy Assessment, derived from a sample of nearly 2,500 Minnesota high school students from eighth grade and 11th grade, reveal that female students' performance on a standardised computer literacy test can be significantly enhanced by expressing the problems verbally rather than mathematically (Anderson, 1987).

Our assertion that the nature of the images and metaphors deployed in the software can be crucial determinants of performance on computer-based problem solving tasks is perhaps not surprising given the findings of Jakobsdottir, Krey and Sales (1994) who studied school children's preferences for computer graphics. Based on design guidelines derived from the research literature, these authors devised different categories of image representing high female interest, high male interest, and equal interest pictures. Over 310 students, from grades 2, 4, and 6 rated all the pictures, which were presented to them on a TV monitor. The results showed that girls rated high female interest pictures highest and high male interest pictures lowest. Precisely the opposite was true for boys. Since there were no significant gender differences in the ratings of equal interest pictures the authors conclude that it is possible to design computer graphics to appeal to both boys and girls and hence to establish a balance between conflicting preferences due to gender. As Huff and Cooper (1987) make clear, however, gender stereotypes in educational software often arise as a direct result of the designers' expectations about the potential users. The authors conclude "that it is not the computer, or even the software, that is at the root of sex bias in software, but the expectations and stereotypes of the designers of the software" (p. 519).

The use of gender-stereotyped software not only impacts on the cognitive performance of girls. Work by Cooper, Hall and Huff (1990) provides a striking demonstration of anxiety arising in young girls as a result of working with software designed for boys. We thus concur with the design guidelines offered by Culley:

> The software that teachers use should be carefully selected and examined for the way it constructs males and females. Teachers could collaborate to develop checklists for use in software evaluation to ensure that it reflects their educational objectives in a non-sexist way. For example, how are males and females presented? Is the language sexually inclusive? Is the software motivating for all pupils? Do examples build equally on the experiences of girls and boys? Does the software provide the opportunity for group work and co-operative learning and is it likely to develop the confidence of pupils? (Culley, 1993, p. 157).

But, as we have seen with software which appears to have no obviously gendered connotations, the description of the activity, for example, as a game as opposed to a task, can result in gendered perceptions of computer use which demonstrably colour performance and become, in effect, self-fulfilling prophesies. This is an issue of some significance, just at the moment, because of the growth of the edutainment industry, aimed at both home and school markets. Increasingly we see computer game formats used as vehicles for educational tasks. The findings reported here suggest that these might not be universally appropriate vehicles if we are concerned about fostering girls' enthusiasm for computers in schools.

Our work on the social context of computing taps into a broader debate concerning the segregation of girls from boys in the context of computer-use. Observations of male dominance, together with evidence that girls from single sex schools are typically more experienced, enthusiastic computer users than girls from co-educational schools (e.g. Funken, 1993; Jones & Clarke, 1995), have been used by some to argue that when it comes to computing, girls and boys should be educated separately. Our findings are a striking testament to the potential costs associated with a reduction in opportunities for interaction between boys and girls when they are working with computers.

The issue of whether single-sex computer classes in co-educational schools is an appropriate way of fostering females' competence and enthusiasm for computer related activities is a hotly debated topic. Kreienbaum and Metz-Goeckel (1992) argue that gender-specific role stereotypes seem to develop much more readily in co-educational settings, hence the need for all-girl computer classes. Others, however, point to the inherent dangers in such a course of action, ". . . separating out girls from boys (boys from girls) will not, in itself, change the perceptions of the teachers and administrators in the schools, any more than it will automatically change the attitudes of the students themselves" (Willis & Kenway, 1986, p. 145). Willis and Kenway also comment that whilst the strategy of segregating boys and girls for computer-based work possesses some superficial appeal, it could result in resources being diverted from the important issue of curriculum reform. Thus, it is argued, if schools are to consider adopting such a strategy it is vital that they ensure that a change in the gender structure of the classes is not a substitute for an examination of curriculum content, teaching methodologies and assessment practices.

Culley (1993), mindful of the comments of Willis and Kenway, is tentative about advocating segregation as an effective intervention strategy for lesson-based computer work. She sees the need for teachers to develop an understanding of classroom dynamics and the ways in which boys may come to dominate the computer classroom. Classroom management strategies, she argues, need to involve girls more centrally and pupils need to be involved in discussion of patterns of gender interaction. Culley does, however, believe that there is a much stronger case to be made for ensuring that girls have access to free-time use of computers in girls-only settings where they are free to develop skills and tinker in a supportive environment. She adds that any such girls-only sessions should be supervised by teachers who are both competent in computing and sensitive to gender issues.

Culley (1993, see also Clark 1989) further stresses that gender stereotypes are socially constructed within a cultural community, and that pupils are active in the construction of gender and their own gender-identity. In some circumstances, then, the mixed-gender computing class could afford an appropriate context in which to challenge existing gender stereotypes in schools, with change being affected through interaction.

The studies we have discussed here clearly raise many issues. One such is the question of why the girls we studied appeared to be more sensitive to the contextualisation of a task than boys. One possibility is that these differences arise as a function of the differential confidence levels of the boys and the girls with computers. Robinson-Stavely and Cooper (1990), for example, have shown with adult students that the performance of women with low familiarity/confidence with computers is greatly affected by subtle manipulations of their expectations about how well they will do. Women with higher levels of experience of computers are largely unaffected by such manipulations. On this analysis, lability of response is more a factor of experience and confidence rather than of gender per se.

Whilst we have focused on educational computing, the context of children's computer-use is not limited to the classroom. Rather, it comprises the many different environments in which the individual is embedded. Thus, in addition to the context of educational computing use the contexts created by the family, the media and the leisure industry have to be considered. In her analysis Reisman (1990) claims that males receive greater support and encouragement in all of these four cultural institutions and that: "Far from revolutionising society, the computer has conformed to society, becoming another element of the status quo" (Reisman, 1990, p. 45). We acknowledge that technologies inevitably arise in the context of existing social relations and for this reason are highly likely to result in the reproduction of these forms of relationship. However, the same technologies may open up possibilities for the transformation of these social relations. In the case of IT it is imperative that we seek out and create the conditions for achieving such a transformation and a clearer understanding of the contexts of computer use may do much to afford such a transformation.

Note: This paper was written whilst Karen Littleton was an Economic and Social Research Council Funded Fellow, working at the Open University.

14

Information Technology and the Culture of Student Learning

Charles Crook and Paul Light

1 Introduction

The reform of universities is prominently placed on the political agenda in many countries. In the UK, there is considerable pressure to extend the reach of higher education. This has come to mean more than simply increasing numbers. It means a greater concern for the participation of less traditional students: those whose circumstances might require more flexibility from the curriculum and timetable. Yet such goals have to be attained in a stringent economic climate. So universities must be more versatile and more accommodating but, at the same time, they are expected to be more efficient and more economical. In seeking to resolve the tension inherent in such a set of demands, many policy-makers (e.g. Hague, 1991) have appealed to developments in information technology (IT). Perhaps, it is argued, a greater deployment of computers in the educational environment will allow teaching and learning to be both more flexible and more affordable.

Here is a quotation from an article in a European Union magazine concerned with IT development. The article anticipates a computer-driven reform of higher and vocational education. We are invited to:

> Picture a system where every citizen would have on-line access, via individual workstations, to terabytes of information...Individual students would have access at any time to material precisely geared to their needs, and would themselves determine the pace of their studies . . . Of course these developments would not dispense with the need for teachers: instead they would free them from routine tasks and leave them more time for individual supervision of self-guided study, at a higher level of the coaching process. (*I & T*, 1995 p. 13)

This quotation will have a familiar ring to anyone who follows the political and promotional literature surrounding educational technology. Typically the rhetoric of this literature invites us to picture a system of the future. The author will then helpfully conjure up a careful selection of features to populate this picture. Emancipation is a common element. So, we are assured, every citizen will have on-line access. Individualisation is another element. So, students will enjoy material precisely geared to their needs. Accounts of this kind typically converge on that liberating slogan: self-guided study.

However, students guiding their own study is a sobering idea to contemplate because it demands that we address the fate of teachers in any such learner-centred future. Just in case student autonomy is misunderstood to mean teacher redundancy, it is common to declare that the only impact on teachers will be a process of freeing up. Thus, in the computer-supported educational future entertained above, teachers are (characteristically) promised individual supervision of students and the opportunity to enjoy interventions at a higher level of the coaching process.

Such descriptions are dangerously seductive because they are so easy to construct from a selective collection of ingredients. In particular, the participating students in such visions are conveniently uncomplicated people: usefully equipped with wholesome motives, ambitions and enthusiasms. Yet these sketches of the future lack authority. Certainly, they are rarely derived from present reality in any systematic or convincing way.

Sometimes, the uneasy feeling of contrivance is sharpened by the photographs used to accompany such projective journalism. In the case of the article mentioned above, we see a 20-year-old male dressed in smart denim in a study/bedroom, lounging back on a swivel chair, trainers pushing against the surface of his desk. One hand hovers over the keyboard on his lap, the other moves a mouse. On the computer screen is Leonardo's *Mona Lisa*, with a colour palette tool to one side. The overall image successfully captures something central to what we commonly understand about being in the state of studying: namely, that it involves a reflective and private engagement with some text. However, pictures of studying that is computer-mediated typically conjure up richer associations: the relaxed posture of the user suggests a continuity between recreation and study, while the computer graphics package suggests a distinctly exploratory or constructive relationship between student and subject matter.

New technology can indeed offer learners ready access to scholarly material. It is also true that this technology can set such material in a challenging exploratory environment. These are appealing features, for they resonate well with prevailing constructivist accounts of learning. On the other hand, the image of the student-at-study considered above makes a number of assumptions. In particular, it makes assumptions about motivation. To take the example offered in the photograph a little further: what inspires a student to inspect or manipulate reproductions of renaissance art? How do people come to want to do such things?

One answer is to suggest that, under ideal conditions, motivation to learn arises from challenges encountered in the course of leading everyday life. So if the futuristic student looks highly motivated, this could reflect an assumption that IT will allow educational experience to be geared to personal and meaningful concerns — as they spontaneously arise within the student's leisure or work. Indeed, it is already commonplace to argue that new technology offers the possibility of more flexible learning, in just this sense, because it furnishes an educational infrastructure that may be available when needs arise — so-called just in time learning. Valuable though it may be to have an educational system that is flexible to people's interests in this way, modern society demands skills that surely require a more accelerated or intensive form of educational preparation. Insofar as this remains the case, the attraction of becoming and remaining a student may often need to be generated in us by the inspiration and efforts of well-structured educational practices.

Another view about the successful management of motivation in an IT-rich learning setting might be that the resources offered by computers are so vivid, so interactive and so comprehensive in their scope that the student will inevitably be seduced into engagement. That computer-based learning materials can be compelling in these ways is not in doubt. However, they may not ever be so compelling that they will simply dispose of the problem of motivation. Those of us fortunate enough to have jobs that make multimedia learning materials readily accessible do not seem to have been transformed into obsessive students! Cuban's (1986) careful history of predictions for teaching technologies cautions us not to expect too much, as arguably we have done in every case in the past.

We are arguing that IT-led visions of an educational future take a lot for granted. Particularly where societies cultivate a period of dense learning in early life, there will remain inevitable problems of focusing and sustaining student engagement. Our own view is that this has traditionally been achieved through the socially-organised character of educational practice. Arguably, futuristic scenarios often seem to recognise this — for they typically acknowledge a continuing role for teachers. However, the form of that role is debatable. We do need to reflect more carefully on the organisation of our current social and cultural context for teaching in order to notice how it works. In the example used earlier, the student-of-the-future encountered teachers as administering individual supervision of self-guided study and offering higher levels of the coaching process. In short, this model abstracts rather particular social features of current instructional practice: the setting of goals (to self-guiding students) and the management of coaching at higher levels (by implication, only at those points that can not be reached with some current state of educational technology).

How does our knowledge of current educational practice help us to judge the detail and workability of such visions of the future? In the next section we shall consider what psychological research can offer on the question of how the learner, the information technology and the teacher might engage together creatively and productively.

2 Psychological Perspectives

The traditional theoretical preoccupations of psychology have not helped it in addressing questions about teaching and learning in higher education, and this is perhaps particularly true in relation to challenges arising from new educational technology. A key feature of cognitive–psychological theorising is its analytic focus on the (often isolated) individual actor. In relation to educational practice, this interest has encouraged the characterisation of learning styles — individual differences that learners bring to the circumstance of being a student. Indeed this has been one of the few active areas of psychological research on learning in the settings of higher education (e.g. Säljö, 1984).

Though individualistic in its origins, such theorising can adopt a more situated flavour. Thus, recent work in this tradition has acknowledged that such learning styles are not inherent characteristics of people but rather they arise from learners encountering particular institutional cultures of educational practice (Ramsden, 1992). Such a perspective orientates us towards understanding the dynamic of educational institutions themselves, and how that dynamic is experienced by participants in the teaching and learning enterprise. The learning styles tradition of research has not yet probed this territory very deeply, but the situated cognition approach, recently introduced to the discipline via cultural psychology, suggests some directions for such an analysis.

Central to this approach is the idea that what someone is said to know must be understood in terms of "a micro-culture of praxis" (Bruner, 1996, p.132). Knowledge is distributed in the material and discursive environment: in the tools, symbol systems, institutional rituals, physical spaces and ways-with-words that make up any micro-culture of intellectual life. Educational practice becomes an orchestrated encounter with such an array of mediational means. Ideally, it should involve organising for the individual learner an intensive exposure to the resources associated with particular disciplinary activities.

The concept of learning style might indeed emerge from an analysis of educational practice in these terms, to the extent that different forms of orchestration may lead participating individuals to adopt distinctive and persistent styles of approach to their exploration of some scholarly discipline. Thus the classic cognitive perspective on learning (learning style) does not necessitate empirical analysis at the level of de-contextualised learners: it actually invites an ecological approach. Analysis can best focus on the *activity of being a learner*, seeking to reveal the character and quality of students' situated interaction with disciplinary resources.

A more ecological approach inevitably takes the social organisation of instruction seriously. However, the study of learning, conceived as a situated achievement in this sense, must entail more than a consideration of the localised circumstances of teacher–learner dialogue. Thus, any evaluation of the possible future impact of IT should not be limited to consideration of the prospects for this form of exchange alone. Our traditions of institutionalised higher education have evolved various forms of *community structure*. This is rarely acknowledged in visions of the future based on educational technology such as the extract cited at the start of this chapter. Typically, these visions preserve the social dimension of educational practice in only the most narrow sense: in terms of the need for tutors to supervise the self-guided learner or to pursue increasingly high levels of coaching.

Educational practice may depend on more than this — for example educational communities may provide crucial ingredients for motivating and sustaining the activity of study. Likewise the creation of common knowledge (Edwards & Mercer, 1987) within such communal settings may provide important resources for intellectual reflection and development among community members. So, this cultural context may both motivate and resource learning. The neglect of such considerations is hazardous, given that IT has the potential to transform, and perhaps to undermine, educational communities as we currently understand them. In the following section we shall explore some of the types of research that are needed to rectify this neglect.

3 Researching Student Learning in Context

What kinds of research are necessary to expose and evaluate more of the culture of undergraduate learning as it takes place in traditional full-time, residential settings? In addressing this question we shall allude briefly to some findings of our own work. Our research has looked at just two university settings, one a relatively old university and the other a metropolitan more recent one each having about 10,000 students. Much of what we can report is probably typical of current practice in the UK, but we do not claim that it is a fully representative snapshot. Rather we have been attempting to identify prevailing patterns of educational life in selected (but unexceptional) higher educational settings, and then exploring the manner in which new technology interacts with them.

Adopting a situated perspective invites a particular analytical stance for research into teaching and learning. In particular, it requires recognition that educational practice, as we find it now, has arisen through a long period of historical development. Resources, institutional structures and interpersonal rituals have *evolved* to their present forms, and researchers cannot take them for granted as if they simply provided a neutral backdrop to the interactional business of teaching and learning. Once the situated perspective is adopted and learning is viewed as an activity, we must attend to how today's students enter an arena in which established materials and practices heavily circumscribe the very ways in which it is possible to act.

The conceptual vocabulary of activity may conjure up associations with apprenticeship models of learning (e.g. Rogoff, 1990), but we mean something more general. Learning as mediated activity might describe the concerns of our research more adequately. To learn about, for example, chemistry or geography is to participate in structured relationships both with a particular set of materials (texts, equipment, etc.) and with a particular set of other people (tutors and peers in particular). The encounter with cultural resources (texts, technologies, rituals) may sometimes be in intimate relation to the actual practice of the relevant discipline, as in apprenticeships, but that encounter may also be a relatively arid experience, perhaps quite remote from actual disciplinary practices.

In either case, the business of unravelling the learning experience demands research in an ecological tradition. A start was made on such a research agenda more than 30 years ago (e.g. Becker et al., 1961; Wallace, 1966), though this work was largely undertaken from a sociological standpoint, and made limited reference to any psychology of learning. Our own research starts from the rather particular purpose of locating the impact of new technology. But, since we suppose that the impact of any new mediational means on education should be evaluated by reference to the existing culture of learning activity, it has been necessary to try to understand more about this setting as we find it.

In this spirit, we asked students to keep diaries that document a wide variety of study-related activities to a fine temporal resolution. We carried out self-report surveys of study practices during a period of revision for examinations. We created sociometric representations of students' collaborative relations with peers. We made direct observations of students working together at academic problems and we organised focus group discussions concerning the experience of local educational practice. The research was in both universities, some of it involved a few students and some of it many. Very little was duplicated on both sites.

In the diary study, 60 students were recruited by post, from a randomly selected list of second year students drawn from six disciplines. Each respondent was required to record their study activity on one day per week, over a period of five successive weeks. In the survey of study and revision practices, the respondents were identified by compiling a random sample from the main student data base representing 20% of the second year student body. Twenty student helpers were recruited to conduct the survey, with some respondents living on campus and some in town (1:3 ratio living in/out). The final number of returns was 159. For the focus groups, four sets of three group meetings were held between November 1996 and July 1997. In each group there were six students drawn from each of three faculties, Science, Social Science and Arts. The first set of meetings focused on the broader culture of student life, for example, college life, departmental life, being at university, relations with one's course peers, relations with staff. The second set of meetings focused on extracurricular study, student–tutor relations and students' experience and attitudes to information technology.

This research has revealed a good deal about the texture of studying in a full-time, residential setting, and has helped to provide a framework within which we may interpret the ways in which computers are entering the teaching and learning system. In reviewing what research commentators say (and illustrate), we noted a blurring of the distinction between recreation and learning, and between teaching, coaching and self-directed study. More generally, the formal and the informal curriculum are made to seem less distinct from each other. With this in mind, in our own analysis of the culture of undergraduate learning we have paid particular attention to the informal, out-of-class experience of study. For example, we have focused upon the possibilities for informal collaboration with peers and informal contact with tutors that arise from participating in a full time community of study.

In fact, however, our various empirical techniques all tend to reveal a strikingly low level of study-related informal collaboration with either peers or (out of class time) with tutors. For example, student diaries recorded very little casual conversation about work with fellow students. Similarly, sociograms of interaction amongst students following the same courses showed few connections based on any form of collaborative study. Revision for examinations tended to be an almost exclusively solitary enterprise.

Some of the ways in which aspects of the local institutional culture support or inhibit informal collaboration emerge from this research. One theme that emerges is that of place: institutions provide spaces in which out-of-class study can go on. In particular, such work tends to be concentrated in libraries and in study-bedrooms. Rules of conduct in the former and room allocation policies in the latter both tend to militate against the effective use of these study spaces for collaborative learning.

How might new technology fit in to this? Where it takes the form of powerful workstations provided in the student's own (private) room it might further militate against face-to-face interaction and collaboration; although, at the same time, it might afford possibilities for computer-mediated interaction. On the other hand, provision of computing resources in so called public computer rooms might afford new possibilities for face-to-face interaction. We conducted (in the older university) an observational study of activity in campus computer rooms, with a view to establishing their significance in supporting informal collaborative exchanges. This study was facilitated by a PC-based program (written by Simon Clark) which allowed both the schematic representation of a room with the location of PCs marked, and an indication of persons, male or female, via appropriate icons. Eight rooms were observed, including one library. Snapshots of the computer room were taken at 90 second intervals and 10 snapshots per session, 33 sessions being recorded. Although such spaces do not demand that users are quiet (like libraries), we found little evidence of casual collaborative exchanges mediated by this arrangement of computing. Instances of productive interactions 'around' the technology (Crook, 1994a) were few and far between.

Colleagues from other universities have expressed surprise at this finding. To some extent this may simply reflect selectivity in what the undisciplined observer of such spaces notices. However, it may indeed be the case that aspects of institutional organisation such as timetabling, or the location of public rooms on the campus, influence what happens in these spaces. The point then is not that our observations expose an inevitable pattern of using computer rooms but they do remind us that collaborative interactions around computers are not easy to provoke. Perhaps, their mediating function in this respect is moderated by other factors in the physical and social context. To this extent, the effect of IT provision needs to be understood in terms of the niches that are created for it within the learning community. The way in which computers work for the students we have observed, for example, may well reflect prevailing traditions of their learning practice such as their relatively modest experience of peer-based collaboration.

One feature of the computer rooms we have studied is that any informal collaboration is likely to depend upon chance encounters among course peers using these facilities. What is the impact of IT-based resources in situations where the joint activity is more planned? To investigate this, we have observed pairs of students who volunteered to work together out of class, revising a course either from their own (paper based) lecture notes or from a (computer based) hypertext document developed for the purpose and covering the same lecture course (Crook, 1996). The interaction within these pairs was very different. Students working around their own notes developed a deeper exploration of the core ideas. Those working around the computer-based document frequently became more enmeshed in discussion about the hypertext itself and the possible motives of the author. More work needs to be done to establish how far this is an issue of private vs. public resources for revision, and how far it relates to the use of new technology per se.

Collaborating involves negotiating common knowledge, and the prospects for that enterprise will typically depend on the shared resources towards which the collaborators can orient. Our survey data indicate that, first, this kind of joint activity may be surprisingly rare in higher education communities but also that, second, computer-based resources may not necessarily have an empowering impact. Yet this need not reflect something intrinsically limited about the medium. How students collaboratively approach on-line access to terabytes of information may depend upon prevailing attitudes to the curriculum and towards texts; attitudes that have been cultivated through extended participation in their existing learning communities.

The examples above have concerned the support of collaboration through interactions around and at computers. Arguably, IT might be expected to lend itself better to supporting asynchronous collaboration; interaction 'through' the technology (Crook, 1994a). In this connection we have been interested in the use of email and electronic bulletin boards. This mode of exchange is not so radical a departure from existing practices within a residential learning setting as it might seem at first sight. To some degree participants have always created shared meanings from events or contributions that are scattered within the time and space of institutional life (via notice boards, pigeonholes, and so on). Computer-mediated communication (CMC) formalises and reconfigures some of the possibilities for this kind of interaction. However, our own observations of its use suggest that the resource does not merely reproduce existing practices in another medium.

One approach we used was to conduct a user-managed survey of email usage among students. This indicated few spontaneous study-related exchanges with peers or tutors. A more proactive approach involved building an email facility (to contact the tutor) and an anonymous bulletin board facility (for comment and discussion) into a set of hypertext course materials. Despite the fact that the course materials themselves were heavily used, the students hardly made any use of the CMC facilities at all (Crook, 1994b).

When writing to email lists is made a more central course activity, the picture can be different. In such skywriting (Light, 1996; Light and Light, in press) students write to a list consisting of all fellow students on the course and the tutor(s) of that course. On one largish 9-week lecture course for first year students (108 students, half registered for psychology and half taking the course as an elective subsidiary to other degrees) that we monitored, the lecturer was highly responsive to direct questions. Both questions and answers were broadcast to the full list. In this instance, skywriting soon became entirely a question-and-answer exchange, supplementing the opportunities usually provided for asking questions of the lecturer (at the end of lectures, in tutorials, etc.).

About half the students made use of the opportunity to ask questions, with more than half reading the exchanges. The medium seemed to provide for a kind of vicarious learning (Mayes & Neilson, 1995) and, importantly, allowed the students to gauge their own level of understanding relative to that of their colleagues. Interestingly, in this situation skywriting exchanges were less dominated by male students than were the traditional face to face tutorial meetings that accompanied the course, and levels of participation were independent of the students' attitudes towards, or experience with, computers generally.

By contrast, where the lecturer offered skywriting as an opportunity for the students on another large lecture course to enter into debates with one another about the issues covered in the course, participation levels were extremely low, and almost all participants were male. Students reported great self-consciousness about their own contributions and, sometimes, ambivalence bordering on antagonism concerning the contributions of others. It seemed that skywriting, in this context, cut across some of the conventions of this student group (as regards criticising one another in public, for example), and it fairly rapidly died out.

Skywriting proved more successful on the two smaller courses we observed. These were followed by only five to ten students who all knew one another personally. In this setting, the skywriting medium came to be used for mutual support, to the extent that students would copy materials to one another and (given encouragement to do so) even share work assignments. All of the students involved felt that the small group size and the fact that they knew each other well were important to the way the skywriting developed.

The lesson we take from these observations is that the established culture of learning can greatly influence the prospects for new CMC initiatives. Existing institutional practices can equip students with particular experiences of their relationship to both their disciplines and their peers, and these experiences will dictate the way in which new technology is appropriated — or, indeed, whether it is appropriated at all.

4 Actual and Virtual Community in Higher Education

We have argued that learning is best conceptualised as arising from the participant structures created by the formal and informal curricula of the learning institution. In our own research we have been trying to interpret the impact of educational technology with reference to the community contexts in which such innovations are set. In general, our results confirm us in the view that students are enculturated into particular communities of learning, and the resulting practices will offer resistance to the bolting on of new educational technologies.

One area in which the results of our surveys of student routines surprised us was that they suggested that the quantity and quality of informal (out of class) collaboration is often extremely limited. Since full-time residential education is often defended in terms of the possibilities afforded for such collaboration, this finding is at first sight rather alarming. In particular, it might seem to question whether (residential) community is so precious that it needs protection from computer-intensive educational futures.

However, face-to-face familiarity amongst students may be important to their learning in less direct ways. Thus, for example, the very effective use to which skywriting was put in the small groups was seen by the students themselves to depend critically upon the fact that they knew one another well. Other research on CMC in the context of distance education, where the students are not close companions, has indeed tended to paint a fairly discouraging picture (e.g. Grint, 1992; Kaye, 1995).

More generally, the fact that full time residential institutions may not support a great deal of informal collaborative learning does not imply that the cultural supports they do provide are irrelevant to their success as places to learn. For example, various aspects of institutional culture (assessment, peer comparison, timetables and other rituals) may be important in motivating learning without drawing students into particularly collaborative relations. Becker et al. (1961) illustrate this in their ethnography of medical undergraduates. Our own focus-group discussions indicate that, although informal peer collaborations may be unusual, there remains a strong sense in which studying is framed in terms of such community structures and demands.

The radical potentialities of information technology for higher education are frequently seen in terms of freeing such education from the constraints of time and space, creating a distributed community of teaching and learning. The technology may indeed be available to free teachers and learners from the constraints of time and place, but our analysis suggests that conjunctions in time and place may as often enrich as constrain interaction — even if the present culture fails to take full advantage of this. For example, in our studies with full time undergraduates, none of those we interviewed would have willingly substituted electronically mediated teaching–learning interactions for face-to-face exchanges. This applied to even the most enthusiastic users of the skywriting facilities. A common observation was that while real-time conversation provides openings, triggers responses, and stimulates engagement, sitting at a computer composing a query or comment for an email list discussion does none of these things.

Contemporary experiments with virtual classrooms, even virtual universities, offer limited guidance here, since they tend to involve highly self-selecting clienteles. Certainly any demonstration that it is *possible* to support learning in this way should not lead us to suppose that the virtual classroom will be an inevitable and unproblematic substitute for present institutions. Relatively small numbers of people do presently learn in this way, just as relatively few academic researchers choose to join the communities of usenet or listserv electronic mailgroups.

Nonetheless, the success or otherwise of such virtual environments for supporting learning activities is eminently worthy of study. For present purposes, it is interesting to see that constraints of time and space tend to be reinvented virtually within such environments. Evolving from collaborative computer games, for example, MOOs (multi-user object-oriented environments) offer the option of creating both personal and public virtual places, resourcing particular kinds of interaction. Frequently, such environments for learning incorporate virtual spaces conceived only for informal interaction. Oldenburg's idea of 'third places' is often cited; places distinct from home and work but within which "conversation is the primary activity" (1989, p. 42).

However, in comparison with communities of traditional full-time study, these virtual settings are relatively impoverished. The potency of interaction in learning does not arise simply from the availability of contexts for synchronous talk (or text). Productive interactions amongst learners depend upon a backlog of common experience and a mutual recognition that experience is indeed held in common. In other words, the context for collaborators in learning extends beyond the immediate resourcing of conversational exchange; it involves those institutional structures that allow individuals to be, and to know themselves to be, participants in the same events.

More generally, a situated approach to evaluating specific IT interventions may serve both to illuminate their impact and to suggest what may be necessary to make them more powerful resources. Both the possibilities and the limits of what can be expected from, for example, email or hypertext or computer rooms may be better recognised when the culture into which they are being introduced is more fully appreciated. At the same time that culture itself may merit more attention, lest we find that we have inadvertently thrown out the baby with the bathwater.

15

Assisting Child–Computer Collaboration in the Zone of Proximal Development (The Vygotskian Inspired System (VIS))

Rosemary Luckin

1 Introduction

This chapter investigates the features of the situated approach to education which must be recognised and incorporated into the design of any software which aims to provide this type of learning experience. The problems which need to be addressed are highlighted and a possible design rationale is presented. The investigation is presented in three sections. Firstly, some of the most relevant features of the situated approach, in particular the theory of Vygotsky (Vygotsky, 1978, 1986, 1987) are considered. The particular implications of this approach for software design are discussed and summarised. The second part of the chapter looks at the approach adopted by VIS (The Vygotskian Inspired System), a piece of educational software designed in line with the implications previously identified. The structure of the system and the educational experience it provides for the child are described. Finally a brief summary underlines the conceptual issues inherent in this approach. The research underpinning this investigation is concerned with the design of educational software for use by 10–11-year-old children in a primary school. Discussion is therefore limited to this setting and endeavour.

2 Situated Learning

Situated learning has been described as a "contextualised way of teaching abstractions" (Clancey, 1992). The constituent themes within this phrase: the context, the teaching and the knowledge (abstractions) provide a starting point for the identification of the key factors in this approach. These three themes can also be found in Vygotsky's work and are expressed in the "general law of cultural development" (Vygotsky, 1978). The higher mental functions of the child, which require the use of abstract scientific concepts, originate in her social context. Acquisition, which is the aim of the instructional process, is through the internalisation of the tools and symbols that mediate interactions within this context. A brief look at each of the three themes in turn will help to illustrate the particular characteristics which must be built into the construction of a design framework.

2.1 Context

A situated approach to the design of a learning environment or context necessitates authenticity and consistency with the educational culture of the child. As Mercer (1992) highlights, despite the increased usage of the term context, it is rarely defined; its current use will therefore be clarified before going further. Within the current investigation, context is used to refer to the environment in which the learning activity is taking place. This term encompasses both the environment of the school and the environment which can be provided by the computer. The learning environment must provide and support opportunities for interactions between the child or others, in recognition of the importance of "the social dimension of consciousness" (Vygotsky, 1987). Different environments afford different forms of mediated experiences and require the use of different tools or signs. Throughout different cultures the nature of the school environment varies, as does the nature of the child's learning. The UK government's current concerns with how other countries achieve their academic results illustrates the practical implications of this recognition. Information technology, and in particular the use of computer software, is becoming an integral part of the primary classroom learning environment. This is recognised by the inclusion of information technology in the National Curriculum as a subject to be employed by all children educated within the English state educational system. It is recommended that IT should be integrated with and support the other National Curriculum subject activities (DFE, 1995). The reality of the extent to which computers are a part of this educational culture is indicated elsewhere in this volume in the chapters by Bannert and Littleton and by Crook and Light.

2.2 Knowledge

The word abstraction refers to a particular type of knowledge: a generalised knowledge which is flexible and available to the user in different contexts. The differentiation of disparate types of knowledge is clearly expressed by Vygotsky (1986), who adapts and develops Piaget's distinction between spontaneous and non-spontaneous concepts. The spontaneous concepts are deemed to be everyday concepts, familiar to the child through her interaction with the world. The abstract or non-spontaneous concepts are dubbed scientific; these concepts are the domain of the child's school education. In order for the child to become conscious of the concepts she is using they must be integrated into a system. The child's contact with the scientific concepts of the school curriculum provides the system into which the everyday concepts, already within the child's mind, can be linked. It is the interrelationships between all of these various concepts within this system that should be the main concern of the instructional process (Vygotsky, 1986). The teacher or learning partner (whether human or computer) must have knowledge of the subject to be taught in a flexible form which is culturally consistent with the child's current knowledge, if it is to integrate with her already existing knowledge structure (Evans, 1993). The implications for the design of the knowledge base of educational software are clear. Careful specification of the knowledge of the domain which is to be taught is essential, as is the availability of opportunities to build links between the to be learnt and the already experienced. The importance of a careful analysis of the domain to be taught should not be underestimated. In their recent evaluation of scaffolding in schools, Bliss, Askew and Macrae (1996) found that before scaffolds can be constructed, or even planned, a careful analysis of the domain, indicating the potential links to the child's existing, intuitive knowledge, is essential. Likewise, Norman and Spohrer (1996) acknowledge the benefits of authentic problem solving activities *but* believe that a structured analysis of the curriculum content is still necessary.

2.3 Teaching

The zone of proximal development (ZPD) is the linchpin of Vygotsky's approach to the clarification of the relationship which exists between instruction and development. It proposes a particular type of instruction; a particular approach to teaching. The ZPD is a means of providing a dynamic assessment of the child's potential through her collaborative performance ability (Vygotsky, 1986, 1987). It is also something which must be created through social interaction. It is the essential contact point between the scientific concepts of the school curriculum and the everyday concepts already familiar to the child. The only good learning is that which is the result of the creation of a ZPD which awakens the internal developmental processes of the child (Vygotsky, 1978). Teaching and learning is an interactive enterprise in which the tools and symbols of the interplay between the different parties shape their subsequent development. The teacher's language and actions are not the only form of mediation. The child's acquisition of the scientific concepts requires mediation by the spontaneous concepts which are already within her mind. It is therefore the role of the teacher to ensure that the child is introduced to scientific concepts for which there are already spontaneous concepts available within her mind to play this mediational role. The recognition of what the child already knows is a problem which must be addressed.

A fundamentally important aspect of the ZPD is the necessity for collaboration or assistance from another more able partner. Within Vygotsky's organisation the teacher acts as a partner, enabling the learner by providing appropriate activities and amounts of assistance. In this way the learner can be inducted into the culture of her society and empowered as an autonomous learner (Becker & Varelas, 1995). This partnership approach needs to be neither exclusively teacher-directed nor exclusively learner-directed. To be effective, both must participate; there must be mental effort on the part of the child as well as assistance from the teacher.

The use of any theory as a basis for software design requires a precise specification and various difficulties arise. For example, the dialogue between student and tutor is restricted by the format of the system's knowledge base and interface. There are however examples of human-to-human instructional methods which have been shown to generalise to human-to-computer instruction (Wood et al., 1992). The issue of whether transfer to a wider context is possible is an open question.

Adaptability to the needs of the child is possible through the use of a combination of domain, pedagogical and learner knowledge. This adaptability is increased with the amount of information about the child which the system can access and use (Laurillard, 1993), but student modelling is not a precise science, it is still an active and contentious research area.

In line with the attention paid to Vygotsky's work in psychology and education, research into intelligent tutoring systems (ITS) and computerised learning environments (CLE) design have recognised the potential usefulness of a Vygotskian approach. Both aspects of the ZPD have been acknowledged within this work. The ZPD as a means of measuring a child's potential has implications for the importance and nature of the student model within an ITS. Gegg-Harrison (1991, 1992) for example used the ZPD as the basis for enhancing the student model to include measures of cognitive potential as well as cognitive ability. The other angle on the ZPD, that of its creation through instructionally effective collaborative interactions, can be seen within varying approaches which consider the importance of providing assistance or collaboration tuned to the needs of the learner. The scaffolding approach of Wood was adapted from face to face tutoring to ITS design (Wood et al., 1992). Whilst originally the computer in an ITS was viewed as the tutor, the acceptance that the computer can play different or additional roles is also evident in Self (1985) who saw the computer as a collaborator or co-learner, Chan and Baskin (1990) who saw it as a learning companion, and Palthepu, Greer and McCalla (1995) for whom the computer was the learner and the student the teacher. More recently, attention has turned to unpacking collaboration, for example, Burton and Brna (1996), Baker and Lund (1996) and Dillenbourg, Traum and Sneider (1996).

3 The Approach Adopted by VIS

VIS offers instruction about food chains and webs to children in Year 6 (aged 10–11 years). It is part of a larger system called TRIVAR (Luckin, 1996a). The approach taken to software design is in line with the Vygotskian framework presented earlier. A child using VIS creates a mini-ecosystem by selecting the animals and plants which will share a common habitat. Through the activities that she is asked to complete these organisms are built into a food web. The level of abstraction can be increased as the food web is constructed so that the child is required to complete increasingly complex activities, using increasingly abstract concepts. Throughout her use of the system the child is offered help of varying qualities and quantities. The decisions made by the system about which activity she should be offered next and the amount and type of assistance are based upon the system's representation of her ZPD. Below, the system is described in terms of the way each of the implications identified in Table 1 have been addressed. Once the structure of the system has been explained the nature of the educational experience it provides is described.

Table 1: The Implications of a Vygotskian Design Approach

Requirement	Implications for software design
The learning environment Attention to the culture of the child's education.	Recognition of the role played by computers in the classroom culture.
Knowledge abstraction A flexible and culturally consistent knowledge representation of the domain to be taught.	Knowledge representation
Potential links between the concepts in the curriculum to be taught and the concepts already experienced by the child.	Knowledge representation
A fabric of systematically organised scientific concepts and systematicity in the introduction of these scientific concepts.	Content sequencing
The role of the teacher A means of representing the child, in particular their ZPD.	The systems' beliefs about learner's ZPD boundaries
Collaborative support	Quantification of assistance
Challenging the child to ensure 'strenuous mental activity'.	Content sequencing
Semiotically mediated interaction	Interface

3.1 The Structure of the System

3.1.1 The Learning Environment

VIS is designed to promote interaction between child and computer, it respects the role currently played by computers within the classroom environment. The software provides assistance and is, in this sense, a collaborative learning partner for the child. The system is designed for use within this classroom context by a single user. The benefits of the one-to-one student-to-teacher interaction have been recognised (Bloom, 1984; Laurillard 1993; Wood, Wood & Middleton, 1978). However, the limitations of this in terms of possible social interaction are also noted. Future work may be enhanced by considering the possible benefits of designing for more than one user. The subject area of ecology and specifically feeding relationships is part of the National Curriculum for children in Key Stage 2. It will be a part of each child's educational experience, with or without VIS. The design of VIS aims to ensure that the requirements of the IT and Science curriculum which are part of current educational culture are addressed.

3.1.2 The Knowledge Base

Within the knowledge representation described here the scientific concepts are deemed to be those which constitute the curriculum to be taught. The representation used recognises that each concept has a level of abstraction with links to sub- and super-ordinate concepts and a position within the world with links to other concepts in terms of the role each plays within the world. The result of this approach is that the links between different elements are divided into two main categories:

L1. Vertical dimension links connecting concepts within a taxonomy in terms of their level of abstraction. For example, specific instances of concepts such as 'rabbit' are linked to the more general concept 'herbivore' which in turn is linked to the concept 'prey'. Trying to clarify instances of concepts within the world and the levels of abstraction of their encompassing concepts is not easy. It is recognised that the taxonomy (illustrated in Figure 1) may not be entirely based upon the abstraction relationship. However, as the level increases the concepts are those which are less

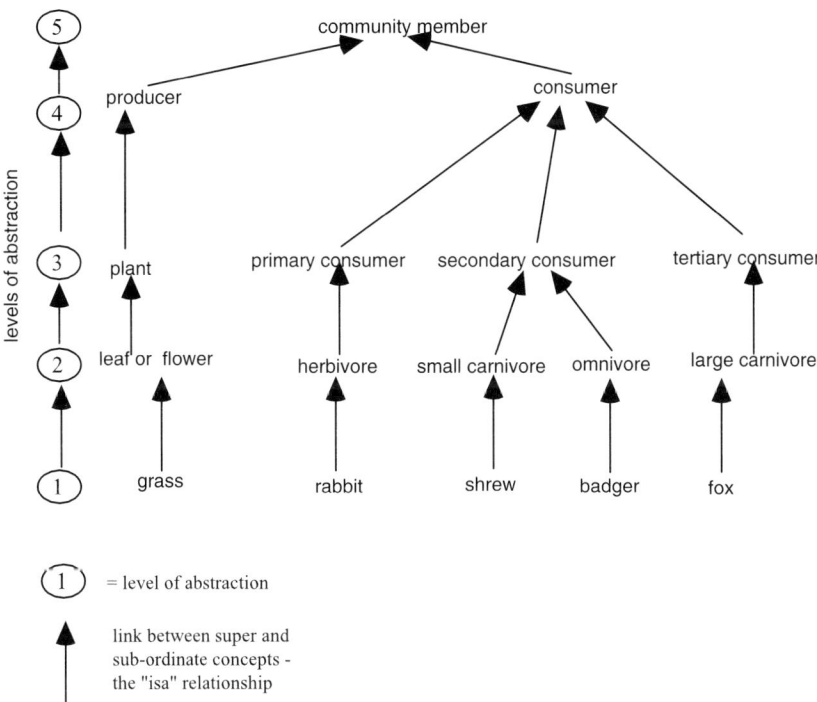

Figure 1: Knowledge representation: vertical dimension.

familiar to the child and more inclusive of the subordinate concepts. Those at the lowest level of abstraction are those most likely to be already part of the child's everyday experience.

L2. Horizontal dimension links which exist between the ecosystem members. These describe the concept's position within the world and the relationships which concepts bear to each other. An example of this type of link would be an 'eaten by' link connecting a predator and a prey. These relationships themselves form a hierarchical structure constructed in terms of the complexity of the particular relationship. For example, the eaten by relationship existing between grass and rabbit is simpler and needs to be understood before the relationship which exists between grass and fox which are non-adjacent members of the same food chain. The relationships which exist between individual organisms within a food web, whether described in terms of rabbit or herbivore, i.e. whatever their level of abstraction within this system, are arranged into a food web rule hierarchy.

The rules which underlie and specify the relationships which exist between members of food webs need to be understood by the child in order for her to understand the dynamics which exist within the ecosystem she creates. The design of the rule structure within VIS has been informed by various sources of information, for example, research by Griffiths and Grant (1985) and Lumpe and Staver (1995) into the teaching of food webs. The relationships within the food web have a rule-like nature which lends itself to a hierarchical style of representation. In this instance the hierarchy is not based upon the relationship of abstraction, but upon the complexity of the relationship which underlies the rule. Relationships further up the system require an understanding of those lower down. Reciprocally, understanding of a later rule is not possible without an understanding of the rules which are a pre-condition for it. The knowledge included is simplified in some instances to respect the age of the children for whom the system has been designed.

The food web rules have been used in the construction of a type of genetic graph (Goldstein, 1982). The original genetic graph was a method of representing knowledge genesis which was used in a game playing coach. It has subsequently been extended and used in two quite different domains by Brecht (1988). In VIS individual rules are represented by nodes in the graph, those which have a common purpose are grouped together into collections, or in Goldstein's terminology islands. In VIS the nodes within an island are linked together as are the islands themselves; the islands are grouped into four phases, with phase 4 containing the most complex rules. Figure 2 illustrates the network of rule islands which make up the horizontal dimension of the domain knowledge representation.

This method of knowledge representation is not without its problems. The domain knowledge which constitutes the curriculum can be carved up in various different ways. In the current food web example the categories of small carnivore and large carnivore could be subordinated to the general category of carnivore, but then of course the division into secondary and tertiary consumer would be difficult. Decisions as to which strategy for categorisation should be adopted have been reached in consultation with teachers who are currently involved in the delivery of this area of

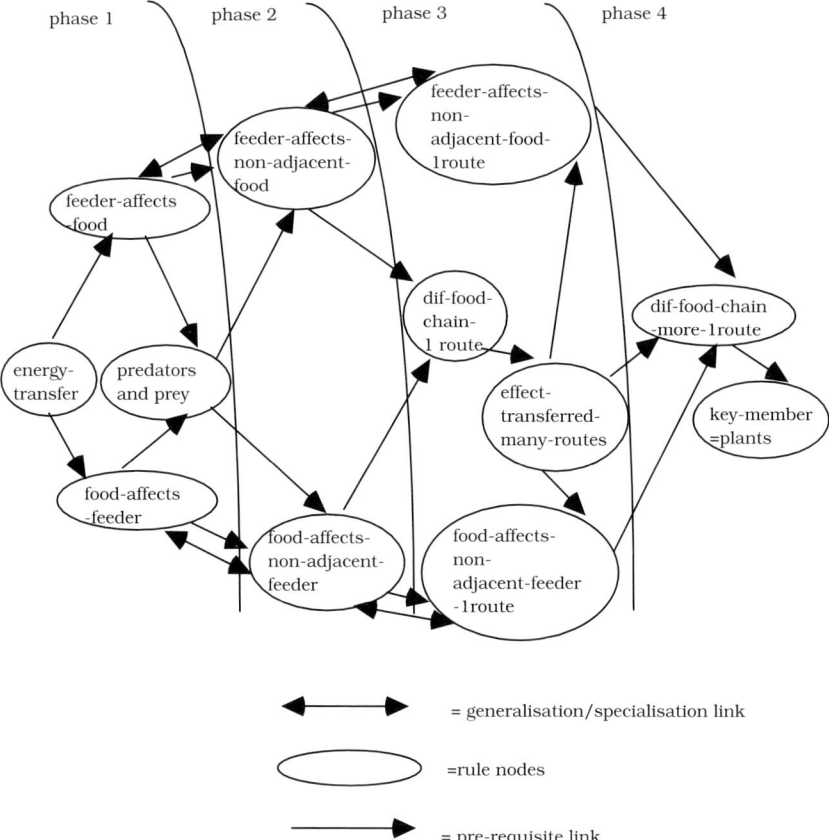

Figure 2: Network of food-web rules which represent the horizontal dimension of the domain knowledge.

the curriculum within the primary school system. In this way the final representation of the curriculum information should be consistent with the context of its delivery. In addition to questions about categorisation the exact nature of the scientific and everyday concepts presents a dilemma. The knowledge representation has been described as constituting the scientific concepts of the curriculum to be delivered to the child. However, the objects at the base of the taxonomy in Figure 1 are basic level concepts (Rosch, 1978) and may be instances with which the child is already familiar. In this sense they are everyday concepts. Indeed, in order for there to be a link between what is already understood, with or without conscious awareness, and the new systematically introduced concepts of the school curriculum, the initial concepts introduced need to be everyday or at least so close to those of the child's everyday experience that she can feel comfortable with their use. Identifying which are these all-important already acquired everyday concepts for each child is a quandary faced by any teacher, human or computer. Within the knowledge representation adopted the concepts which are likely to be familiar to children of the particular age and culture for which the software is designed need to be identified and made explicit.

3.1.3 The Role of the Teacher

Collaboration within the ZPD provides the contact point between the everyday and the scientific (Vygotsky, 1987) and it is this contact point that VIS attempts to recreate. VIS aims to act as a more able learning partner for the child and to promote the creation and maintenance of a ZPD between child and system. Within Vygotsky's discussion of the ZPD there is little detail about the sort of assistance required to create a ZPD. For example: .

> Different experiments might employ different modes of demonstration in different cases: some might run through an entire demonstration and ask the children to repeat it, others might initiate the solution and ask the child to finish it, or offer leading questions (Vygotsky, 1978).

> the first step in a solution, a leading question, or some other form of help (Vygotsky, 1986).

The nature of the collaboration required is therefore very much left open (Wood & Wood, 1996). Collaboration in VIS consists of:

(i) The selection of activities which are just beyond what the child can achieve alone.
(ii) The provision of help of different qualities, ranging from general encouragement to a demonstration of the activity being attempted.
(iii) The differentiation of activities through adjustments to the style of their presentation.

For each activity the child is invited to complete there are hints available. The system can offer the child hints which can be selected from five levels of control (Wood et al., 1992). The lower the level of hint the more the control of the task is maintained by the child. At level 5 the system completes the activity for the child. There is no notion of failure in VIS, only variation in the levels of support offered to ensure success. In addition to offering the child specific hints to ensure that the activity is completed the presentation of the activity itself can be adjusted or differentiated. There are 18 possible combinations of help and activity adjustment (collaborative support). Each carries with it a certainty value. The higher the value the stronger the system's view or belief that this activity is within the child's independent capability. The measure of collaborative support used by the child in the past is used to calculate the amount of support to be offered in the future. This information is weighted so that the most recently tackled templates contribute most to future decisions.

Within VIS, decisions about the content and delivery of the curriculum depend upon the system's knowledge of the child. The creation of such a model requires that care has been taken in the design of the curriculum to ensure that what the child needs to learn is represented precisely (Corbett, Anderson & Patterson, 1988). It also requires that sufficient information about the child's performance is recorded. Whilst the child's performance at a particular activity can be directly observed, her ability itself cannot. A human teacher would interpret her knowledge of the child's past performance to assess that child's capability and form a view about it. This view would be used as the basis for the selection of an activity and a collaborative approach. Likewise, VIS must rely upon the student model as its representation of the student's past performance and find a means of interpreting this representation to form a view about her future potential. To complicate matters, the situation is constantly changing. As the child completes more activities, the system must be able to adjust its view accordingly.

In the domain knowledge representation (described earlier) each node represents something which the child needs to understand: a relationship or a concept. Associated with each node is an activity which deals with the particular element of the curriculum represented by that node. Common misconceptions are recognised within the knowledge representation and the activities. In the student model there are two tags associated with each node. The first tag is the system's belief about the child's independent ability, the second is the amount of collaborative support which the child needs to complete the activity associated with this node. One conception of the ZPD is as the difference between collaborative and independent success. The first tag value is the system's belief about the value of the lower limit of her ZPD. The upper limit which relates to the child's collaborative capability entails decisions about:

- which nodes are within her independent capability and, therefore, outside the lower bounds of her ZPD;
- which nodes are outside her collaborative capability and, therefore, outside the higher bounds of her ZPD;
- which nodes are within her collaborative capability and, therefore, within her ZPD.

The levels of help and activity adjustment which comprise the amount of collaborative support used for a particular activity are used to assess the probability that the particular node defining that activity is within the child's independent ability. If the node is considered outside the child's independent ability then the measures of collaborative support can also be used to assess the extent to which the node is within the child's collaborative capability and therefore within her ZPD. There are prerequisite links between the nodes of the knowledge representation structure; this partial ordering allows the use of conditional probabilities. As has already been stated these values are uncertain; they are the system's best guess based upon the evidence available. In all they add up to the system's opinion about the child's capability in this domain. The student model must be constantly updated to take the changing situation into account, not just for one particular node, but for the nodes influenced by that particular node also. Conditional probability is a basic concept in the Bayesian probability calculus (Jensen, 1996) opening up the possibility of using the conditional probability relationships between nodes as the basis for the construction of a Bayesian Belief Network (BBN). BBNs provide a form of representation for uncertain knowledge. Algorithms for computing probabilistic inference can be associated with the BBN representation. In this case the probability in question would be that a particular node is a member of the set of nodes which represent the child's ZPD, or put another way that the set of nodes which are within the child's ZPD has a particular composition.

The prerequisite relationships within the domain knowledge allow a partial ordering of the curriculum elements. They can be viewed as the guiding goals of the instructional process. The system uses its beliefs about the child's ZPD and the level of collaborative support required to adjust and react to the current state. Decisions about the next activity and the amount of collaborative support to be offered for its completion are based upon the most up to date beliefs in the student model.

4 The Educational Experience

When the child first starts up VIS she is told that she is going to construct a simulated community of plants and animals. These plants and animals each play a particular role within the community and react with other members in certain ways. Through using the computer the child can discover the nature of these roles and relationships and how different combinations of organisms can live together. The screen is introduced and possibilities for interaction are specified. The elements within the domain knowledge system and the food web rules are represented by objects which the child can manipulate on the screen. The behaviours which are possible with these objects are specified in the object's description. The interface is designed to be consistent with the windows type software with which the children for whom the software was designed are familiar. Figure 3 illustrates the nature of the interface. The mouse can be used to select from a tool palette or a menu across the top of the screen and to manipulate the plants and animals via clicking or clicking and dragging. The world of the community which each child will create can be viewed in different ways. It can be seen as:

- an animated picture of the plants and animals in an appropriate landscape;
- a series of energy meters each of which describes the energy level of a particular community member;
- a puzzle or diagrammatic representation of the feeding relationships which exists within this child's world.

The child can switch between these views by clicking in the tool box. There is also a notebook facility; this is a simple editor where she can record information of her choice. In addition to these different views there are two modes in which interaction can take place:

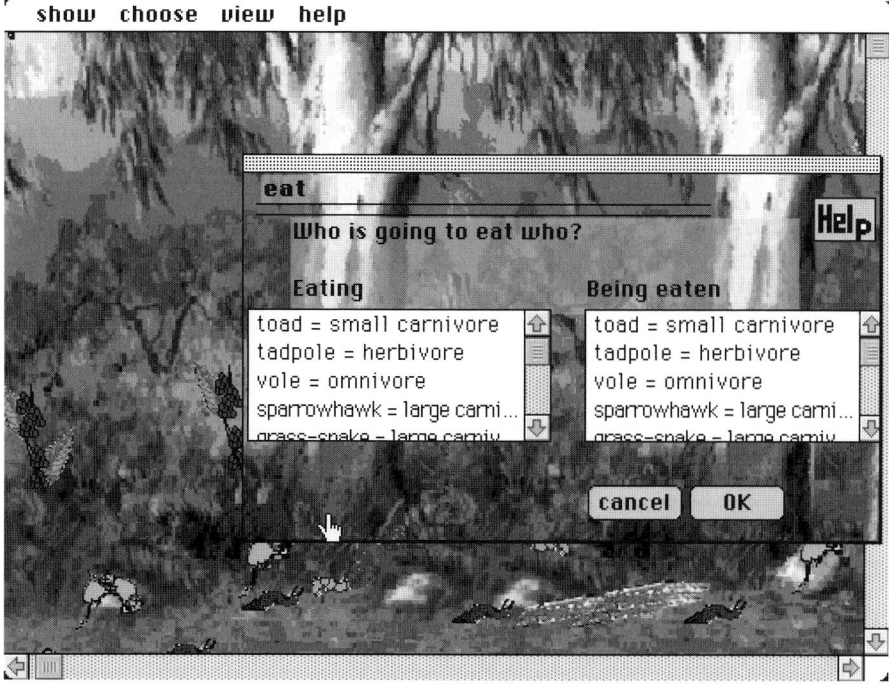

Figure 3: Ecolab World View showing the box which appears when the 'eat' command is selected.

Building mode, during which the world is static and organisms can be added or removed. This mode allows the child to consider the current world situation before any actions within it take place.

Running mode, during which the world is dynamic and the creatures within it act and react in accordance with their nature. The nature of the reactions which can occur is restricted in accordance with the phase of the domain knowledge of the current node. There are four phases (as indicated in Figure 2): the first deals with simple feeding relationships, the second with food chains, the third with food webs and the fourth with the whole community of relationships within this world. Running mode is further divided into *Step* in which a single action can be specified, *Program* in which a series of actions can be linked and *Free* in which the user has no control and the organisms fulfil their natural roles. When each organism is placed in the world it is allocated a certain amount of energy, which, if not replenished via the intake of food, will reduce to nil and the organism will die. If the appropriate food is present it will be eaten and the consequences can be observed.

Each node in the knowledge base has a goal. This goal is either that the child will be able to illustrate that she has understood the rule which underpins a relationship within the food web, or that she has understood a concept's definition. The activities presented when the child is at each node are designed to achieve that node's particular goal. Activities can be of the following types:

- *Investigation*: this encourages the child to use the simulation to answer questions about the particular relationship or concept definition currently being presented.
- *Definition*: this requires that the rule or concept definition appropriate to the current node be specified by the child. Once again she is encouraged to use the simulation to observe relationships and behaviours in order to reach a conclusion.
- *Integration*: this type of activity is designed to link what the child has learnt previously with what they have just learnt.
- *Explanation*: this is an untutored activity, the purpose of which is to encourage reflection and provide information for subsequent evaluation of the system.

There are templates for each of these categories of activity. The content of the activity is stipulated by the nature of the current node. There is also an introductory template which allows an initial explanation of what the ensuing activities will be about.

The following scenario describes an interaction between a child, Helen, and a VIS prototype. The system's student model reveals that Helen is a novice, both in terms of her use of this system and in terms of her knowledge of food webs. The initial task which faces Helen is the selection of some plants and animals for her community. She is offered a selection of organisms and as each one is chosen it is added to the screen representation of the world. The availability of a choice of organisms is designed to promote the possibility that those chosen will be familiar to Helen and a part of her informal everyday knowledge of ecosystems. For example, Helen's first choice is a sparrowhawk, from a woodland ecosystem. The sparrowhawk now appears in each of the three perspectives of the world: as a picture, as an energy level and as an element of the food web puzzle. Further selections are made, the system offering guidance as to when there are sufficient organisms to allow dynamic interactions.

The first activity which Helen is offered is one which relates to the first node: energy transfer. The underlying basic rule is: energy is transferred from food to feeder when the food is eaten; there is then the possibility of a qualifying refinement to this rule which is that not all the energy of the food is transferred to the feeder, some is dispersed. The level of abstraction is the lowest and the organisms will be described accordingly. The investigation activity template asks Helen to use the world to answer questions of increasing difficulty for example:

- What happens to the energy level of the sparrowhawk when it eats the thrush?
- Why does the thrush eat the snail?
- How does the snail get enough energy to live?

Helen can change the mode of the system to run in order to investigate the answers to these questions. The Definition activity template asks her to construct a rule which describes what she has observed while answering the questions. The constituent parts of various rules are available on the screen, but she must build them together to form the correct rule. At this point in the interaction an integration ACTIVITY template would normally be appropriate, but on this occasion this is the first node Helen has tackled and therefore there is no previous experience to consolidate. To conclude this node Helen is therefore asked to explain, using the notebook, what she has learnt.

The above brief description illustrates how Helen can interact with VIS, and explains the types of activities that she encounters. However what is not clear from this text is how the system acts as a more able collaborative learning partner. This will now be clarified. As has already been identified there are two basic types of assistance which VIS can offer. First, there are specific help statements which can be of five different levels. For example, if Helen has difficulty in answering the investigation questions the system can prompt generally with "try setting the world to run and see what happens" or more specifically with "try programming the sparrowhawk to eat the thrush and look in the energy view to see the changes". The most specific help, i.e. that in which the system takes the greatest control would be a demonstration of the programming of the system for the sparrowhawk to eat the thrush and then a switch to energy view to see the outcome.

The second category of assistance is the manner in which the activity presented to Helen can be adjusted. The adjustments possible in the current implementation of VIS are organised into two levels. The first level incorporates two types of adjustment. First, the number of organisms used in the activity is restricted to those which exhibit the relationship of features which is currently the subject of instruction. Second, when the child is required to select an answer, or construct an answer from its constituents, the number of possible wrong answers or constituents is reduced. The second level of adjustment also encompasses two kinds of alteration in addition to those utilised for level one. First, some of the elements of the activity are already completed and second any rule refinements are ignored.

In the current example this would mean that Helen would be introduced to the rule that energy is transferred from food to feeder, but not that some of the energy is dispersed. Once Helen has completed further activities there will be another form of adjustment available. The selection of which node is to be the subject of the next activity can also be viewed as an adjustment. Shifting the level of abstraction to a higher level will mean that in the next and future activities Helen will be required to manipulate concepts such as herbivore rather than snail. She will need to understand the relationships which exist between the organisms in terms of their roles rather than in terms of category instances. Such roles are likely to be less familiar to Helen and therefore more difficult. If however such a move presents Helen with an unreasonable degree of difficulty, as would be indicated for example by her requiring a high level of help to succeed, then the level of abstraction can be lowered again.

These adjustments comprise the zone of available assistance (ZAA) for VIS, i.e. its overall scope for adjustment (Luckin, 1996b). Collaborative support entails the tailoring of these features of the ZAA to be in line with the requirements of the child who is currently using the system. This adaptation requires that the system create a zone of proximal adjustment (ZPA) (Murphey, 1996) from the most appropriate elements of the ZAA. This ZPA specifies the quality and quantity of the system's collaboration, and should mirror the child's ZPD. The student model described earlier dictates the nature of the collaboration which will comprise the ZPA: the selection of the next activity, the level for the adjustment of the selected activity and the level at which help should first be offered.

The TRIVAR system has been implemented in the Ecolab software. An empirical study of Ecolab was undertaken with a class of Year 6 (10–11-year-old) children in a local primary school. Limitations of space restrict the current discussion of this evaluation, which is fully reported in Luckin (1998). The study indicated that children who used the VIS variation of Ecolab demonstrated the greatest learning gain between pre-interaction and post-interaction test. Both learning gains and the nature of the interactions which occurred between child and system were strongly related to the child's ability at the start. The ZPD is not a concept which allows an easy definition or diagnosis. However, the interactions between the Ecolab system and child which targeted the available resources of the system to its model of the child's ZPD were most commonly characterised in the interactions of the VIS users. These results are promising and indicate that the design of Educational Software can benefit from attending to the ZPD.

5 Conclusion

This chapter has considered the requirements for a learning situation that will promote the contextualised teaching of abstractions, highlighting the implications these requirements have for the software designer. VIS offers a possible approach — designed for use in the school environment, which is a culturally influenced context. Information Technology is now a part of the current UK primary school culture and the ecological subject matter is in line with the science National Curriculum. The instruction provided by VIS aims to mediate the child's experience of the curriculum in the classroom. The provision of help through assistance and through differentiation of the activities addresses the need for a balanced partnership approach to support the learner to a higher level of achievement. The system needs to be able to adjust to the child dynamically; because the child's ZPD alters so must the system's ZPA.

Throughout this chapter the requirements of this approach have been highlighted. The need for software designers to respect the culture which exists within the educational environment has been recognised. However, culture is a dynamic entity and any design methodology will need continuous review. The system's knowledge of the domain needs to be structured and flexible with explicit potential links to the child's current understanding at any point during instruction. The structuring of the domain is not straightforward and decisions between alternative categorisations, for example, need to be made with respect to the context of delivery. The provision of potential links to the already understood is also problematic. It requires not only flexible knowledge of the domain but also a means of eliciting and evaluating information about what the child does already understand. This information about the child is important if the selection of the most appropriate assistance is to be effective and the system's ZPA is to match the child's ZPD in a manner which results in balanced collaboration. The ZPD is dynamic and so must be the system's ZPA. The constant challenge faced is the provision of collaboration which will ensure that the child is both supported and challenged. This is a demanding role for any learning partner whether computer or human. Computers are a recognised tool for the delivery of the school curriculum. Children gain experience when they enter school at 4 and commonly enjoy computer-based activities. However, the interactions which are possible between computer and child are still limited. The social interaction available within the school culture is important and an interesting development for VIS would be in its further adaptation for use with more than a single user.

16

Learning and Discovering with Computational Aids

Vincent Corruble and Joan Bliss

> The reasonable man adapts himself to the world; the unreasonable man persists in trying to adapt the world to himself. Therefore, all progress depends on the unreasonable man.
>
> George Bernard Shaw

1 Introduction

The research on human learning and that on scientific discovery have for a long time entertained a rich and mutual cross-fertilisation through what researchers speculate the two processes have in common. It is often considered that the process that leads the scientist to a discovery has some similarities with the one that takes the learner from one level of knowledge to the next. One of the advantages of such a process is that it enables an individual to think in relatively simple terms even though the content of the knowledge involved can be too complex for the non-specialist to grasp. This article investigates the parallel between learning and discovery from a specific angle: we are interested in the study of the use of computational aids to support these two processes. In particular we focus, in two case studies, on how these methods provide an original way to study some aspects of the way in which the context of each of the two activities is often critical to the outcomes.

One of the most common causes of frustration for computer programme users is that the programme does *what they say* instead of *what they mean the computer to say*. This is indeed usually an annoyance, but in the context of this article, it has actually proved to be a benefit. Since computers do what they are told, they provide a way to study the *meaning of what they are told*, and hence reflect directly on the user's original ideas. How is that done? Computer programs perform operations on their inputs (data, or knowledge) according to a number of rules, expressing a form of rationality. The inputs to the program as well as its rationality are essentially disconnected from the user's own context and will therefore often behave unexpectedly. This surprise element is a tremendous opportunity for users to reflect on the situation and its background and how this context helped to generate some of the ideas related to the situation. There are also dangers with this approach, as we will see in our first case study: users can place too much faith in the computer program, and hence try to find in their own understanding of the world an explanation for this surprise element, instead of reconsidering either their own expectations, or indeed the inputs to the system.

We are now going to see two applications of these types of method in two very different case studies. The first study addresses the use of a computational aid for reasoning about everyday complex systems, whereas the second one deals with the use of a program to aid discovery in medicine by suggesting hypotheses on the causes of a disease. In these two studies, the use of computational tools helps to question the user's initial assumptions, and also illustrates how users rely on knowledge they already have about the situation to respond to the task.

2 Aiding the Modelling of Dynamic Systems

IQON is a dynamic modelling tool developed within the Economic and Social Research Council's Initiative, Tools for Exploratory Learning Programme, which investigated students' reasoning between the ages of 11 and 14 using computational modelling tools (Bliss 1994, Bliss and Ogborn 1992, Mellar et al., 1994). We had two underlying questions about modelling:

(i) Can working with modelling tools which contain models of a domain facilitate students' reasoning in that domain?
(ii) Are learners helped to reason about a domain by using modelling tools to express their own ideas about that domain?

SMALL TALK was used to develop a direct manipulation tool, IQON, in which the variables created increase or decrease in value, and affect one another. Essentially the user needs no mathematics when creating relationships between variables. IQON is suited to expressing models of unquantifiable ideas such as the effect — on enjoyment of holidays — of weather, food, quality of accommodation, expense, etc.

An IQON model is built by making and defining boxes to represent relevant variables, and linking them to show their mutual effects. A plus-link says that a

Table 1: Two case studies and the tasks they address

	Dynamic system modelling	**Inductive modelling**
Domain	Common knowledge	Medical research
What people start from (what they mean)	Expectations about system's behaviour	Expectations about system's behaviour, experimental data
What they tell the machine	How different things affect each other in the system	Description of cases Some causal relationships
What the machine provides	Simulated behaviour of the system studied according to the model given as input	Causal hypotheses based on the inputs

variable which is high causes the one to which it is linked to slowly increase; if low, the effect is reversed. A minus-link says the opposite, that is, a variable which is high causes the one to which it is linked to slowly decrease, and vice versa. All variables have a middle normal level at which they have no effect on others. Links between variables can be made stronger or weaker. One of the novel features of IQON is that it allows the user to build a model expressed in a semi-quantitative graphical language, as shown in the example model below (Figure 1).

Students between the ages of 11 and 14, mainly 12–13, were asked to build models to think about the following situations: (i) how to keep fit with a sensible diet; (ii) how to keep a shop in profit; (iii) how to control traffic and congestion. In the study, we distinguished between expressive learning activities where students expressed their own ideas with the modelling tool, and exploratory learning activities where they worked with another's model and made judgements about its worth after using it in several trials. All students worked individually for several hours with a researcher two weeks before the task, learning both to use an Apple Macintosh ™ and how to use IQON before carrying out a task on one of the above topics in either an expressive or exploratory mode.

IQON provides opportunities for students to learn more about some fundamental aspects of modelling. They can find out what exploring a model is all about, either through the experience of expressing some ideas themselves or by trying out and criticising others' models. In both cases they have the opportunity of thinking about the relation of models to reality or about how a model or modelling system looks at the world in its own special way.

The findings discussed in this chapter focus essentially on students working in an expressive mode. Before considering the main focus of this section — the relation of

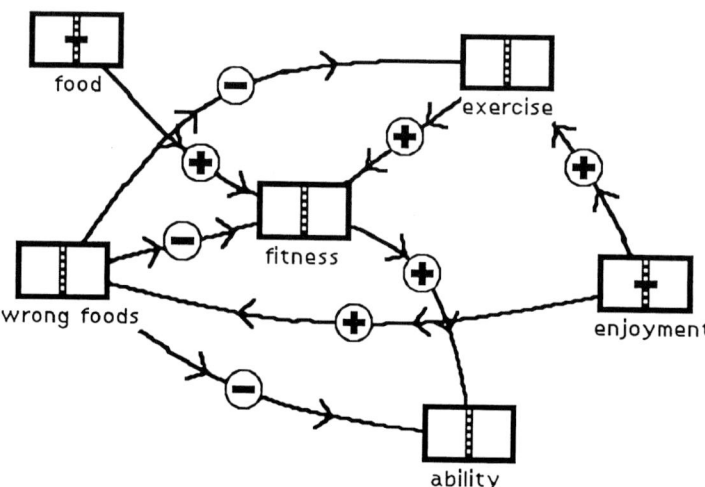

Figure 1: Thirteen-year-old Anthony's IQON model for fitness and health.

models to reality — we mention one or two of the most important results to give the reader a taste of what these young students could achieve (see Bliss 1996, 1997 for further details). All students who undertook the expressive task created a model. They could think of a good number of relevant variables for the problem and most models contained between four and six variables, a few being considerably larger, and only one smaller. Linking variables was not a problem. Students were aware of ways in which the model variables affected one another, and no model contained any variable not linked to some other. About half the models showed some considerable degree of interdependence of variables, with variables affecting one another in multiple ways or with feedback.

Nonetheless, students had difficulty with the necessary simplification or idealisation involved in the creation of any model. Thus in such a learning activity, the relationship students perceived between model and reality is of obvious importance. We can ask:

- When is the model thought of as just itself, independent of reality?
- When is reality thought of as giving sense to the model?
- When is the model to be thought of as accounting for reality?

Three criteria were used to judge whether or not students used their own knowledge and experiences of the world in their reasoning about the relationship between the model and reality:

(i) Does students' reasoning stay close to what is happening in the model, simply describing the variables and links as prescribed by the model on the screen?
(ii) Does students' reasoning, while respecting the variables and links in the model, elaborate on the meaning of the links?
(iii) Does students' reasoning use both knowledge outside the model and also call upon the students' own experience?

When a model was running, students focused on how the model was functioning and rarely if ever made reference to anything outside it. When asked to predict what the model would do there was also little reference to reality. It seems that because students were thinking about how to manipulate their models and at the same time working out how their models would function — but without yet running them — their thinking stayed close to the model.

However, reality and students' own knowledge and experience were crucial in a number of other situations. The more students sought to explain how the model worked, providing sophisticated causal explanations, the more they tended to make reference to reality or to their own experience. When models produced puzzling or unexpected results students would also introduce the real world into their explanation to try and make sense of what was happening. The passage from the complexity and richness of the real world of students' knowledge and experience to a simplification to be represented in the model was sometimes difficult. We turn now to give some examples of these dilemmas, dilemmas which reveal the importance of the context of the situation and students' prior knowledge of such contexts.

In our first example, to do with the task of how to keep a shop in profit, Nesta was trying to keep profits normal in the model. In doing so, she put her model (see Figure 2) into an unexpected oscillation by decreasing both the independent variable (*helpful staff*) and the dependent variable (*prices*). She first tried to make sense of this situation with a causal argument, using simple and multiple connections (many-to-one):

> I've put the prices down a tiny bit, and also down (decreases helpful staff) a tiny bit so there's less helpful staff, so they (the customers) would come because of the prices — but maybe not come because of the staff (it is implicitly understood that customers affect profits).

She then attempted to explain the oscillation only in terms of the model:

> Profits went down a tiny bit and now it's gone up because the prices have gone down but now they're going up again. The prices and the profits are — it keeps going up and down. Now the profits are going down and the prices are up and the prices are going down. It's just going to keep going up and down probably.

However she was not satisfied with this description and tried to make sense of the situation, calling on reality to help her:

> I suppose sometimes the customers would come and sometimes they wouldn't because it's not anything special, like they could go somewhere else that's cheaper — that's why it keeps going up and down — sometimes they might come and sometimes they might not.

Here, faced with a difficulty, the pupil uses her own knowledge of shops and shopping, not represented in the model, to explain the unexpected result, and in some ways to better understand her model.

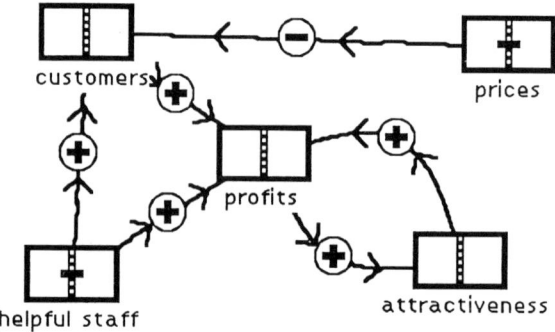

Figure 2: Nesta's model of a shop making profits.

In yet another example (Figure 3), Brenda, in the Fitness task, discussed the issue of keeping the variable of fitness high: "You should do a lot more running, keeping the fat off, you're eating better food, it's giving you more energy".

Neither keeping the fat off nor more energy were variables in her model, though they could be seen as re-interpretations of the variable fitness, incorporating her experience and knowledge into the model and helping her to understand the problem better.

Anthony built a model for fitness and sensible diet, as shown in Figure 1. Twice he increased the variable enjoyment to achieve a rapid fall in the fitness variable. His understanding of the variable enjoyment was explained with the glee of an adolescent who loves junk foods despite adult insistence that they are not healthy:

> Enjoyment is a little more up — quite a bit more up he's very happy in other words! Because food's very bad, exercise is very bad, his ability is poor, he eats all the wrong foods but he enjoys everything a lot — he enjoys the wrong foods.

This statement implies links that were not in his model, though when he came to reason in detail about the model Andrew knew exactly how the links worked. His knowledge and experience allowed him fun in experimenting with the model.

There were cases where the students' knowledge and experience did not help them with the modelling task, and where it tended to dominate over the task. Emma had four variables positively linked to profits: clean shop, earning money, being friendly, how good things are. She saw the cleanliness of the shop affecting the customers. Although customers did not feature in her model they were part of her thinking. She argued:

> A clean shop is better than a dirty one. People won't want to go into a dirty one because they've got to buy all their food and or something from those — or um — you, or if it's a clothes shop, they don't want dusty old clothes bought from a shop.

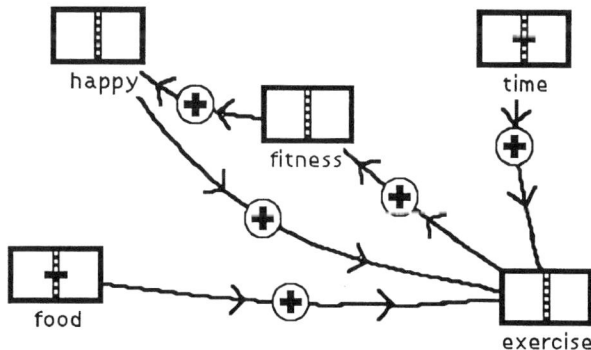

Figure 3: Brenda's fitness model.

Emma also imagined a new link between the variable being friendly and that of profits which would result in less profits. Her explanation proposed that, ". . . if the shopkeeper is too friendly then someone might take advantage (steal), thinking they could get away with it", because of the friendliness of the shopkeeper. She elaborated on the consequences, suggesting that the persons taking advantage must be punished, since otherwise they might do it again. Here, real-world sequences of events wholly replaced model variables and links. In much of her reasoning Emma used her own knowledge of people, shops and shopping, not represented in the model, to explain how things functioned.

2.1 Interactions of Topic and Type of Reasoning

In semi-quantitative expressive tasks, models built for the Traffic task were simpler than those for other topics. The semi-quantitative Traffic models did not contain fewer variables than the others, but had fewer links. Over half of them were star, or star-like, that is, a collection of variables linked to one central variable, with no interaction between the variables. Star models were produced less frequently in the other tasks. Fitness expressive models were both complex and the best understood of all models on all topics. Models for the Shop and Fitness tasks were very alike in sophistication, both sets of models being more complex than Traffic models. But the Shop models were the less well understood.

It seems likely that such differences are due both to differences in students' knowledge about and experience of the topics and to differences in the nature of the tools. The transcripts suggested that the Fitness topic was close to students' immediate experience, and had often been the subject of personal reflection. IQON helped them to represent this knowledge. Shop knowledge was extensive but one-sided, experienced as the customer and not as the seller. Students believed they knew a lot, but when they tried to articulate their knowledge from a different perspective, the lack of sureness about the ideas often interacted with the difficulties presented by the formalism of the tool being used.

Knowledge of traffic seemed the least well articulated in students' thinking. Certainly the problems posed, of local authority control of congestion, were outside their experience. And while they knew about traffic jams as an immediate experience, they were not able to see them from a whole town perspective. This is consistent with the fact that expressive semi-quantitative Traffic models tended to be a collection of things that might affect congestion, with no inter-connections (star models).

An advantage of having tasks based on everyday knowledge was that it permitted and valued the use of students' own everyday knowledge and experience. The more they wished to explain and interpret models the more they tended to seek reasons in the real world. This was not without problems. Sometimes students would suggest a real world event, not in any way represented in the model, as a reason for an effect of the model. There were times when knowing too much proved to be a problem. A few students who were personally involved in the issues raised in the tasks would import very substantial knowledge into the task, which could distract them from the modelling activity. However, overall the majority of students broadly used reference to the real world and to their own knowledge and experience in a reasonable and appropriate manner.

3 Aiding Inductive Reasoning in Science

3.1 Background to the Work

The system presented in the previous section is primarily addressed to the study of children learning/ reasoning with dynamic systems. In the field of scientific research, another important task is the articulation of hypotheses (candidate theories) from observational data. Logicians, philosophers of science, statisticians, and more recently data analysts and the artificial intelligence community (within the field of machine learning) have addressed this problem under the term of induction. The inductive inference has been problematic for the field of logic because it is not a valid inference: usually, its conclusion, a hypothesis, is not entailed by its premises (a number of cases or examples). Hence, it is necessary to define an inductive bias that bridges the gap between the information provided by experience (the examples) and the commitment constituted by the hypothesis.

It is even more problematic to study the inductive inference if the interest is in how people actually perform this type of reasoning. Finding categories and classifying objects are processes central to human cognition, yet they remain largely mysterious because humans are not aware of them as processes. Hence the reasoning steps carried out by humans, even in a scientific setting, are not usually formalised, at least at the hypothesis generation stage. Moreover, the inductive bias is usually left implicit. Thus the focus of research in formal induction, over the past century, though sound from the formal standpoint, is only remotely related to the human inductive process.

This difference is not a problem in the absolute, when the interest is in a normative study of inference; it does however become a problem when the focus is on aiding the inductive process of the scientist. In this latter case, it is essential that the two processes (in the machine, and in the scientist's mind) are strongly related so that a fruitful interaction can be obtained. Artificial intelligence, in the field of machine learning, has contributed to this research goal in realising that:

- The careful definition of the inductive bias is sufficient to specify the behaviour of the inductive system, independently of any algorithmic detail;
- This definition can be carried out at the knowledge level in an explicit way, thus specifying a declarative bias (Utgoff, 1986).

Two inductive systems have been used in a number of experiments aimed at the reconstruction of historical medical discoveries. The first system, CHARADE (Ganascia, 1987), induces rule-systems from a set of examples described in an attribute/value representation. The second one, PASTEUR (Corruble, 1996), induces theories with exceptions which are, in turn, modelled themselves. CHARADE has been useful in the reconstruction of the discovery of the causes of scurvy (Corruble & Ganascia, 1997), and PASTEUR for the discovery of the causes of leprosy (Corruble & Ganascia, 1996).

The study of the performance of induction from observational data, independent of pre-existing knowledge, is the question tackled in these experiments. This question has been previously studied within the fields of history of science (history of medicine in our case) and philosophy of science. It is important because, even though modern medicine claims to base its theories on facts and observations, evidence seems to show that, at least historically but also maybe during this century, it has not always been the case.

One of the main contributions of computational methods is that they provide the means for an experimental study of this theoretical question. The possibility of expressing candidate background theories as knowledge given to the system is a powerful way of testing the historical plausibility of various knowledge contexts for induction from cases. A pure empiricist approach, which considers that science can abstract itself from previous theories, would challenge the ideas developed in the situated movement according to which no knowledge or experience (such as observations) is independent from the cultural context in which it has developed.

The computational approach has also a practical advantage. If hypothesis generation in science is the result of an interpretation process situated in a cultural context, it is essential that this context is made as explicit as possible. Through computational modelling and computational simulation, previous knowledge can be externalised. Any inconsistency or incompleteness in this externalisation process is likely to show up by causing the system to behave unexpectedly. Hence, a refinement process is made possible in which a better account of contextual knowledge is given to the system and tested. Second, as a result of externalisation, this contextual knowledge can be questioned since it is brought back into the field of investigation.

4 Experiments on the Discovery of the Causes of Scurvy

4.1 Motivation

Our ideas about the role of implicit contextual knowledge on hypothesis generation in science have been tested and refined by some experimental studies of historical medical discoveries. In order to illustrate our approach, we present here work done on the discovery of the causes of scurvy. Because of the important role of contextual knowledge in our work, we first give a short overview of medical research on scurvy. We will then present our computational experimentation, putting the machine in the data context of the 19th century and then experimenting with different knowledge contexts. For a more detailed account of the technical aspects of this work, readers are invited to refer to Corruble and Ganascia (1997).

4.2 History of Scurvy Research

There is a rich history of research on scurvy. As a disease of seafarers (not exclusively), its impact on European naval powers was enormous, both in human terms (over a million deaths), but also politically (Carpenter, 1986). Consequently, research on the causes of scurvy attracted the brightest medical minds of the time (Lind, 1753). From the 17th to the 20th century (when the actual cause of the disease, i.e. lack of vitamin C, was discovered), dozens of theories on the origins of the disease were put forward. Many of these were totally disconnected from the real cause, for example, those referring to the psychological effect of being at sea, far away from home. Other theories were quite close to the real cause, especially the one that became widely accepted, that is, that scurvy was the result of the conjunction of the humidity in the air and of the lack of fresh fruits and vegetables in the diet.

The lack of fresh fruits and vegetables was eventually accepted as the only cause of scurvy in the early 20th century. A first explanation for this late discovery is the lack of a concept necessary for a global understanding of the disease. The concept of vitamin (the idea that a small quantity of a chemical has a great influence on the functioning of the human body) is a key to the comprehension of the mechanism leading to scurvy. However, given that seamen were acquainted with the importance of fresh fruits and vegetables as early as the 15th century, it is surprising that a practical cure was not widely accepted earlier. To put it into more formal terms, though the lack of an explanatorily adequate theory of scurvy is easily understandable due to the necessity of a new concept, the reason for the absence of a descriptively adequate theory (a theory establishing only the conditions of development of the disease) is unclear.

This last question provided a good motivation to try modern inductive techniques on data available in the 19th century. This attempt is described in the following sections.

4.3 Induction from Historical Data

Some descriptions of cases of scurvy were collected in the medical literature of the 19th century (hence, prior to the actual discovery). Each case was directly translated from the original, natural language description to a simple language defined by a number of attributes (the ones found in the original descriptions), each with a domain of possible values. For example, attributes describing a scurvy case were: year, location, temperature, humidity, food-quantity, food-variety, hygiene, type-of-location, fresh fruits/vegetables, disease-severity (note that for some cases, only some of the values are known). These new descriptions were made available to an inductive system (CHARADE, Ganascia, 1987) to produce rules linking the description of cases to the severity of the disease.

CHARADE is an inductive system which, in this study, explores a search space of hypotheses relating descriptions and disease severity. Following this method, 12 rules were obtained. An analysis of the rules was then done in a number of stages:

- regrouping of the rules in coherent subsets, each a proto-theory on the causes of the disease;
- evaluation according to the number of cases explained;
- comparison with explanations proposed in history.

After this analysis, it would seem that the system was putting forward mainly three factors to predict the disease: the diet (in terms of variety and quantity); the hygiene; and the climate (through the temperature attribute). A first observation is that these three factors were all proposed as possible explanations in the medical literature until the 19th century (for example, in the medical encyclopaedia from which the case descriptions were taken). This is a good sign of correspondence between simulation and history.

A second, more surprising, result is obtained by looking at the relative strength of the theories put forward by the machine. The system, to which no prior knowledge was given (except for that encoded in the data), outputs as the strongest theory the one corresponding to the real cause of scurvy, i.e. the lack of fresh fruits and vegetables in the diet.

Despite the encouraging sign noted in our first observation, there is a major problem in this reconstruction for the researcher interested in the correspondence with history: the main theory of scurvy proposed in history, which referred to humidity as the main cause of the disease, does not appear at all in our reconstruction. This concern motivated a second experiment.

In this second experiment, some concepts and knowledge which could have influenced 18th and 19th century reasoning were given as theoretical background knowledge to our inductive system. These concepts and knowledge are the necessary building blocks to articulate the blocked perspiration theory. This theory, inherited from Ancient Greece (Hippocrates) and developed further by Galen, became very popular from the 16th century onward. It is centred around the idea of the essential role of invisible perspiration in preserving health: perspiration is the only way for the body to get rid of some poisons naturally produced by the friction of internal fluids on our solid tissues. Hence this theory introduces two abstract concepts (quality of fluids, and quality of perspiration) which are not directly observable, and not mentioned in the historical descriptions of scurvy cases. In our second experiment, we defined as prior knowledge the existence of these two concepts and how they can be derived from the available descriptions.

With this background knowledge, the rules obtained by induction contain a new proto-theory which puts forward the role of the two new concepts and of humidity as causes of scurvy. An important remark is that the number of cases explained by the new abstract concept of fluid quality is significantly higher than the number of cases explained by the theory putting forward the role of fresh fruits and vegetables. This reconstruction, therefore, provides an explanation for the following puzzling phenomenon. Even though the data necessary to reach a satisfactory theory were available as early as the 16th century, and certainly by the 19th century, the medical community favoured a competing theory putting forward the role of humidity until the end of the 19th century. Our explanation, i.e. the role of pre-existing knowledge influencing inductive reasoning, had been suggested by the history of medicine, but our computational reconstruction provides empirical evidence of its plausibility.

Analysis of Results

Because a computational simulation needs to have all its inputs explicit, and since the system used for induction (here CHARADE) has a large part of the inductive bias as input rather than coded within the algorithm, a main advantage of the computational approach is to have most of the inductive bias explicit. This is a first major difference between this inductive process and the process followed by scientists or others in common scientific practice. The consequence is that the context of the induction is made explicit through this approach and, hence, becomes subject to investigation and experimentation.

There is also a major difference with other formal approaches to induction such as the one defined in Bayes' theory. The inductive bias can be expressed as domain knowledge (in our experimentation the blocked perspiration theory) that directly makes sense to the user. This is a necessary condition so that the expert can benefit directly from the interaction with the computational system.

5 Conclusion

Earlier we noted that the two types of tasks presented in this chapter could be viewed as radically different. Yet our aim has been to show that the types of aid provided by the use of computational tools on these two tasks have strong similarities. A common feature of both is that they are to some extent situated. Because children are asked to work on problems encountered on an everyday basis they have many ideas about them which are rooted in an individual's experience and cultural context. Because the scientist, and maybe even more the physician, have had to deal with similar problems, a good account of their inferences needs to take into account the rich cultural context of their research, and the theories and observations made during their education and practice.

In both cases, the tasks are generally too complex for humans (student or scientist) to accomplish at a formal level, and therefore they usually rely heavily on the use of heuristics. One limitation in the use of these heuristics is that they often work under a number of assumptions that are not made explicit. Such an assumption can be, for example, that one can ignore the influence of a given variable on an event. There is nothing wrong with these assumptions, until they need to be reconsidered. The major help provided by computational tools in both tasks is therefore to help the user, student or scientist, to externalise these assumptions (knowledge or beliefs) about the problem they tackle. This is by no means an easy or straightforward process. The computational tool does not spell out explicitly the nature of an assumption, yet it reveals its existence through the behaviour of the system being modelled. Because each model is expected to behave in a certain way by its creator, it is through the interpretation of the unexpected results that one can be led toward (i) externalising the assumptions; (ii) questioning their validity; (iii) refining the model.

The computational tool offers therefore the possibility of being used simultaneously as a medium and as a mirror. It is a medium because it lets users externalise their thinking through modelling at the appropriate level of formalism. It also acts as a mirror through the simulation of the model (or more generally through the execution of a programme), which reflects back to the user the true face of what has been externalised.

17

Situated Cognition: A Challenge to Artificial Intelligence?

Dolores Cañamero and Vincent Corruble

> A man, viewed as a behaving system, is quite simple. The apparent complexity of its behaviour over time is largely a reflection of the complexity of the environment in which he finds himself.
>
> Simon, *The Sciences of the Artificial*

1 Introduction

One of the most challenging problems faced by researchers in artificial intelligence (AI) is having our systems interact in the real world — a dynamic, complex, and partially knowable environment — and learn from this experience. Traditional AI approaches have attempted to tackle this situation by endowing the systems with more complete and flexible models of the world. The different issues involved in modelling change when dealing with the real world constitute what has come to be known as the frame problem. Over the last decade, an alternative view has arisen that advocates situatedness as a more plausible solution to this problem. By this term, we mean considering the interactions of a system with its environment as the global unit that has to be modelled. This is in opposition to the traditional view that models isolated input/output systems.

Situated cognition has its roots in cognitive psychology and has been the object of long debates in the cognitive sciences, as illustrated for example by the 1993 special issue of the journal *Cognitive Science* devoted to situated action. In AI, we can distinguish two different attempts to put into practice the general philosophy of this movement, which has undergone some reinterpretation in the light of problems specific to the field. One has emerged within the symbolic AI community, made explicit in particular in the field of knowledge acquisition, and has its roots in Newell's knowledge-level hypothesis and Simon's bounded rationality principle. The other has developed within the behaviour-orientated AI community and takes ethology, evolution, and biology as its inspiration sources. In this chapter, we examine landmark work in both lines of research and analyse their underlying assumptions according to criteria that allow a better definition of their commonalities and differences. In particular, we focus on how each community approaches the following hypotheses that we see as a core manifesto of the situated movement in AI:

- Knowledge does not exist physically somewhere but is dynamically (re)constructed through interactions between the agent and its environment.
- Knowledge cannot be stored in static representations, and therefore it cannot be transferred, only reconstructed.

- Problem solving is a constructive process, as opposed to a process of search.
- Intelligent systems behaviour has to be understood within a particular environment and task.

2 Situatedness and Knowledge Level

Knowledge acquisition (KA) is the process of eliciting and structuring the knowledge necessary to build a knowledge-based system (KBS). During the late 1980s, the KA community experienced a switch in paradigm. KA was not perceived any more as a process of transferring the knowledge possessed by a human expert to a computer system but as the process of modelling a system situated in the world. This resulted from the application to knowledge engineering of Newell's knowledge-level (KL) hypothesis.

2.1 Hypotheses

Newell (1982) proposed the KL hypothesis in an attempt to clarify the role played in AI by the key notions of knowledge and representation. His solution was based on a functional view of intelligent systems (Newell & Simon, 1972) coupled with the physical symbol system hypothesis (Newell, 1980). This introduced a new level of analysis for intelligent systems, which was above the symbol or representation level, and where knowledge was to be defined.

Functional decomposition of physical symbol systems. Newell and Simon (1972) proposed a theory to explain (general) human intelligence and in particular problem-solving behaviour, taking the computer as a metaphor. According to this, a human operates as an information processing system (IPS), or physical symbol system (PSS) that transforms symbolic expressions by performing elementary processing operations. These systems are embedded in a world populated by other objects besides symbols. Therefore, the notions of denotation or reference, and of interpretation, are crucial to establish a relation between the system and the external world. Such a system possesses the necessary and sufficient means for general intelligent action. Since an IPS is an instance of a universal machine, human intelligence can be realized by a universal computer. This hypothesis has its roots in what could be termed a behaviouristic view of intelligence, in the sense that intelligence is considered as a behaviour that can be recognized by its effects. The expression general intelligent action is meant to reflect a fact observed in real-world problem solving: the behaviour of humans is appropriate with respect to their goals, and adaptive to the environment. It is rational not in the sense of an ideally perfect rationality or optimization but in the sense of a bounded rationality (Simon, 1947). The environment, coupled with the goal that motivates the agent, constitutes the task environment. Since the behaviour of an agent is adequate to the situation where it happens, it says more things about the task environment than about the agent, about which we only know that its behaviour is motivated by a problem, i.e. goal-oriented.

The IPS model provides a functional explanation (in the sense of different cognitive functions) of this problem-solving behaviour, and a decomposition of the agent in different functional modules, such as perceptual, memory, information processing, and motor systems, according to it. The agent, embedded in a task environment, processes external information and builds an internal representation that constitutes the problem formulation — the identification of its components — and delimits the problem space — the set of possible behaviours. Problem solving is performed within this representation. The system first selects a particular problem-solving method, that is, a sequence of information processes that establish a rational relationship between the different means defined in the representation to attain the desired solution. This method is then applied, controlling the behaviour, both internal and external, of the problem solver.

The knowledge level. The KL hypothesis analyses an intelligent system, human or machine, as an agent seen from an intentional viewpoint, i.e. as composed of goals, actions, and a body of knowledge. The system's physical body allows it to interact with its environment. It can be any physical system of any complexity, but this remains external to the KL description, only concerned with function; therefore, the physical body is described as consisting of a set of actions. The body of knowledge can be seen as the system memory to which actions can add knowledge. No structural requirements constrain the organization or addition of knowledge, or access to it. Goals are bodies of knowledge about a state of affairs in the environment that the agent wants to achieve. Knowledge is the medium that the agent processes to achieve its goals, following the principle of rationality, that is, it selects actions that it knows will lead to one of its goals. Accordingly, Newell characterizes knowledge in functional terms, "... whatever can be ascribed to an agent, such that its behaviour can be computed according to the principle of rationality" (Newell, 1982, p. 105). Knowledge is used in a situation involving an observer and an agent. The observer treats the agent as a system at the KL, ascribing knowledge and goals to it so as to be able to make predictions about its behaviour, as the environment is accessible to both. The observer is itself an agent that can be described at the KL. However, what both really have is a PSS that allows them to compute actions to be taken according to the knowledge and goals attributed to the other agent. The observer thus determines what the agent can know and do from what it can compute itself (functional equivalence rests thus on structural similarity, which somehow questions the usefulness of Newell's analysis). There is no physical structure corresponding to knowledge itself. Knowledge is not in any physical place but is dynamically created in the situation where it is used. It can only be imagined as a result of applying interpretative processes to symbolic expressions.

As a model of description for intelligent systems, the KL is radically incomplete, it does not allow an agent's behaviour to be fully predicted since some behaviours can only be defined at lower levels. It is useful to rationalize an observed behaviour based solely on information about the external environment. However, an intelligent system can only be fully described by taking into account both the knowledge and symbol levels. Newell ends up coupling even more tightly his new hypothesis to the PSS one. A question remains open, though: Could the knowledge level be decoupled from the PSS hypothesis and be realized by other architectures and physical systems? And, would it be as useful?

2.2 Engineering the Knowledge Level

Clancey pioneered the analysis at the KL of a number of existing KBSs (Clancey, 1985) in an attempt to characterize their competence, the type of problems they can solve, at an adequate level of abstraction. He noticed the existence of common reasoning patterns across different problem-solving activities, and came up with a functional methodology for KBS design. The experience gained from this work led him to propose a research program for knowledge engineering based on a reformulation of the KL (Clancey, 1989). KBSs contain models of systems situated in the world and the expert serves as an informant of how these systems behave, not as a repository of the knowledge that has to be transferred to them. The knowledge engineer is an observer that elaborates a KL model of the problem-solving behaviour of a situated agent, expert or KBS. KL descriptions themselves are representations of explanations that the knowledge engineer gives of a situated system and cannot be associated with structures present in the head of the expert. As any other representation, they have an open interpretation that depends on the point of view of their observer. Finally, Clancey seems to decouple the KL from its realization by a PSS, saying (p. 290):

> The idea that human-like intelligent behaviour could be generated by interpreting stored programs that predescribe the world and ways of behaving must be abandoned, for this view confounds descriptions an observer might make with physical mechanisms inside the agent.

The adoption of the systems-modelling perspective, he says, suggests that methods for modelling situated systems which have been largely disregarded by the knowledge engineering community, such as cybernetics, general systems theory, and chaos theory, need to be incorporated.

The model-based KA community retained three main ideas out of these principles: (i) The elaboration of an abstract functional KL-model of the desired KBS in terms of the different types of components put forward by the KL hypothesis; (ii) The independence of this model from the symbol level; and (iii) The search for generic functional knowledge components that can be reused across applications and domains. The situatedness aspect was not so successfully grasped. Three new elements were also introduced: (i) the application of KL analysis to specific (types of) problems or tasks rather than as a model of general intelligence; (ii) the introduction of structuring elements at this level; and (iii) a top-down analytic approach to problem/ system decomposition. The main modelling methodologies, such as those proposed by Chandrasekaran and Johnson (1992), Musen (1992), Steels (1990), Wielinga,

Schreiber and Breuker (1992), can all be seen as organizing models around three main types of components: knowledge models containing domain knowledge necessary to solve a problem, which are structured according to one or several ontologies; tasks or (types of) problems to be solved; and problem-solving methods or descriptions of strategies or procedures for solving a problem, i.e. how domain knowledge can be used to solve a task. Some methodologies also use inference structures which specify the inferences (primitive reasoning steps) that the system will be able to perform, its competence, the types of input and output knowledge that each inference requires, and the data flow among them.

Throughout the years, a number of generic model components have been characterized and stored in libraries of reusable components (Breuker & Van de Velde, 1994). Generic components provide support to knowledge engineers at different stages of the development cycle. In modelling, an existing component can be adapted to analyze the problem that a new KBS must solve. Knowledge elicitation is eased as these models point to the types of knowledge that must be acquired from the expert. System design can also be guided by the structure of the KL model. However, as pointed out among others by Rademakers and Vanwelkenhuysen (1993), the application of generic components to real-world situations is problematic since they do not give enough pointers about what sort of knowledge to use for this. It is difficult to identify the most adequate model for analysis of a problem when several candidates exist. Due to their intentional nature, KL models shape what is and what is not perceived, and therefore some elements that are essential when reusing the model for other tasks can be missing. Finally, their underlying view of problem-solving, as an observable behaviour modelled as an input/output process, disregards the processes of problem formulation and solution construction. As a consequence, they cannot adequately deal with problem-solving tasks with no visible effects, such as comprehension or intention recognition (Cañamero, 1994).

A constructivist view of problem solving that takes into account problem formulation and makes transparent the process of solution construction was proposed by Van de Velde (1993). Constructivist approaches to the KA process itself have also been proposed, stressing the fact that KL models are not fixed abstractions of behaviours that can be readily instantiated for their application to particular cases but the outcome of many interacting circumstances (Vanwelkenhuysen, 1993). This view has an impact on the way we have to build our modelling libraries. Libraries must also include reusable components that take into account the genesis of the model so that their validity can be better assessed (Van de Velde, 1994).

3 Complete Creatures Situated in the World

The behaviour-based or autonomous agents approach to AI is concerned with the study and development of complete intelligent agents situated in their environment, as opposed to the simulation of separate cognitive subsystems. Its philosophy advocates the design of systems that are able to interact with their environment without external intervention and to do the right thing (Maes, 1989), i.e. decide what to do next without relying on a designer-provided model of what the right thing is in that situation. The agent or creature must also possess some capability to adapt its behaviour to the situation.

These systems are designed to act in a closed-loop interaction with the world, which they sense continuously (or very often) and on which they can act, thus modifying their subsequent perception. It is the dynamics of the interactions of the agent(s) with the environment which is important in this case as opposed to building a snapshot static model of an agent considered in isolation. This has important consequences for the way intelligent agents have to be designed. First, now we are modelling non-linear dynamic systems that cannot be readily explained by sequential, linear computational models, such as the top-down analysis promoted by functionalist accounts of intelligence. Second, autonomous agents are adaptive: the interactions evolve over time, resulting in a system that operates far from a global equilibrium. This implies that we have to focus on complex structures that undergo progressive modification, and whose global behaviour feeds back to the individual parts, modifying their behaviour. It also means that the notions of improvement and satisfaction are usually more relevant than that of optimization. Third, situatedness can be seen as a way to overcome the frame and the symbol grounding problems, traditional in symbolic AI, and both instances of the ungroundedness problem (Harnad, 1995). The former is concerned with what changes and what remains the same after an action, and how to represent it. The latter raises the issue of the infinite regress in symbolic representations: if the symbols in a PSS always have an interpretation that is dependent on an external observer, then the mind of the observer cannot be a PSS that must be interpreted by another external observer. (Note, this criticism would equally apply to KL-models of situated agents.) Finally, as argued by Smithers (1995), the input/output information processing metaphor seems to be inadequate to model the relationship between the dynamics of an agent and that of its environment, which are better thought of as a whole entertaining a relationship of structural coupling, that is, a history of recurrent interactions leading to the structural congruence between two or more systems (Maturana & Varela, 1987 p. 75).

The notion of situatedness itself is interpreted in various ways by different researchers. For some, being embedded in some world is enough to speak of situated agents, whereas others argue that embodiment is a necessary condition for this, the requirement of physical embodiment being at the extreme of the spectrum.

3.1 Hypotheses

The main working hypotheses and key ideas that led to this new paradigm of AI research are made explicit in Brooks (1991):

Situatedness. The unrealistic assumptions that traditional AI makes concerning perfect and direct perception and action through designer-provided symbolic descriptions of the world pose big problems of brittleness in the agent behaviour. Such agents are unable to perceive, learn or do anything for which they have not been programmed. These assumptions also pose epistemological and practical problems when the system is put to work in the world. These systems need to build models of the world that show a one-to-one correspondence between the representation and the external reality. On the conceptual side, this view is loaded with all the problems of a strong objectivism, such as identifying the necessary/sufficient features to describe an object, or how to assess truth in terms of correspondence. On the pragmatic side, many years of work on robot planning provide a good example of the inadequacy of using the world model rather than looking at the world itself to reason about and go around in. Also it shows how time-consuming it is for a robot to build such a complete representation, leading to unrealistic and useless execution times. Robots that continually refer to their sensors avoid all these problems and allow for much easier and richer perceptual processes. Thus, there is a significant increase in overall performance and speed on activities classically performed by robots, involving navigation, path planning, or object identification and manipulation.

However, being purely reactive to the world is not enough on many occasions. An investigation of types of representations that avoid the problems of symbolic ones has also been undertaken: deictic representations (Agre & Chapman, 1990; Maes, 1989) take into account the spatial and functional relationship of entities to the agent instead of representing individual entities in the world. Active-constructive decentralized, non-manipulable representations (Mataric, 1991; Cañamero, 1997) themselves do the computations necessary for the relevant task. The key idea stemming from situatedness is: the world is its own best model.

Embodiment. Physical embodiment is thought to be critical for two main reasons. First, only embodied intelligent agents are fully validated as systems that can deal with the real world. Physical embodiment forces the designer to face all the issues posed by being in the real world, such as imperfect perception and action. Also, physical experiments leave no (or much less) room for self-delusion, thinking that perceptual issues are trivial, that a simulated perceptual system is fully realistic, etc. Second, only physical embodiment can give meaning to any internal system (symbolic or other) and to the processing within the agent, providing a solid ground that breaks the infinite regress loop posed by the grounding problem. Additional support comes from biological systems, which lead us to question the possibility of a disembodied mind, and suggest that the connections of the human mind, the only form of intelligence we actually have access to, to the world are so strong and complex that symbolic abstractions might not be plausible at this level. The key idea arising from embodiment is: the world cuts off regress.

Intelligence. Taking the very short life span of intelligent activities in humans within the evolutionary time scale as an argument, and ethological studies of animal behaviour as an inspiration source, this approach proposes to look at simpler animals as a bottom-up model for building intelligence. Activities having to do with perception and motor action in a dynamic environment took a much longer time in evolution, and have also proven to be the most time-consuming tasks in robots built following a traditional symbolic approach. As pointed out by Simon in the quotation at the beginning of this chapter, the dynamics of the interactions of the agent with its environment determine to a great extent the complexity of the agent's behaviour — of (the structure of) its intelligence. The key idea here is: intelligence is determined by the dynamics of interaction with the world.

Emergence. Contrary to homuncular models that look for unique elements to explain (parts of) intelligence, and in the spirit of Minsky (1985), intelligent behaviour is seen as a functionality that emerges from the interactions of primitive components among themselves and with the world (Steels, 1991). Symbolic AI has also focused on the generation of intelligent behaviour out of the combination of simpler components, although both the components and their interactions are of a very different nature. In behaviour-based AI the components are task-achieving, behaviour producing — versus information-processing — modules, each one interacting with the world and directly contributing to the global behaviour of the system. It is not possible to exactly identify what the contribution of each module is or to analytically decompose the system. These modules can be seen as implementing diverse goals, if we, as external human observers, wish to describe them in terms of intentional systems (Dennett, 1978). However, there are no explicit goals in the system; they are just in the eye of the beholder. This strongly reminds us of KL descriptions of agents, but contrary to the use that the symbolic community has made of KL-models, the behaviour-based approach does not see or use intentional descriptions as tools that can guide the analysis and design of intelligent systems. The key idea arising from emergence is: intelligence is in the eye of the observer.

3.2 Engineering Complete Creatures

The above principles have direct consequences on the design of intelligent systems. The top-down functionalist approach is substituted by a bottom-up design methodology that stresses the development of distributed, minimal architectures. The intelligent agent is designed as composed of a set of activity- or behaviour-producing subsystems, each connecting perception to action in a different way. An activity or behaviour is a pattern of interactions with the world. Behaviours compete to get the control of the agent's performance. They can be either hierarchically organized, as in the 'subsumption architecture' (Brooks, 1986) or locally linked to each other in a flat network, such as Maes (1989).

In flat networks, all the behaviours have the same status from an evolutionary or developmental perspective. In the subsumption architecture, the behaviours that are considered more fundamental from the viewpoint of survival are implemented first, forming the lower-level behavioural layers. A layer consists of a fixed-topology network of more primitive behaviours (implemented in this case as networks of finite-state machines augmented with a timing element and running asynchronously) and implements an activity that the agent can autonomously perform in the world. Layers run in parallel without any internal communication. The modules are data-driven by the messages they receive, with no possibility of access to global data nor of dynamically established communication links, as no central locus of control exists. In the same way, there is no central representation of the world resulting from the creature's perceptual system. Running in parallel, each activity processes sensor data independently and in a different way. Layers are incrementally added on top of the existing ones, i.e. without modifying them, as we want to move from simpler to more complex agents. The interactions among activities do not emerge out of nothing — chaos is not considered to be a necessary ingredient of intelligent behaviour — but have to be carefully engineered. They are combined through mechanisms of suppression — a higher-level module can suppress the usual input of a lower-level one and provide a replacement — and inhibition: a higher-level module inhibits any output signal from a lower-level one for some time. A simple example of layers are those of the robot Allen, described in Brooks (1986):

- A basic competence layer makes sure that the robot does not come into contact with other (dynamic or static) objects, through the interactions of a network of modules such as Turn, Forward, Feelforce, Runaway, etc.
- A second layer combined with the previous one has the effect of having the robot walk without hitting obstacles mainly through the combination of the Wander and Avoid modules, the output of this latter suppressing the output of the Runaway module from the previous layer.
- A third layer adds an exploratory behaviour to the robot, using visual observations to select interesting — distant and empty — places to visit.

Following an evolution-inspired design, the robots are fully tested in the real world each time a new level is added, each design step corresponding to an autonomous creature on its own. This principle is aimed at ensuring robustness, i.e. the robot must be able to show some kind of behaviour when some of the modules fail.

Many types of robots, some of them described in Brooks (1990), and simulations have been built using this paradigm. However, they implement very simple forms of intelligence, and therefore are still a long way from the complex tasks that symbolic AI attempts to model in the first place. Researchers in this field are well aware of the main challenges posed to them (Brooks, 1991), mainly:

- How many activities can be added in this type of architecture before their interactions become too complex?
- How far can we get in behavioural complexity without central and symbolic representations?
- Can higher-level functions such as learning (See Chapter 18 for a survey that analyses this question) emerge out of the local interactions of simple components?

Only attempts at practically realizing higher-level intelligence can provide answers to these questions, as several projects on humanoid robots, such as Cog (Brooks & Stein, 1994), intend to do. This attempt has led to a second shift in viewpoint, from behaviour-based to cognitive robotics (Brooks, 1997) that has different concerns at many levels. Early behaviour-based AI was too much of a reaction against deliberative planning systems, and therefore remained to a great extent at the same level of abstraction, not fully suited to realize other kinds of tasks. Contrary to early behaviour-based robots, the morphology of the body is of paramount importance because a human-like body is needed both to have humanoid perception and interactions with the world, and to have humans interact with the robot in a human-like manner.

Another crucial issue is endowing the robot with a good set of motivations and emotions that allow it to deal with the very complex action selection, focus of attention and problem-solving (among others) scenarios that humans encounter. Self-adaptation — self-calibration — is not less important, the complexity of this task requiring the use of learning types very little used in robotics so far, such as habituation and sensitization. How many of the capabilities desired from the robot should the designer implement, and how much should be left for the robot to acquire? A developmental approach seems the best choice but at the same time the most difficult one, not only because of our still limited knowledge and the existence of diverse alternative explanations about human development, from both neurophysiology and psychology, but also because of the complexity of the engineering task itself. Taking inspiration from the brain seems essential, though, be it only in the form of adopting similar functional decompositions. Finally, an old philosophical problem, which in this case becomes an engineering one, is to single out what is essential and what is accidental in the human-plus-human-environment system.

4 Two World Views in Debate

In an already classic paper, Vera and Simon (1993) attempted to present behaviour-based AI ideas within a symbolic framework and challenge some of their ideas about symbolic AI that they consider as a caricature of the field. This is done essentially by presenting a number of working programs performing tasks which can be considered as situated (the Phoenix Project, and the Navlab System), involving complex, real-time interactions with the environment. Additionally, they analyse some behaviour-based systems, such as Pengi (Agre & Chapman, 1990). They consider that the claim that these systems cannot be interpreted within a symbolic framework results from a misunderstanding of the notion of symbol. In another section, they question the idea that the symbolic approach locates cognition exclusively in the head, and acknowledge the fact that learning is essentially a social process. They claim, however, that this complex process can theoretically be simulated within a symbolic framework, and that some important parts of this process involving only one individual have already been the subject of cognitive simulations (for example with the system KEKADA simulating some aspects of scientific discovery).

Reciprocally, Clancey (1993) attempts to present the symbolic approach within a situated framework, and considers that traditional AI will manage to simulate only a subset of all intelligent tasks.

Both points of view have interesting arguments (Slezak, 1994), and one possibility is, as Agre (1993) suggests, to consider that the two approaches belong to two distinct world views which cannot directly interact in a scientific fashion: each has its own coherence and internal validity, and is not touched by external arguments. It is important to realize that this is a low-key version of the situated movement, which otherwise can be considered as questioning the whole enterprise of symbolic AI.

Yet following further Agre's argument, one can use Lakatos' ideas about the (internal) evaluation of scientific research programmes (Lakatos, 1978): how much new knowledge is generated by each programme? A research programme is alive as long as it manages to produce new knowledge, explaining more classes of facts. This argument is also internal in some way because each approach has its own means to evaluate new knowledge. Within the Lakatos framework, the symbolic approach, anchored in the view of AI as an empirical science (Newell & Simon, 1976), evaluates its successes by looking at the production of programs performing tasks usually thought of as requiring intelligence. On the contrary, the major problems they have recently encountered, compared with the initial promises, can be seen as showing its limitations in production of new knowledge. This anchoring into empirical science is not as strong in the more varied situated action community.

As far as behaviour-based AI is concerned, this approach starts with a very different hypothesis about what intelligence is, which has led it to focus on tasks very different from those which constitute the main concern of symbolic AI. In a sense, behaviour-based AI can be seen as a different world view that cannot be assessed by the same standards. As pointed out by Brooks (1990), the need for criteria to measure success (e.g. generality) that arises when building isolated modules of intelligence that are not grounded in the world can be a counterproductive approach when building physically grounded systems; this leads to much more complex systems in order to meet a constraint that is very unlikely to happen in the world. When a physically grounded system is interacting in the real world, the way in which it does it becomes immediately apparent.

A second point of view can be adopted, however, to see the physical grounding hypothesis as empirical in the same sense as the PSS one (Brooks, 1990). In order to show that it has uncovered the fundamental foundation of intelligence, the strategy adopted by behaviour-based AI has been to build robots that are more autonomous and can do more things in the world. The measure of success would in this case be twofold: in the best case, the ability to achieve (more and more) higher-level complex behaviour starting from the bottom up; in the worst case — or in the meanwhile — the engineering achievements. As is the case with symbolic AI, incremental work is needed to come up with both systems that are closer to the final goal and a better analysis of domains and tasks that will eventually lead to a theoretical understanding of the dynamics of the interactions of a complex system with its environment. However, the fact that this approach deals with human intelligence at a different level makes it difficult to compare both fields. An external theory would be needed for this task, but again the empirical nature of research in symbolic AI and the pragmatic, task-orientated approach of behaviour-based AI, which still lacks an adequate understanding of many theoretical issues, prevent us from doing it in the near future.

5 Conclusions

During the last decades, AI has become more concerned with the study and implementation of systems situated in the world, thus approaching the philosophy of the situated cognition movement in psychology and cognitive science. Traditional symbolic AI has adopted a functional and analytical perspective towards situated systems, focusing on the simulation of high-level cognitive capabilities. Situatedness is seen here, to a large extent, as constructivism. The strong functionalistic nature of the PSS hypothesis underlying this approach, and its dependence on the architecture of the von Neumann computer, lead to several problems of ungroundedness and of knowledge representation. As for the KL hypothesis, being tightly coupled to the PSS one, it inherits most of its problems. The question of whether the KL can be decoupled from the PSS hypothesis, and whether it could then be useful as a blueprint for the design of situated (embodied) autonomous agents seems a natural one. A negative answer is provided by Smithers (1995), who argues that the problem with the KL is not the incompleteness of this hypothesis itself but the fact that it leaves out the dynamics of the interactions between the agent and its environment. Therefore, adding to a KL description everything that is necessary for a specification of an autonomous agent to be useful and realizable will result in a radical reconstruction of that description.

The usefulness of finalistic or intentional descriptions of autonomous agents is however emphasized by Prem (1996), in particular when the generation of some internal model comes into play. According to Prem, for embodied systems the subjective component in their perception of the world is so strong that functions (instead of the physical properties retained by purely causal and mechanistic models) become the key epistemological and ontological primitive. Contrary to traditional AI, however, functions are not to be seen as additional properties attached to objects but as the elements at the heart of the nature of things. Knowledge about the agent's world requires the previous understanding of the agent's actions. The main engineering task is thus the design of the functional world of the autonomous agent.

Behaviour-based AI has adopted a synthetic, bottom-up approach to situated agents, where situatedness is mainly rephrased as embodiment. The way in which this view captures the situated aspect has also undergone some criticisms many of which are summarized in Clancey (1995). First, the robots built up to now are too reactive, lacking any significant ability to remember and reuse/adapt past experience of their interactions with the world, as well as many types of learning which are relevant for situated agents to adapt to their environment in complex ways. Second, there are still some hidden information-processing notions underlying the design of situated robots, such as the use of predefined categories for modelling the world, while more constructive mechanisms should be investigated. Another important element that has not been extensively explored yet is the social aspects of agent interaction. Finally, the engineering design approach adopted in the early days of behaviour-based AI makes a scientific appraisal of some of the contributions of this field difficult. An important prerequisite will be to find a common theoretical framework and engineering vocabulary in order to compare both the symbolic and behaviour-based approaches to situated agents if we are to have a better understanding of both approaches to situated artificial intelligence.

Acknowledgements

While writing this chapter, Dolores Cañamero was a postdoctoral fellow at the MIT Artificial Intelligence Laboratory. She wishes to thank Rodney Brooks and the Zoo group for their support, and for very valuable discussions on this and related topics. Her research was funded by the Spanish Ministry of Education under grant PF9500410164. Vincent Corruble is currently Marie Curie Research Fellow at the University of Aberdeen, Scotland (TMR: European Union Programme for Training and Mobility of Researchers).

18

Situated Learning in Autonomous Agents

Bart de Boer and Dolores Cañamero

1 Introduction

In Chapter 17 we started the discussion about autonomous agents as part of the situated cognition approach in AI, and we develop this further here, focusing on learning.

An agent is autonomous when it is able to function on its own in a complex and changing environment. We will call an agent adaptive when, in addition, it is able to improve its behaviour to make it (more) appropriate for its environment and task. Through interactions with the environment, the agent learns to improve its behaviour on the basis of its own appraisal of the perceived external or internal effects of its actions. We will thus focus on learning as a form of adaptation, where learning is an integral part of the agent's behaviour that cannot be disconnected from its performance, as opposed to a separate cognitive function that can be studied and modelled on a stand-alone basis.

In biological systems, long-term adaptation to the environment is obviously crucial for survival as is the ability to accommodate to incremental changes and to react to unexpected events. What is the significance of learning as adaptation for artificial agents? From an engineering point of view, adaptation is a design issue. When the agent is to be embedded in a dynamic, unpredictable, or insufficiently known environment, endowing it with the capability of adapting its behaviour to the context and of improving its performance over time greatly simplifies the design and construction of the system because it is not necessary to pre-wire from the outset all the knowledge the agent will be able to use.

From an epistemological point of view, autonomous agent learning can be seen as a way to cope with the grounding problem (Harnad, 1990), i.e. how the semantic interpretation of the agent's internal representations is made intrinsic to the system, rather than dependent on the interpretation that an external observer attributes to them, as happens in physical symbol systems (Newell, 1980). When representations are learned while acting in the real world, they automatically have a meaningful relation to it.

From the biological, ethological, and psychological perspectives, the study of learning in artificial autonomous agents can also act as a testbed for existing models. The analysis of learning in humans and animals can be complemented with a synthetic approach that produces or simulates learning behaviour in computer models and robots.

The major challenges encountered in autonomous agent learning stem from two main facts. First, the agent cannot stop acting in the world in order to start a learning phase when confronted with some novelty. Second, a teacher that guides or scaffolds the agent in its learning task is seldom available in dynamic situations. At every moment, the agent will have to exploit the information it can obtain from its interactions with the environment to improve its behaviour over time and to generalize what has been learned across situations and tasks. In particular, it will have to: (a) observe the effects that its own behaviour has on/in the world; and (b) assess them as something positive, contributing to its own interests and therefore worth remembering and repeating, or as something negative and avoidable in the future. This class of learning situations and the set of techniques developed to deal with them are known as reinforcement learning in the machine learning literature (Kaelbling, Littman & Moore, 1996) due to their resemblance with this paradigm (both classical and operant conditioning) in psychology. Two main ways of applying these techniques have been used in the modelling of autonomous agent learning behaviour. The first is based on statistical analysis of the agent's observations, the other is based on Darwinian evolution of the agent behaviour.

The problems that have to be solved by these learning mechanisms are manifold. We can consider them as caused by two factors: the need for the agent to survive and the complexity of the (interactions with the) world. The drive for survival prevents the agent from spending all its time exploring the world — it also has to efficiently exploit what it already knows about the world. A mechanism is thus needed to continuously arbitrate between exploration and exploitation (Wilson, 1996). Since the agent cannot stop the world, learning must take place while it simultaneously works on survival. Furthermore, the learning mechanism should be flexible enough to allow the agent to react to sudden changes in its environment. A learning mechanism that can learn one task very well, but that is not able to unlearn it, is not sufficient. Also, the agent needs to be able to distinguish between what is useful and what is not useful to learn.

The environment in which artificial autonomous agents usually operate — partially observable, noisy, real-time, changing — also poses a number of problems. If the agent performs an action usually the reward (or punishment) arrives after a certain time. This reward then has to be correctly assigned to the right action — the credit assignment problem. This becomes more complex when a number of different actions have been performed either sequentially or simultaneously. Since the agent's perception of the world can be extremely inconsistent, and as the agent cannot stop performing in order to learn, forming a useful world model becomes a difficult task.

Two types of situated agents, synthetic and robotic, are now discussed, as there are a number of interesting problems and possibilities particular to these two different embodiments.

2 Learning in Software Implementations

Software implementations of autonomous agents avoid some of the specific problems posed by physical (hardware) embodiment as well as those arising from having to deal with the physical world (see next section). For this reason, they usually offer more freedom to experiment with novel architectures and learning techniques. In particular, they allow for an easier integration of symbolic and nonsymbolic representations and learning models. In this section, we will examine three landmark architectures that can be considered hybrid in the sense that they attempt to combine the best of both worlds: classifier-system-based animats, behaviour networks, and a schema-based simulation of the Piagetian development theory. Although the three integrate a (more or less structured) symbolic representation within a neural network computational model, each of them illustrates a different type of learning technique. They are all instances of synthetic agents — autonomous agents that inhabit computer-simulated environments — or, in other words, simulated robots inhabiting artificial worlds.

2.1 Animats

Animats are simple artificial animals created with the aim of studying intelligent behaviour at a primitive level where, ". . . intelligent behaviour is to be repeatedly successful in satisfying one's psychological needs in diverse, observably different, situations on the basis of past experience" (Wilson, 1985). They are adaptive systems designed to operate in a closed-loop interaction with their environment, from which they receive signals in the form of binary feature vectors and on which they can perform a predefined set of actions that change the signals. Only part of these input signals are relevant at each moment, and some of them (or their absence) have a special status for the animat, being perceived as a reward. Past experience is taken into account by keeping an internal record of earlier interactions with the environment and their results.

The basic problem of animats is to generate rules that associate sensory signals with adequate actions so as to maximize the amount of reward, and therefore survive and move around more efficiently in their world. This problem is decomposed in three parts: (1) to come up with efficient rules; (2) to discard inefficient ones; and (3) to generalize the rules that are kept in an optimal way. Different animats have been implemented using classifier systems (Holland, 1992; Holland & Reitman, 1978). These are parallel message-posting systems that learn binary condition–action rules, called classifiers, which compete to control the behaviour of the system. Learning classifier systems are composed of three main elements: a performance module, an apportionment of credit system, and a genetic algorithm. The performance module is responsible for action selection. It consists of a set of finite-length classifiers and a finite-length message list of binary strings. Classifiers are defined over a binary alphabet plus a wildcard or 'don't-care' symbol that allows for some generalization. At each time step, incoming environmental messages — perceptions — are posted to the message list, where the messages activate those classifiers whose condition part matches the message; these classifiers may in turn activate others or may cause an action to be taken by the animat. Conflict among competing classifiers is solved by the outcome of an activation auction, which depends on the evaluation of the weight (usefulness) of each classifier. This value — the classifier's strength — is assigned by the apportionment of credit module according to each classifier's role in obtaining reward from the environment. This module is thus a form of reinforcement learning that stores stimulus–response (S–R) associations in their simplest form, i.e. for purely reactive systems, or S–R–S associations for systems that learn more complex models of the world that predict not only rewards, but also world states, allowing therefore, for some anticipation, such as in Riolo (1991).

At a certain number of time steps, the genetic algorithm introduces new, better rules into the system by applying a tripartite process of reproduction, crossover, and mutation to the existing ones. Rule strength is used to select the best classifiers that will generate the new ones. The genetic algorithm usually comes up with more general classifiers — those containing more wildcards — that allow the system to use fewer rules, possibly organized in default hierarchies, to model more complex worlds with (relatively) many states. However, transfer across tasks remains difficult, since once a rule has been assigned a given strength for having solved a task, current classifiers are not able to use it for a different task, perhaps due to the fact that their condition part is not sensitive to the animat's goals or motivations. Also, the choice between exploitation, done by the performance module, and exploration, carried out by the genetic algorithm, is done in an ad hoc manner, since it is the system designer, based on experimentation over many trials, who decides how often the genetic algorithm comes into play.

2.2 Behaviour Networks

This architecture was proposed by Maes to model action selection as an emergent property of a parallel process in such a way that the system can show various trade-offs between reactive and more deliberative behaviour.

In this approach, an autonomous agent, a creature, consists of sets of fixed behaviours in which it can engage, which constitute the nodes of the network. For example, the creatures presented in Maes (1991) inhabit a two-dimensional computer-simulated world containing obstacles, food and water sources which they sense at very short time intervals, and in which they can perform behaviours such as feeding, drinking, sleeping, exploring, approaching or fleeing from other creatures, etc. They also have motivations to act: safety, curiosity, fear, hunger, etc. Behaviours are implemented using STRIPS-like operators which, in addition to preconditions (stimuli), add- and delete-lists, consist of a set of processes implementing the behaviour, an activation level that reflects their relevance for the perceived situation and an activation threshold that must be surpassed for the behaviour to become active. Behaviours are linked through three types of causal relations — successor, predecessor, and conflictor links — defined according to the relations between propositions in their precondition, add- and delete-lists.

Behaviour selection is the emergent functionality of a homeostatic process involving the behaviours and motivations (goals) of the creature and the state of the environment. This process spreads both activation and inhibition between behaviours using the above links so that the activation energy accumulates in the behaviours that represent better action choices for the current situation.

In Maes (1992) weights or probability estimates of the reliability of causal relations between the behaviours were added in order to exploit environmental feedback which improves the network as the creature accumulates more experience about the actual effects of its behaviours in the world. Weights are used in both the action selection and learning mechanisms, performance and learning being fully integrated. The learning task is, for every behaviour, to learn its successor, predecessor and conflictor links together with their respective reliability, when the creature is acting in a noisy, dynamic and non-deterministic environment. This learning algorithm presents many interesting features. The probabilistic behaviour selection mechanism leads to an elegant trade-off between exploration and exploitation, since the system reduces the amount of experimentation as learning progresses and the links take values approaching 0 and 1. It also uses a priori knowledge that allows the designer to test how different factors bias learning by controlling the degree to which the creature's knowledge is innate versus learned. It also allows for some transfer since the links learned in the context of one motivation can be applied in the context of another. As for its relation with human (and animal) learning, this algorithm partially models both instinctive learning and instrumental conditioning. In return, the algorithm makes a set of simplifying assumptions that limit its applicability, although some possible extensions are sketched in Maes (1992). For example, it assumes that the effects of behaviours are very immediate, a behaviour being completed before the next one is activated, which is seldom the case in many real applications, especially involving robots. The creatures are supposed to know what behaviours achieve the different motivations.

Finally, the repertoire of innate behaviours is assumed to be sufficient and granular enough to allow the creature to learn how to achieve the motivations, and therefore it does not support the learning of new behaviours. This latter type of learning is developed in the architecture we examine next.

The Schema Mechanism

This general learning and concept-building mechanism constitutes an explicit attempt to replicate key aspects of human cognitive development with AI techniques, taking Piaget's theory as a source of inspiration (Drescher, 1991). Following Piaget's constructivism, the mechanism uses almost no a priori knowledge of the world. As a consequence, the system should be able to learn in a broad range of environments and transfer its knowledge from one to another more easily than a strongly biased one; it should also have powerful exploration capabilities, as it will have to make discoveries far beyond the knowledge it has built in. In return, such a system faces two main problems that can make learning very slow: interpreting its experiences in a non-deterministic world without the guidance of a bias in such a way that it can learn from them, and inventing new concepts suited to the different problems it faces. Drescher (1991) implements this mechanism in a simulated robot embedded in a discrete bidimensional microworld that very roughly replicates the world as perceived by an infant younger than 8 months. The cells of the world can be either empty or occupied by one object. The robot consists of a one-cell body/head that cannot move and a one-cell mobile hand disconnected from the body. It also has a mobile eye with a square visual field, the centre of which — the fovea — can see additional details in objects so as to determine their identity. The mechanism uses three kinds of data structures: items, actions, and schemas. Items convey sensory information — either proprioceptive, tactile or gustatory — in the form of propositions about the state of the world, and can be on, off, or in an unknown state. Actions designate events that can affect the state of the world. Schemas are complex units of knowledge, both declarative and procedural, reflecting regularities. A schema comprises a context, an action, and a result, stating that if the action is performed when all the context conditions are satisfied, the result conditions will hold (i.e. these items will be turned on) with a certain degree of reliability that the schema maintains. The system is initially provided with some primitive sensorimotor data structures of every category. However, this built-in knowledge is encapsulated to each sensory system and is not available to the schema mechanism, primitive structures being featureless entities for it. Based on its experiences while randomly interacting with the world, the system learns: new items to designate novel aspects of the world; new schemas that connect items and actions to express discoveries about regularities in the world; and composite actions that achieve a result independently of which schemas are used for that, therefore allowing it to represent events at levels of abstraction higher than the sensorimotor one, and to designate externally caused state transitions.

Some of the typical learning achievements of the system are the elaboration of intramodal schemes about body-related positions adjacency; schemas for intermodal coordination such as visual effects of hand motions that allow the system, for instance, to anticipate tactile contact when the hand is seen in motion beside an object, or the elaboration of synthetic items that begin to designate objects as distinct from their perceptions (object persistence). This system presents two main drawbacks. First, it is not coupled to a performance system, and therefore the closed-loop interaction with the world is missing. Second, learning is too slow — it is not even guaranteed to happen — because there is nothing to guide the learning, neither *a priori* knowledge nor internal motivations. Foner (1994) has attempted to overcome this second problem by integrating an attentional system into the schema mechanism that makes learning more selective by having the system focus its attention on what is important for it.

3 Learning in Robotic Agents

The problems of autonomous learning are amplified by making a physical embodiment, or, in other words, a robot implementation of the learning system. Not only does the robot have to cope with the general problems of autonomous learning, such as the lack of a teacher, the assignment of credit and the problem of performing and learning at the same time but it also has to process complex sensor data, plan complex actuator actions and operate in real time (Smithers, 1996).

Let us first make clear what kind of robots we are talking about. The robots that are used are of course not at all like the humanoid robots known from science fiction, nor are they like the robot manipulators found in car factories and the like. They are usually small vehicles, between 10 cm and 1 m in size, usually with three or four wheels, sometimes with simple legs and equipped with a number of sensors, such as a camera, touch sensors, light sensors or infrared sensors. Generally they have a small, but powerful, computer on board, together with a rechargeable battery, so that they can operate completely autonomously for a reasonable amount of time.

At the moment a number of projects are underway that try to build more human-like robots, with complex sensors and manipulators. Note that in the design of these robots a number of important trade-offs must be made. If one equips the robot with limited sensing abilities, it will be able to process more sensor data per unit time but it will have a less accurate view of its environment. If it has more powerful sensors, it will need more time to process the sensor data. If one makes a big robot, one can equip it with big batteries, a big computer and lots of sensors, but it will also be more difficult to do quick experiments or to build more robots in order to do experiments with multiple robots. If one builds a small robot, it is easy to do quick experiments, and easy to build many of them, but the computing and sensing abilities will be less. Different laboratories have made different design decisions (Donnet & Smithers, 1991; Mondada, Franzi & Ienne, 1994; Brooks & Stein, 1994).

The learning paradigms that are implemented on autonomous robots are the same as the ones in autonomous synthetic agents. There is on-line learning, in which individual robots try to improve their behaviour over time and there is evolutionary learning, in which a population of robots tries to improve its behaviour over time. Both are trying to solve the problem of assigning the reinforcement they get to the right behaviours. The first (Gaussier & Zrehen, 1994; Lambrinos, Scheier & Pfeifer, 1995; Mahadevan & Connell, 1992; Marjanovic, Scassellati & Williamson, 1996) does this by estimating which actions of a robot were responsible for the reinforcement and by making it more likely that these actions will be performed in a future similar situation. The second (Cliff, Harvey & Husbands, 1993) assigns reinforcement by looking at a large number of robot control systems, considered good if they cause a robot to survive long enough. A new generation of robots is then equipped with small variations of the best control systems. These control systems are tested in the same way and the cycle is repeated.

The fact that a robot has to operate in real time places time constraints on the learning mechanism but this can usually be overcome by using a more powerful on-board computer. The real-time requirement is much more of a problem for the sensing of the environment because in a limited amount of time only a very limited amount of knowledge about the environment can be acquired.

The problems with sensors and actuators are often underestimated in learning robot research or are considered not to be central to the issue of learning. In most experiments, learning tasks and environments for robots are constructed to provide just the right sensor cues for the learning system so it is not necessary to worry about the sensor processing. For practical applications, however, and in order to achieve results that are applicable to learning in humans and animals, one would like to have a learning robot that is able to extract the necessary information in a realistic environment instead. In fact, in order to understand learning in humans and animals and in order to be able to build learning robots for realistic applications, an understanding of sensor processing and actuator planning seems essential. No known learning system can cope with a continuous stream of data so it has to partition the continuous stream of sensor data into events, both in time as well as in space. Examples of such events can be the detection of a charging station, the touching of a wall, or the performance of some useful task. These events can then be associated with actions and with increasing or decreasing fitness. The partitioning of continuous data streams into events is very hard and rather task-specific.

A related problem is that of sensor fusion. Sensor fusion is the use of different sensors in combination in order to be able to distinguish certain events better than the individual sensors could. Robot control systems that can fuse sensors autonomously operate better than ones that cannot. The robot must also calculate its own reinforcement from what it perceives through its sensors, as there is no external teacher to provide it with the reinforcement signal. Often, given a certain learning task, it is not directly obvious how the reinforcement should be calculated and sometimes it is necessary to have different reinforcements for the different sub-behaviours of the robot.

Furthermore, new actuator actions should be made up. This is also no trivial task as, given a certain goal the agent wants to achieve, it is not always easy to calculate the exact actions of the robot's motors that have to be performed. Usually some kind of feedback mechanism is necessary where a sensor is used to monitor and guide the movements of the actuator. A robot arm can serve as an example here: given the task of grasping a certain object at a certain location, how should the motors be activated? If a camera to sense the position of the robot arm is used it is easier to guide it to its destination.

Another difficulty is that of the inherent unreliability of real-world sensors and actuators. A sensor never gives exactly the same signal to the same stimulus, nor does an actuator always give the same response to the same command. Sensors and actuators also change over time: motors get hot, sensors get displaced, actuators wear, and usually performance is dependent on the amount of energy in the robot's batteries. Of course, one could try to compensate for this by using better sensors, actuators and batteries but, in reality, it appears to be necessary to tune the sensors and actuators constantly.

These problems are generally tackled by designing task-specific subsystems for the different levels on which learning is necessary. These are meant to function in the environment for which the robot is designed. The subsystems are for partitioning the sensor-streams into events, for inventing and planning new actuator actions, for fusing sensors, for calculating the reinforcement and for tuning the sensors and actuators. The actual interesting learning is performed by the same types of general learning mechanisms that can be found in autonomous synthetic agents. However, because the inputs of these learning systems depend on the many environment-specific subsystems that process sensor data, the overall learning control system of the robot is environment-specific as well.

The problems can be illustrated by an example of a simple robot learning experiment which has been performed at the AI laboratory of the Vrije Universiteit Brussel. The robot is equipped with two wheels and two motors and steers by running one motor faster than the other (differential steering). Its sensors are touch sensors and infrared sensors. It has been programmed to retract if its touch sensors are touched, thus effectively avoiding obstacles. The robot's performance would be better if it could use its infrared sensors to avoid obstacles since it would then be able to avoid physical contact with them. However, it is rather hard to predict what kind of signals the infrared sensors will produce if the robot gets near an obstacle, so learning (or rather, self-tuning) is necessary.

Although this task is rather trivial, it illustrates all the problems of learning in robots. It has to happen on-line (while the robot is running) and it has to be self-reinforcing (there is no external teacher). The robot has to filter events from the stream of sensor inputs because it should only learn when it is near an obstacle. Furthermore the explore/exploit problem has to be tackled: in order to learn, the robot needs to encounter obstacles but, in fact, the learning goal is to avoid them. Also, the task is an example of sensor fusion: infrared and touch sensors are used in co-operation to avoid obstacles.

This particular problem was solved by designing an ad hoc learning mechanism. The performance part consisted of connections with a certain strength between the infrared sensors and the motors. If the infrared sensors gave a strong signal, and if the connection with the motor was strong, the motor would run fast. If there was little signal, the motor would run slowly. The connections were initialized with a very low strength. But whenever an infrared sensor and a motor were highly active at the same time and a touch sensor had just been touched, the connection between that motor and that infrared sensor would be strengthened. This learning mechanism is called Hebbian learning (Hebb, 1988). This algorithm solves the learning task. Every time the robot has just hit an obstacle, the motors are running in such a way as to avoid the obstacle. The infrared sensors are close to the obstacle so they give a strong signal. After a while the robot will start making the same steering actions through the signals of the infrared sensors as when the touch sensors were touched.

One problem is that of overtraining: if the connections become too strong, the robot will stay too far away from obstacles which would prevent it from navigating efficiently. Therefore the strength of the connections has to decay slowly. One could say, with the risk of taking a human metaphor too far, that the robot forgets how to avoid obstacles. This also solves the explore/exploit problem. If the robot bumps into obstacles very often, its connections are strengthened until it does not bump into obstacles any more. But because the strength decays slowly, every once in a while it will bump into an obstacle again, so it makes sure it does not stay too far away from the obstacles.

This particular learning mechanism is completely situated. It only works because in the robot's environment, infrared signals and touch sensor signals correlate. It can only touch an obstacle if it is close by and then the infrared sensors will also give a strong signal. Also the explore/exploit problem is solved through the environment;

connection strengths decay until the robot bumps into an obstacle again. If the obstacles were to be removed, the strengths would decay to zero.

Learning in autonomous robots crucially depends on the ability to partition a robot's sensor streams into meaningful events, and to derive information on how well it is performing from its sensor inputs. Humans and animals are very good at this, and research into human and animal learning often assumes this ability as a given. Research into autonomous software agent learning can avoid these difficulties because, with synthetic agents, the environment is usually designed to deliver exactly the right data for the software agent to perform (although one tries to simulate more and more complex environments). The research into robot learning has pointed out that handling the stream of sensor inputs is no trivial task, and that a lot of the apparent intelligence of a robot resides in its ability to process sensor data. As the sensor data which the robot receives is completely dependent on its environment the learning such a robot can do is inherently situated. Indeed, it would be impossible to build a learning robot (by hand or through evolution) that can operate independently of its environment.

4 The Role of Situatedness in Learning as Adaptation

From the machine-learning perspective, what does the situatedness aspect contribute to learning, as opposed to disembedded symbolic approaches? From constructing artificial autonomous learning systems we have learned that situatedness can present both advantages and problems. From the engineering viewpoint, perhaps the most obvious advantage comes from the fact that having an agent learn from its experience and adapt to the environment can dramatically reduce the design element. Moreover, a situated learning system avoids being faced with the grounding problem, and is more truly autonomous. Situatedness also allows for efficient learning — indeed, the most efficient and effective learning and behaviour are often very situation-specific because the learning algorithm for a given environment is specialized (either by design or through self-adaptation), in the same way as a behaviour is specialized. This means that a certain learning mechanism can be highly adapted and useful in a given environment, while in a different environment it can be almost useless. The negative side is that transfer becomes a hard problem, and it would seem that situatedness and transfer are in contradiction with each other — the more situated the system, the more difficult transfer is.

Transfer can be seen at several levels in autonomous agent learning. The first is transferring the agent across environments: as with biological agents, this is usually possible — the agent adapts to the new situation — as long as the new environment is not significantly different from the previous one, i.e. it is the same type of ecological niche. The second is transfer of learned behaviours across tasks. This type of transfer is not easy for most reinforcement learning algorithms, in particular when (a) the goals or motivations that guide the agent's behaviour are implicit, and (b) what is learned is just S–R associations based on sparse reinforcement signals coming from the environment. Once an association between a (set of) behaviour(s) and a goal or task is learned it is very difficult to associate those same behaviours to a different task. One typical solution is to have the system forget by using a decay function that progressively decreases the strength of the association so that it can relearn a new one (or the same) when needed. However, in this case we cannot say that the agent is transferring previously acquired knowledge, but rather re-adapting to a new situation.

Apart from the difficulties mentioned above, there are two other partial reasons for the difficulty these algorithms have in dealing with transfer: their very limited use of internal states and the absence of variables in the traditional sense which makes it hard for the agent to keep a good memory of its actions. In particular, we are dealing with distributed and non-linear systems whose global behaviour is an emergent property of local interactions and, therefore, it is usually not possible to single out which elements and structures are responsible for a certain behaviour and can be usefully reused.

The complexity of the environment, usually very dynamic, uncertain and noisy in non-trivial learning situations, highly increases the complexity of the learning task. This poses problems concerning the processing of sensory data and the division of the sensory stream into meaningful events, in particular for robotic agents. However, although these problems complicate the learning process, they affect the sensory processing and motor contol aspects more than the learning mechanism itself. It is a non-negligible problem, though, especially if we take into account that the environments and tasks that have been used for autonomous agent learning are relatively simple, and therefore scaling up becomes a very hard endeavour.

From the perspective of the designer, transferring the experience acquired from building one system to another is also problematic since the results of research on autonomous agent learning are often rather hard to interpret and evaluate. This is in part due to the task-driven, pragmatic approach adopted to design learning systems and the experiments created to test them.

Until now, the field has been more concerned with building working (and demonstrable) agents than with following a rigorous scientific approach, which is still a long way from the current state-of-the-art. For example, not enough is known about the different environments to come up with a good classification of them, although some attempts have already been made in this direction (Littman, 1993; Todd & Wilson, 1993). As a consequence, there are no standard environments in which to perform experiments that would allow better comparisons of results. The same is true for the learning tasks. As far as learning performance is concerned, neither do we have good standards, nor are there standard means (or sets of criteria) to compare and evaluate the performance of agents. Even more importantly, we have been lacking good theories and models about fundamental concepts and issues, and only now do we start to see the first steps: the relationships between learning and other not fully understood problems such as action selection, adaptation, evolution, development, or emergent functionality. Also, researchers in this field have taken for granted a simplified version of only one learning paradigm — behaviourism — disregarding other models that only now start to be acknowledged as relevant. As Maes puts it, ". . . the agents built using this approach end up looking more like 'a bag of hacks and tricks' than an embodiment of a set of general laws and principles. Does this mean that the field will evolve into a (systems) engineering discipline, or will we find a path towards becoming a more scientific discipline?" (Maes, 1994). We are not in a position to answer this question yet, so we let the reader guess where our hopes and efforts are directed.

5 Conclusion

In this paper we have examined and illustrated how AI approaches the problem of learning in artificial autonomous agents. Situatedness is seen to be essential to this kind of learning, although it also poses a whole range of problems which are very difficult to solve given the current state of the art. Even though the results in the field cannot be considered spectacular, we believe that the study of learning as adaptation in situated agents is the right approach for achieving meaningful learning in artificial systems.

At this point, one might ask what is the significance of our research for research in human learning. We see several important contributions. The study of learning in artificial autonomous agents offers a synthetic approach to learning and intelligence that can nicely complement the work on human (and animal) learning, which traditionally adopts an analytic perspective, in the sense that it is concerned with the study of existing systems. In this respect, the study of learning in autonomous situated agents is more relevant to human and animal learning than the study of non-situated learning since humans and animals are also agents that must continuously operate in their environment.

The operationalization of psychological theories in autonomous agents can also help to improve understanding of these theories, to test and analyse their predictions, and to contribute to their usefulness or validity in a direct way. Some interesting observations have already been made in this direction concerning, for example, the tight coupling between learning and performance, the development and use of dedicated and environment-specific learning subsystems, or the cruciality of knowledge about the task and the environment for some inference processes. Some projects, such as the humanoid robot Cog (Brooks & Stein, 1994), explicitly advocate the use of ideas from developmental psychology in their design. This has attracted the interest of psychologists and we begin to see studies about the parallelisms between infant development and the behaviour-based approach to AI and learning and its implications for the psychological study of human development (Rutkowska, 1994). Many important technical and methodological problems in autonomous agent learning have yet to be solved. Among the latter, perhaps the most important ones stem from the ad hoc manner in which we have approached our designs and learning algorithms. We believe that a close collaboration with researchers from other disciplines, such as neurosciences and psychology, can be the first step to take in order to overcome them.

Acknowledgements

We are indebted to Joan Bliss for her very valuable suggestions and her continuous support, and to the VUB AI-Lab for providing an excellent research environment. Bart de Boer was financed in part by the Belgian federal government project on emergent functionality (FKFO contract no. G.0014.95) and the IUAP Construct project (no. 20). While writing this chapter, Dolores Cañamero was a postdoctoral fellow at the MIT Artificial Intelligence Laboratory. She wishes to thank Rodney Brooks and the Zoo group for their support, and for very valuable discussions on this and related topics. Her research was funded by the Spanish Ministry of Education under grant PF9500410164.

19

Situated Learning at the Threshold of the New Millennium

Yrjö Engeström

1 The Agenda for the 1990s

The idea of situated learning was forcefully and programmatically formulated by Jean Lave and Etienne Wenger in their 1991 book. Throughout the 1980s, elements and early versions of the idea had been brewing in a number of contexts: these included the Laboratory of Comparative Human Cognition at San Diego (Laboratory of Comparative Human Cognition, 1983) and its *Quarterly Newsletter* (see Cole, Engeström & Vasquez, 1997), the Laboratory of Cognitive Studies of Work at City University of New York (Scribner, 1984), as well as a group of researchers on street mathematics in Recife, Brazil (Carraher, Carraher & Schliemann, 1985), and the Institute for Research on Learning in Palo Alto. The volume *Everyday Cognition* (Rogoff & Lave, 1984) was an important predecessor, as was Lave's (1988) book, *Cognition in Practice*.

Situated learning is not a unified theory; it is a broad and relatively loose theoretical platform, informed by a number of contextual and practice-oriented theories and schools of thought. Prominent among these is the cultural–historical theory of activity initiated by Vygotsky (1978) and Leont'ev (1978); for a representative collection of recent research, see Engeström, Miettinen & Punamäki (1999). In *Cognition and Practice* and in *Situated Learning*, Lave and Wenger draw also on the sociological theories of Bourdieu (1977) and Giddens (1979). The notion of "situatedness" itself is closely related to central ideas in Garfinkel's ethnomethodology (Garfinkel, 1967; Suchman, 1987). One could add practice-oriented variants of symbolic interactionism (Strauss, 1993) and the late Wittgenstein (see Schatzki, 1996) as intellectual relatives – if not direct forebears – of situated learning.

Under the umbrella of situated learning and situated cognition, an important alternative agenda was formulated for studies of learning and thinking in the 1990s. James Greeno (1989, pp. 134-5) condensed this agenda into three points:

(1) the locus of thinking and learning is not in the individual's mind; thinking and learning are situated in physical and social contexts;
(2) processes of thinking and learning are not uniform across persons and situations; they are situated in contexts of personal and social epistemologies that differ between individuals and social groups and determine fundamental properties of thinking and learning;
(3) thinking and learning are not built up from simple components transmitted through school instruction; thinking and learning are activities in which children elaborate and reorganize their knowledge and understanding rather than simply acquiring cognitive structures and procedures.

Greeno's version of the agenda was quite vague and ecumenical. Lave and Wenger took a stronger stance.

> "The notion of situated learning now appears to be a transitory concept, a bridge, between a view according to which cognitive processes (and thus learning) are primary and a view according to which social practice is the primary, generative phenomenon, and learning is one of its characteristics. There is a significant contrast between a theory of learning in which practice (in a narrow, replicative sense) is subsumed within processes of learning and one in which learning is taken to be an integral aspect of practice (in a historical, generative sense). In our view, learning is not merely situated in practice – as if it were some independently reifiable process that just happened to be located somewhere; learning is an integral part of generative social practice in the lived-in world." (Lave & Wenger, 1991, pp. 34–5)

In other words, there is a weak version and a strong version of the agenda. The weak version argues that learning is situated in physical and social contexts – thus, context needs to be taken into account as one studies learning. The strong version argues that learning is literally a by-product of participation in any social practice. To understand learning, one must start by analyzing the particular social practice one is interested in. There is an important shift in the vocabulary between the weak and the strong version: the former speaks of contexts, the latter speaks of practices, and of participation in communities of practice.

2 How is the Agenda Pursued?

We may look at the chapters of this volume as a diverse but fairly representative set of attempts at pursuing the agenda of situated learning, whether the weak or the strong version. I will highlight and discuss some of their contributions. To organise my discussion, I will use Greeno's three dimensions of the agenda: (1) individual *vs.* situated locus of learning, (2) uniformity *vs.* diversity of learning and (3) componential *vs.* constructive-holistic nature of knowledge.

(1) I do not think it an exaggeration to say that in their treatment of the locus of learning, the chapters of this volume, like most empirical work on situated learning in the 1990s, are by and large following the weak version of the agenda. This does not mean that the research is weak or ill-founded. It simply means that social practices and communities of practice are not very often taken as the starting point of analysis.

To take the strong version seriously would minimally mean (a) that the study is focused on some relatively durable and socially-important collective practice, and (b) that the researcher presents some sort of ethnographic description and analysis of the collective practice in which the learning is embedded (for previous examples of such ethnographies, see Darrah, 1996; Henning, 1998). In this volume, de Abreu (Chapter 2), Sinha (Chapter 3), and Muller and Perret-Clermont (Chapter 4) each give examples of this stance. Prompted by "real difficulties in giving them access to modern technologies", de Abreu studied the use and learning of mathematical tools among sugar-cane farmers in Brazil. Sinha analyses his observations of pot-making in a Zapotec village in Mexico. Muller and Perret-Clermont examine a French Knowledge Exchange Network as a site of transmission of knowledge.

Lave and Wenger (1991) put forward the notion of community of practice and Wenger (1998) recently explicated it further. These conceptualisations of the locus of learning, or unit of analysis, are not theoretically particularly strong to begin with (see Cole, 1995; Engeström & Cole, 1997, pp. 302–5). Empirical studies, even those pursuing the strong version, tend to be conspicuously silent about the theoretical grounds for the selection, bounding and dimensions of their units of analysis. While not tackling this issue head on, de Abreu, Sinha and Muller and Perret-Clermont each offer some promising openings toward enriching our understanding of practice as unit of analysis. de Abreu points out that the value participants attach to various tools and knowledge plays an important role in their acceptance and acquisition; valorization is an essential aspect of practice. Sinha demonstrates how different macro-frames (e.g., commodity-producing labor *vs.* creative self-expression) give radically different significance and value to ostensibly identical micro-events and products of learning. Muller and Perret-Clermont conclude that context is not something given; it is interpreted and negotiated by the participants in a dynamic fashion. All three chapters point to the crucial importance of issues of identity in situated learning.

Starting from a different approach, the research of Bliss, Ogborn, Cronin, Reader and Tsatsarelis (Chapter 10), through children's accounts of physical phenomena, shows how the physical context can determine the way they reason about it. Further, they argue that learning about the physical world is intimately tied in with learning about the social world, its customs and practices. For them, "social practices are important in that they contain ideas and knowledge which in the end must become taken for granted. That is, there is a level at which human beings – in order to function well in life – must no longer question all things around but must be able to take these as givens."

(2) For Lave and Wenger (1991), situated learning is essentially participation. A fundamental feature of participation is the movement of newcomers from the periphery toward the center of the community of practice. Because patterns of participation change historically and across specific communities of practice, not much more is said about the mechanisms and processes of learning. Diversity of learning processes is taken as foundational, but not explored or conceptualised further.

Again, the chapters of this volume do not tackle the theoretical challenge of diversity *vs.* uniformity head on, in global terms. They contribute to the elaboration of the issue in more subtle, fine-grained and domain-specific ways. The diversity of epistemologies and learning processes is nicely illustrated by Wyndham and Säljö (Chapter 6) and Blaye, Ackermannn and Light (Chapter 9). Wyndham and Säljö focus on the role of sign systems in learning. Diversity of learning processes is largely a function of the different sign systems and intellectual tools employed as resources. Thus, it makes a marked difference whether or not the students are using a calendar in a task of figuring out the number of days between dates: "the calendar metaphorically speaking seems to exert some agency in the situation and it invites the students to count the number of days instead of using subtraction".

Blaye, Ackermann and Light approach the issue through the notion of relevance. They compared children's performance in standard *vs.* contextualised versions of tasks requiring logical reasoning. The authors conclude that logical reasoning is only one amongst several different modes of adaptive reasoning available in a child's mental toolkit: "A critical characteristic of most traditional tasks is that they require not only the activation of logical reasoning but also the inhibition of alternative and highly accessible routines". In other words, diversity is not only benign co-existence; it can also be suppression and exclusion of alternatives. A very similar finding is reported by De Corte et al. (Chapter 5): "the context and culture created by the prevailing instructional practices in today's mathematics classrooms induce in upper primary school children the belief that real-world knowledge is irrelevant when modeling and solving word problems."

(3) For Lave and Wenger (1991), knowledge is something continuously constructed and reproduced in a community of practice. They point toward two interesting aspects of knowledge in situated learning: the importance of transparency of the artefacts, and the importance of stories as resources for problem solving. Beyond that, they left the agenda wide open: what would be the central dimensions and categories for understanding knowledge in situated learning?

Ackermann (Chapter 11) approaches the issue through hybrid representational forms such as enactments and simulations, invoking the crucial question of the role of imagination in knowledge (also discussed by Bliss et al. in Chapter 10 in terms of imaginative denial of rules). Questioning Piaget's distinction between figurative and operative aspects of cognition, Ackermann suggests that we distinguish between object and actions used to transform a situation, and objects and actions used to consolidate a situation: "The capacity to use one's mind to move from black-boxing to glass-boxing – from running a procedure to freezing it – is at the core of the human's ability to learn."

Rizzo et al. (Chapter 8) and Eteläpelto and Light (Chapter 12) discuss knowledge in organisational and professional contexts. Both struggle with the tensions between a more traditional cognitive perspective and an emerging situated perspective for the understanding of knowledge. In organisations, a central issue is the codification and storage of knowledge in shared repositories and databases. For Rizzo et al., the dilemma is that "not only the products but also the processes associated with the production and use of knowledge" need to be represented. For Eteläpelto and Light, the dilemma and challenge is the integration of the level-oriented individual and the situated-contextual aspects of expertise.

Renkl, Gruber and Mandl (Chapter 7) describe yet another tension between established approaches to the psychology of instruction and contemporary "situated" approaches, these latter approaches suffering a number of limitations. However, as they point out, ". . . our focus is not upon the strengths and limitations of attempts to describe learning and problem-solving from a situated learning point of view. Rather, we focus on prescriptive instructional models that aim to implement situated learning in instructional settings."

Part 3 of this book does not fit quite so readily into the above classifications. "Learning with and by machines" is a fairly recent theme within situated cognition (for previous attempts, see Koschman, 1996). This is not surprising, since although micro-computers have been around since the end of the 1970s, they have only gradually become an integral and pervasive part of the social contexts of everyday life. As Bliss and Säljö point out in Chapter 1, part of the task of this book is to consider "the inclusive nature of the link between people and artefacts in various social practices. When technologies change, so does the nature of human thinking and learning and so do our practices." Part 3 attempts to examine this issue in relation to new information and communication technologies.

Nonetheless, over the past 20 years, computers and all that goes with them have gradually become a part of the office furniture in industry and commerce, and largely in homes, too. Similarly schools, colleges and universities have sought, and are still seeking, ways of incorporating them within the learning environment as well as integrating them into the curriculum. But more importantly, as Bliss points out in her introduction to Part 3, knowledge can now be stored and distributed between learning sites in ways that are very different from the past, and access to knowledge is now more widely available than previously through the recognized learning institutions. As Bliss puts it, "ICT challenges traditional boundaries between knowledge communities and the demarcation of knowledge domains. So we can ask: 'What will count as epistemic authority in a society of distributed knowledge and how will this authority be exercised?'"

Three of the chapters in Part 3 (Littleton and Bannert, Chapter 13; Crook and Light, Chapter 14; and Corruble and Bliss, Chapter 16) demonstrate in different ways the importance of the need for deeper and more insightful studies of contexts. At first it may appear that technologies can be neutral but the social practices into which they are being introduced are not. Yet Littleton and Bannert show that software designers bring their own biases to the creation of "innovative software programs" in the same manner as teachers bring their own sets of pre-formed ideas as to what constitutes appropriate computing activities for girls' and boys' computing in classrooms. Technologies do not emerge, they are created and designed by people brought up in a social world with strong sets of expectations.

Crook and Light's chapter approaches the issue of learning in higher education through a study of the physical and social contexts that nurture and enhance learning, particularly with the introduction of new technologies (Greeno's second agenda point). They show that higher education has evolved various forms of community structure which are rarely acknowledged in technological visions for the future and which preserve the social dimension of educational practice in only the narrowest sense. They argue that "the neglect of such considerations is hazardous, given that IT has the potential to transform, and perhaps to undermine, educational communities as we currently understand them".

Corruble and Bliss (chapter 16) rather reverse our ideas of what happens when working with computers. Generally, the advantage of the computer is that it is seen as able "to provide new knowledge outside that of the individual"; as being able to furnish access to new and different knowledge arenas at the press of a button. Yet a common feature of the computer tools studied by these authors (modeling tools and expert systems) was the way in which the young students and the expert physicians imported knowledge from their own experience, their cultural context, or observations made during their practice, into the use of the tool. As Greeno might say, the way in which they articulated the tools was in the contexts of their own "personal and social epistemologies".

Luckin's Chapter 15 looks at the role of the teacher or the activity of teaching where "teaching and learning is an interactive enterprise in which the tools and symbols of the interplay between the different parties shape their subsequent development". In this chapter the teacher is represented by a piece of software whose design features are an attempt to simulate Vygotsky's (1978) zone of proximal development. This attempt is another way of considering the issue of diversity of learning processes. Indeed it can be asked whether or not the computer, within such a Vygotskian design, can genuinely act as a partner by providing "appropriate activities and amounts of assistance".

The chapters of Canamero and Corruble (chapter 17) and de Bart and Canamero (Chapter 18) change the focus from learning with computers to learning by computers. As Cañamero and Corruble state "the most challenging problems faced by researchers in Artificial Intelligence (AI) is having our systems interact in the real world – a dynamic, complex, and partially knowable environment – and learn from this experience." At first sight these chapters seem a world apart from the contexts of the Brazilian sugar cane farmers or of the potmaking in a Zapotec village in Mexico. Yet if we turn for a moment to what these AI researchers quote as the core manifesto of the situated movement in AI, it will be seen that it does reflect that of Greeno's agenda. For the AI community in situated learning, knowledge does not exist physically somewhere but is dynamically (re)constructed through interactions between the agent and its environment. Knowledge cannot be stored in static representations, and therefore it cannot be transferred, only reconstructed. Problem solving is a constructive process, as opposed to a process of search, and intelligent systems' behavior has to be understood within a particular environment and task (see also Agre, 1997; Clancey, 1997).

3 Toward an Agenda for the New Millennium

It seems clear to me that the agenda set for situated learning some 10 years ago, in both its weak and strong versions, needs to be reformulated. This is not so much because the initial questions have been answered. In fact, the agenda consisted mainly of rather global claims, not of questions. Research and discussions published during the 1990s have demonstrated that it is time to ask carefully-focused and theoretically-grounded questions.

In the following, I will present five questions that may serve as one attempt at redefining the agenda for research in situated learning. Some of these questions were prompted by the chapters of this volume, some are more inspired by recent research published elsewhere (for another recent attempt to chart and redefine the agenda, see Greeno, 1998).

3.1 Why do People Learn?

In their overview article, Pintrich, Marx and Boyle (1993) point out that virtually no research has been conducted on the relationship between motivation and conceptual change. In the realm of situated learning, notions of identity, perspective, subjectivity, values, and relevance (see Chapters 2-4 in this volume) are employed to approach this "why" question of learning. The challenge is to go beyond traditional psychological conceptions which place motives inside the individual. Motives need to be constructed as dynamic two-way bridges between the culture, the local collective practice, and individuals participating in it. Recent work by the anthropologists D'Andrade and Strauss (1992), Holland et al. (1998), and Strauss and Quinn (1997) is beginning to provide conceptual building materials for such bridges. Still, following Leont'ev's (1981) pioneering work, I see a need for much more pronounced efforts to analyse the internal contradictions of value, commoditisation and profit motive as central to the understanding of motives of learning.

3.2 What is the Relationship Between Learning and Teaching?

Simplistic conceptions of situated learning have tended to reject teaching as unimportant or outright harmful for learning. Recent work by Gutierrez and her colleagues (Gutierrez, Baquedano-Lopez & Turner, 1997; Gutierrez, Larson & Kreuter, 1995; Gutierrez, Rymes & Larson, 1995) powerfully demonstrates how important it is to our understanding of learning to analyse teaching as well. Gutierrez et al. analyse familiar situations of classroom discourse where the teacher follows her own authorised script while some students pursue their own defiant counter-script. Occasionally these two scripted discourses intersect and merge in unexpected ways, creating a "third space" open to innovation and novel meanings. The authors suggest that learning in the "third space" is fundamentally translation and code-switching, negotiation of the linguistic and sociocultural borders of communities. In a similar vein, Sutter (in preparation) makes the provocative claim that "without instruction, there is no learning!" In other words, negotiation always requires two parties, and the teaching actions are equally important as the learning actions. This also applies to communities of practice where there is no officially designated teacher: there are the instructionally-relevant scripts, actions and affordances of supervisors, colleagues, and approved artefacts (see also Alterman, in press).

3.3 What is the Role of Lateral Transitions in Learning?

People do not move only from the periphery to the center of a practice; they also and increasingly move between communities of practice. In his pioneering work, Bronfenbrenner (1979, pp. 209–36) discussed such "ecological transitions" in a way still to be surpassed by students of situated learning. This sideways movement and boundary crossing is exemplified in transitions between school and work, between school and community, between organisations forming partnerships and alliances, and between producers and users (Engeström, Engeström & Kärkkäinen, 1995). A study of such "consequential transitions" offers a fresh view on transfer as negotiation and hybridization of knowledge and tools across activities (Beach, in press). The negotiated, conflictual and innovative nature of learning and concept formation is accentuated in such lateral transitions, and new forms of collaborative problem solving emerge (Engeström, Engeström & Vähäaho, 1999).

3.4 How to Integrate Learning, Innovation, and Change?

It is increasingly obvious that situated learning in communities of practice seldom happens in a stable, stagnant situation. To the contrary, learning typically happens in the midst of ongoing transformations of the practice. Lave and Wenger (1991, pp. 98–9) were aware of this and suggested that we study the multiple embedded developmental cycles of communities of practice. Few studies of this kind have been conducted, with the notable exception of research coming from the developmental work research school in Finland (see Engeström, 1987; 1994; 1999). The crucial task is to examine and explain connections between small-scale cycles of innovative problem solving and large-scale cycles of organizational and technological transformations and social movements. If situated learning is to be understood as more than mere adaptation to given circumstances, researchers need to study how individuals and groups in their local learning processes contribute to the shaping of history.

3.5 What are the Potentials of Multiple Instrumentalities in Learning?

Social sciences, including education and psychology, have been going through a rather massive linguistic or discursive revolution during the past 20 years or so. Much of the best research in situated learning has been based on analyses of discourse and conversation. However, the privileging of language and talk often means that other means and modalities of mediation – physical artefacts, body movements and gestures, visual images and pictures – are excluded from the analysis. Recent work by Bloch (1998), Keller and Keller (1996; in press), and Lakoff and Johnson (1999) underlines the multiplicity and interconnectedness of the tools of human intellectual functioning. What is needed are studies of the potentials of multiple complementary and conflicting tools and sign systems functioning in concert. Some of this mediational complexity is captured in the notion of "instrumentalities", coined by de Solla

Price (1984) to characterise the knowledge embodied in the use of tools and instruments, the dynamics of technology in use.

All in all, knowledge needs to be understood in terms of "'a micro-culture of praxis" (Bruner, 1996, p.132). Knowledge is distributed in the material and discursive environment – in the tools, symbol systems, institutional rituals, physical spaces and ways-with-words that make up any micro-culture of intellectual life. Educational practice becomes a web of orchestrated encounters with such arrays of mediational means in order to deal with socially-significant objects and tasks. Due to their complex and multi-level nature, such encounters are not easily contained within lessons and classrooms or controlled by curricula and textbooks. As Bateson (1972) demonstrated, the learning of specific bits of knowledge and skills is always embedded in the wider context of learning what it is to learn and be a learner. This wider context of "Learning II" is in itself full of internal tensions and contradictions. When learners resist and question the existing norms of their context, they may embark upon expansive Learning III, opening up possibilities of creating entirely new patterns of activity (Engeström, 1996). In the new millennium, research on situated learning needs to include these poorly understood expansive potentials in its agenda.

References

Chapter 1

Bakhurst, D. (1991). *Consciousness and Revolution in Soviet Philosophy*. Cambridge: Cambridge University Press.
Barnett, M. (1995). Literacy, technology and 'technology literacy'. *International Journal of Technology and Design Education, 5*, 119–137.
Chaiklin, S., & Lave, J. (Eds.). (1993). *Understanding Practice. Perspectives on Activity and Context*. Cambridge, MA: Cambridge University Press.
Cole, M. (1996). *Cultural Psychology. A Once and Future Discipline*. Cambridge, MA: The Belknap Press.
Fairclough, N. (1992). *Discourse and Social Change*. Cambridge: Polity Press.
Giddens, A. (1997). *The Constitution of Society* (Reprint, first published 1984). Cambridge: Polity Press.
Goody, J. (1977). *The Domestication of the Savage Mind*. Cambridge: Cambridge University Press.
Goody, J. (1986). *The Logic of Writing and the Organisation of Society*. Cambridge: Cambridge University Press.
Inglis, B. (1971). *Poverty and the Industrial Revolution*. London: Hodder & Stoughton.
Lave, J. (1988). *Cognition in Practice: Mind, Mathematics and Culture in Everyday Life*. Cambridge: Cambridge University Press.
Ogborn, J. (1996). Science and the made world (Keynote Address). In K. S. Volk. (Ed.). *Bridging Science and Technology Education: Innovations and Experiences. Proceedings of the Science and Technology Education Conference, University of Hong Kong, June 21–22 1999, pp. 2–20*. Hong Kong: University of Hong Kong. ISBN: 962-8093-94-0.
Olson, D. (1994). *The World on Paper*. Cambridge: Cambridge University Press.
Ong, W.J. (1982). *Technologising Orality and Literacy*. London: Methuen.
Resnick, L., Säljö, R., Pontecorvo, C., & Burge, B. (1997). *Discourse, Tools, and Reasoning. Essays on Situated Cognition*. Berlin: Springer.
Säljö, R. (1997). *Learning and Sociocultural Change*. Inaugural lecture, Belle van Zuylen Chair, Faculty of the Social Sciences, Utrecht University, July 1997. Research Papers, Onderzoekschool Arbeid, Utrecht University.
Vygotsky, L.S. (1986). *Thought and Language* (A. Kozulin, Trans.). Cambridge, MA: MIT Press. (Original work published in 1934.)
White, S.H. (1996). Foreword. In M. Cole, *Cultural Psychology. A Once and Future Discipline*. Cambridge, MA: The Belknap Press.

Further Reading

Engeström, Y. (1993). Developmental studies of work as a testbench of activity theory: the case of primary medical care. In S. Chaiklin & J. Lave (Eds.). *Understanding Practice: Prospectus on Activity and Context*. (pp. 64–103). Cambridge, MA: Cambridge University Press.
Säljö, R. (1996). Mental and physical artefacts in cognitive practices. In P. Reimann & H. Spada (Eds.). *Learning in Humans and Machines* (pp. 83–96). Oxford: Pergamon/ Elsevier, 1996.
Wertsch, J. (1991). *Voices of the Mind: A Socio-cultural Approach to Mediated Action*. London: Harvester, Wheatsheaf.
Wertsch, J. (1998). *Mind as Action*. Oxford, New York: Oxford University Press.

Part 1

Bruner, J. (1990). *Acts of Meaning*. Cambridge, MA: Harvard University Press.
Bruner, J. (1996). *The Culture of Education*. Cambridge, MA: Harvard University Press.
Chaiklin, S., & Lave, J. (Eds.). (1993). *Understanding Practice: Perspectives on Activity and Context*. Cambridge, MA: Cambridge University Press.
Harré, R., & Gillett, G. (1994). *The Discursive Mind*. London: Sage.
Keller, C., & Keller, J.D. (1993). Thinking and acting with iron. In S. Chaiklin & J. Lave (Eds.). *Understanding Practice: Perspectives on Activity and Context* (pp. 125–143). Cambridge: Cambridge University Press.
Lave, J. (1988). *Cognition in Practice: Mind, Mathematics, and Culture in Everyday Life*. Cambridge: Cambridge University Press.
Mercer, N. (1995). *The Guided Construction of Knowledge. Talk Amongst Teachers and Learners*. Clevedon, UK: Multilingual Matters.
Resnick, L., Säljö, R., Pontecorvo, C., & Burge, B. (1998). *Discourse, Tools, and Reasoning. Essays in Situated Cognition*. New York: Springer.
Rogoff, B., & Lave, J. (1984). *Everyday Cognition: its Development in Social Context*. Cambridge, MA: Harvard University Press.
Säljö, R. (1996). Mental and physical artefacts in cognitive practices. In P. Reimann & H. Spada (Eds.). *Learning in Humans and Machines* (pp. 83–96). Oxford: Pergamon/ Elsevier.
Wertsch, J. (1998). *Mind as Action*. Oxford, New York: Oxford University Press.

Chapter 2

Abreu, G. de (1988). O uso da matematica na agricultura: O caso dos produtores de cana-de-açúcar. [The use of mathematics in agriculture: The case of sugar-cane farmers].Unpublished Masters Degree Thesis. Recife, Brazil: Universidade Federal de Pernambuco.
Abreu, G. de (1991). Psicologia no trabalho, um enfoque cognitivo: o uso da matematica por agricultores de cana-de-açucar. [Psychology at the workplace, a cognitive approach: the use of mathematics by sugar-cane farmers]. *Psicologia: Teoria e Pesquisa*, 7(2), 163–177.
Abreu, G. de (1993). *The relationship between home and school mathematics in a farming community in rural Brazil*. Unpublished doctoral dissertation. Cambridge: University of Cambridge.

Abreu, G. de (1995a). A matematica na vida versus na escola: uma questao de cogniçao situada ou de identidades sociais. [Mathematics in everyday life versus school: a question of situated cognition or of social identities]. *Psicologia: Teoria e Pesquisa, 11*(2), 85–93.

Abreu, G. de (1995b). Understanding how children experience the relationship between home and school mathematics. *Mind, Culture and Activity, 2*(2), 119–142.

Abreu, G. de, & Carraher, D. W. (1989). The mathematics of Brazilian sugar-cane farmers. In C. Keitel, P. Damerow, A. Bishop & P. Gerdes (Eds.). *Mathematics, Education and Society* (pp. 68–70). Paris: UNESCO.

Childs, C.P. & Greenfield, P.M. (1980). Informal modes of learning and teaching: the case of Zinacanteco weaving. In N. Warren (Ed.). *Studies in Cross-cultural Psychology* (Vol. 2, pp. 269–316). London: Academic Press.

Duveen, G. (1996). *Psychological development as a social process.* Paper presented at the Piaget-Vygotsky 1996 Centenary Conference, April 1996, Brighton, UK.

Fuson, K.C., & Kwon, Y. (1992). Korean children's single-digit addition and subtraction: numbers structured by ten. *Journal for Research in Mathematics Education, 23*(2), 148–165.

Goodnow, J.J. (1990). The socialisation of cognition: what's involved? In J.W. Stigler, R.A. Shweder & G. Herdt (Eds.). *Cultural Psychology: Essays on Comparative Human Development* (pp. 259–286). Cambridge, MA: Cambridge University Press.

Lave, J. (1977). Cognitive consequences of traditional apprenticeship training in West Africa. *Anthropology and Education Quarterly, 8*, 177–180.

Lave, J. (1988). *Cognition in Practice: Mind, Mathematics and Culture in Everyday Life.* Cambridge: Cambridge University Press.

Lave, J. & Wenger, E. (1991). *Situated Learning: Legitimate Peripheral Participation.* Cambridge: Cambridge University Press.

Nunes, T. (1995). Sistema de signos e aprendizagem conceptual. [Systems of signs and conceptual learning]. *Quadrante, 4* (1), 7–24.

Nunes, T., Schliemann, A., & Carraher, D. (1993). *Street Mathematics and School Mathematics.* Cambridge: Cambridge University Press.

Rogoff, B. (1984). Introduction: thinking and learning in social context. In B. Rogoff & J. Lave (Eds.). *Everyday Cognition: Its Development in Social Context* (pp. 1–8). Cambridge, MA: Harvard University Press.

Rogoff, B. (1990). *Apprenticeship in Thinking: Cognitive Development in Social Context.* New York: Oxford University Press.

Rogoff, B. & Lave, J. (Eds.). (1984). *Everyday Cognition: Its Development in Social Context.* Cambridge, MA: Harvard University Press.

Säljö, R. (1996). Mental and physical artefacts in cognitive practices. In P. Reimann & H. Spada (Eds.). *Learning in Humans and Machines: Towards an Interdisciplinary Learning Science* (pp. 83–96). Oxford: Elsevier Science.

Säljö R., & Wyndhamn, J. (1993). Solving everyday problems in the formal setting: an empirical study of the school as a context for thought. In S. Chaiklin & J. Lave (Eds.). *Understanding Practice. Perspectives on Activity and Context* (pp. 327–342). Cambridge: Cambridge University Press.

Saxe, G.B. (1982). Culture and the development of numerical cognition; studies among the Oksapmin of Papua New Guinea. In C. G. Brainerd (Ed.). *Children's Logical and Mathematical Cognition* (pp. 157–176). Berlin & New York: Springer.

Saxe, G. B. (1990). *Culture and Cognitive Development: Studies in Mathematics Understanding.* Hillsdale, NJ: Lawrence Erlbaum Associates.

Saxe, G.B., & Posner, J. (1983). The development of numerical cognition: Cross-cultural perspectives. In H.P. Ginsburg (Ed.). *The Development of Mathematical Thinking* (pp. 291–317). London: Academic Press.
Scribner, S. (1984). Studying working intelligence. In B. Rogoff & J. Lave (Eds.). *Everyday Cognition: Its Development in Social Context* (pp. 9–40). Cambridge, MA: Harvard University Press.
Tajfel, H. (Ed.) (1978). *Differentiation Between Social Groups: Studies in the Social Psychology of Intergroup Relations*. London: Academic Press.

Chapter 3

Bates, E. (1976). *Language and Context: The Acquisition of Pragmatics*. New York: Academic Press.
Bateson, G. (1973). *Steps to an Ecology of Mind*. St. Albans: Paladin.
Bruner, J. (1975). From communication to language: a psychological perspective. *Cognition, 3*, 225–287.
Bruner, J. (1990). *Acts of Meaning*. Cambridge, MA: Harvard University Press.
Cole, M., Hood, L. & McDermott, R. (1978). *Ecological Niche Picking: Ecological Invalidity as an Axiom of Experimental Psychology*. New York: Laboratory of Human Cognition, Rockefeller University.
Elman, J. Bates, E., Johnson, M., Karmiloff-Smith, A., Parisi, D. & Plunkett, K. *Rethinking Innateness: A Connectionist Perspective on Development*. Cambridge, MA: MIT Press/Bradford Books.
Geertz, C. (1974). On the nature of anthropological understanding. *Bulletin of the American Academy of Arts and Sciences 28* (1). Reprinted in R. Schweder & R. Levine (Eds.) (1984). *Culture Theory*. Cambridge: Cambridge University Press.
Harré, R. & G. Gillett. (1994). *The Discursive Mind*. London: Sage Publications.
Johnson, M. (1987). *The Body in the Mind: The Bodily Basis of Meaning, Imagination and Reason*. Chicago: University of Chicago Press.
Johnson, M.H. (1997). *Developmental Cognitive Neuroscience*. Oxford: Blackwell.
Lakoff, G. (1987). *Women, Fire and Dangerous Things: What Categories Tell us About the Mind*. Chicago: University of Chicago Press.
Langacker, R. (1987). *Foundations of Cognitive Grammar Vol. 1, Theoretical Prerequisites*. Stanford: Stanford University Press.
Lave, J. & Wenger, E. (1991). *Situated Learning: Legitimate Peripheral Participation*. Cambridge: Cambridge University Press.
Lock, A. (Ed.). (1978). *Action, Gesture and Symbol: The Emergence of Language*. London: Academic Press.
Lock, A. (1980). *The Guided Reinvention of Language*. London: Academic Press.
McGarrigle, J. & Donaldson, M. (1975). Conservation accidents. *Cognition, 3*, 341–50.
Piaget, J. (1972). *The Principles of Genetic Epistemology*. London: Routledge & Kegan Paul.
Plunkett, K. & Sinha, C. (1992). Connectionism and developmental theory. *British Journal of Developmental Psychology, 10*, 209–254.
Rommetveit, R. (1978). On Piagetian cognitive operations, semantic competence, and message structure in adult–child communication. In I. Marková (Ed.). *The Social Context of Language*. Chichester: John Wiley.
Said, E. (1985). *Orientalism*. Harmondsworth: Penguin.

Sampson, E. (1993). *Celebrating the Other*. Hemel-Hempstead: Harvester-Wheatsheaf.
Shotter, J. (1993). *Conversational Realities*. London: Sage Publications.
Schweder, R. & Bourne, E. (1984). Does the concept of the person vary cross-culturally? In R. Schweder & R. Levine (Eds.). *Culture Theory*. Cambridge: Cambridge University Press.
Sinha, C. (1988) *Language and Representation: A Socio-naturalistic Approach to Human Development*. Hemel Hempstead: Harvester-Wheatsheaf.
Sinha, C. (in press) Grounding, mapping and acts of meaning. In T. Janssen & G. Redeker (Eds.) *Cognitive Linguistics: Foundations, Scope and Methodology*. Berlin: Mouton de Gruyter.
Smolka, A.L.B., De Góes, M.C.R. & Pino, A. (1997). Transforming the canonical cowboy: notes on the determinacy and indeterminacy of children's play and cultural development. In A. Fogel, M. Lyra and J. Vaalsiner (Eds.). *Dynamics and Indeterminism in Developmental and Social Processes* (pp. 153–164). Mahwah, NJ: Lawrence Erlbaum Associates.
Taylor, C. (1989). *Sources of the Self: The Making of Modern Identity*. Cambridge: MA: Harvard University Press.
Tomasello, M., Kruger, A.C. & Ratner, H.H. (1993). Cultural learning. *Behavioral and Brain Sciences*, 16, 495–552.
Vygotsky, L. (1989). Concrete human psychology. *Soviet Psychology*, 27(2), 51–64.
Waddington, C.H. (1977). *Tools for Thought*. St. Albans: Paladin.
Walkerdine, V. & Sinha, C. (1978). The internal triangle: language, reasoning and the social context. In I. Marková (Ed.). *The Social Context of Language*. Chichester: John Wiley.
Wertsch, J. (1985). *Vygotsky and the Social Formation of Mind*. Cambridge, MA: Harvard University Press.
Zlatev, J. (1997). Situated embodiment: studies in spatial semantics. PhD thesis, University of Stockholm.

Chapter 4

Abreu, G. de (1995). Understanding how children experience the relationship between home and school mathematics. *Mind, Culture, and Activity*, 2(2), 119–142.
Bourdieu, P. (1976). Le champ scientifique. [The scientific field]. *Actes de la Recherche en Sciences Sociales*, 2(3), 88–120.
Cole, M., & Scribner, S. (1974). *Culture and Thought*. New York: Wiley.
Donaldson, M. (1978). *Children's Minds*. Glasgow: Fontana.
Dreier, O. (1993). Re-searching psychotherapeutic practice. In J. Lave & S. Chaiklin (Eds.). *Understanding Practice. Perspectives on Activity and Context*. (pp. 104–124). Cambridge: Cambridge University Press.
Elbers, E. (1994). Ground rules for testing. Expectations and misunderstanding in test situations. *European Journal of Psychology of Education*, 9(2), 111–120.
Flahault, F. (1978). *La Parole Intermédiaire*. [The intermediate speech]. Paris: Seuil.
Garfinkel, H. (1967). *Studies in Ethnomethodology*. New York: Prentice-Hall.
Grossen, M. (1988). *La construction de l'Intersubjectivité en Situation de Test* [The construction of intersubjectivity in test situations]. Cousset, Switzerland: DelVal.
Grossen, M., Liengme Bessire, M.-J. & Perret-Clermont, A.-N. (1997). Construction de l'interaction et dynamiques socio-cognitive. [Construction of the interaction and sociocognitive dynamics]. In M. Grossen & B. Py (Eds.). *Pratiques Sociales et Médiations Symboliques* (pp. 221–247). Berne: Peter Lang.

Heber-Suffrin, C. & Heber-Suffrin, M. (1992). *Echanger les Savoirs*. [Exchanging knowledge]. Paris: Desclée de Brouwer.

Hundeide, K. (1985). The tacit background of children's judgments. In J. V. Wertsch (Ed.). *Culture, Communication and Cognition. Vygotskian Perspectives* (pp. 306–321). Cambridge: Cambridge University Press.

Hundeide, K. (1992). Intersubjectivity and interpretative background in children's development and interaction. *European Journal of Psychology of Education*, 8(4), 439–450.

Iannaccone, A., & Perret-Clermont, A.N. (1993). Qu'est-ce qui s'apprend? Qu'est-ce qui se développe? [What is learned, what is developed?] In J. Wassmann & P.R. Dasen (Eds.). *Les Savoirs Quotidiens. Les Approches Cognitives dans le Dialogue Interdisciplinaire* [Everyday knowledge: Cognitive approaches in interdisciplinary dialogue] (pp. 235–258). Fribourg (Suisse): Presses Universitaires de Fribourg.

Jacobi, D. (1987). *Textes et Images de la Vulgarisation Scientifique* [Texts and images of scientific vulgarization]. Paris: P. Lang.

Jurdant, B. (1973). *Les Problèmes Théoriques de la Vulgarisation Scientifique*. [Theoretical problems of scientific vulgarization]. Unpublished dissertation, Strasbourg.

Jurdant, B. (1992). Rapports distance/proximité dans la construction des savoirs. Etude des conditions de mobilisation des connaissances scientifiques et techniques dans les réseaux d'échanges de savoirs [Distance/proximity relations in knowledge construction. Study of the conditions of mobilization of scientific and technical knowledge in the knowledge exchange networks]. Personnal communication.

Jurdant, B. (1993). La science est-elle un bien public? [Is science a public property?]. In D. Borrillo (Ed.). *Sciences et Démocratie* [pp. 43–49]. Strasbourg: Presses Universitaires de Strasbourg.

Kerbrat-Orecchioni, C. (1988). La notion de "place" interactionnelle ou les taxèmes qu'est-ce que c'est que ça? [The notion of interactional "place" or taxèmes, what is that?] In J. Cosnier, N. Gelas & C. Kerbrat.-Orecchioni (Eds.). *Echanges sur la Conversation* [Exchanges on the subject of conversation] (pp. 185–198). Paris: CNRS.

Lave, J. (1991). *Cognition in Practice: Mind, Mathematics and Culture in Everyday Life* (2nd ed.). Cambridge, MA: Cambridge University Press.

Lave, J. (1993). The practice of learning. In J. Lave & S. Chaiklin (Eds.). *Understanding Practice. Perspectives on Activity and Context* (pp. 3–32). Cambridge: Cambridge University Press.

Light, P. & Perret-Clermont, A.-N. (1989). Social context effects in learning and testing. In A. Gellathy, D. Rogers & D.A. Sloboda (Eds.). *Cognition and Social Worlds* (pp. 99–112). Oxford: Oxford University Press.

Mead, G.H. (1934). *Self, Mind and Society*. Chicago: Chicago University Press.

Mehan, H. (1993). Beneath the skin and between the ears: a case study in the politics of representation. In J. Lave & S. Chaiklin (Eds.). *Understanding Practice. Perspectives on Activity and Context.* (pp. 241–268). Cambridge: Cambridge University Press.

Moscovici, S. (1961). *La Psychanalyse, Son Image, Son Public* [Psychoanalysis, its image, its audience]. Paris: PUF.

Muller, N. (1994). *La Transmission du Savoir dans le Réseau d'Échanges de Savoirs de Strasbourg* [The transmission of knowledge in the Strasbourg Knowledge Exchange Network]. Dossier de Psychologie, 44. Université de Neuchâtel, Switzerland.

Perret-Clermont, A.-N. & Nicolet, M. (Eds.). (1988). *Interagir et Connaître: Enjeux et Régulations Sociales dans le Développement Cognitif* [Interacting and knowing. Issues and social regulations in cognitive development]. Cousset: DelVal.

Perret-Clermont, A.-N. & Schubauer-Leoni, M.-L. (1981). Conflict and cooperation as opportunities for learning. In. P. Robinson (Ed.). *Communication in Development* (pp. 203–233). London: Academic Press.

Perret-Clermont, A.-N., Brun, J., Saada, E.H. & Schubauer-Leoni, M.-L. (1983). Learning: a social actualization and reconstruction of knowledge. In H. Tajfel (Ed.). *The Social Dimension* (pp. 52–68). Cambridge: Cambridge University Press.

Resnick, L. B., Levine, R. A., & Teasley, S. D. (1991). *Perspectives on Socially Shared Cognition*. Washington: APA.

Rogoff, B. (1990). *Apprenticeship in Thinking. Cognitive Development in Social Context.* Oxford: Oxford University Press.

Rommetveit, R. (1992). Outlines of dialogically based social-cognitive approach to human cognition and communication. In A.H. Wold (Ed.). *The Dialogical Alternative. Towards Theories of Language and Minds* (pp. 19–44). Oslo: Scandinavian University Press.

Säljö, R. (1991). Learning and mediation. Fitting reality into a table. *Learning and Instruction*, *1*(3), 261–273.

Säljö, R. & Bergqvist, K. (1997). Seeing the light: Discourse and practice in the optics lab. In L. Resnick, R. Säljö, C. Pontecorvo & B. Burge (Eds.). *Discourse, Tools, and Reasoning. Essays on Situated Cognition* (pp. 385–405). New York: Springer.

Säljö, R. & Wyndhamn, J. (1993). Solving everyday problems in the formal setting: an empirical study of the school as context for thought. In S. Chaiklin & J. Lave (Eds.). *Understanding Practice. Perspectives on Activity and Context* (pp. 327–342). Cambridge: Cambridge University Press.

Schubauer-Leoni, M.-L. (1988). *Les Mécanismes de la Communication Didactique* [The mechanisms of didactic communication]. Cahiers de Psychologie, 27, Université de Neuchâtel, Switzerland.

Schubauer-Leoni, M.-L., & Grossen, M. (1993). Negotiating the meaning of questions in didactic and experimental contracts. *European Journal of Psychology of Education*, *8*(4), 451–471.

Schubauer-Leoni, M.-L. & Ntamakiliro, L. (1994). La construction de problèmes impossibles [The construction of impossible problems]. *Revue des Sciences de l'Education*, *20*(1), 87–113.

Schubauer-Leoni, M.-L. Bell, N., Grossen, M. & Perret-Clermont, A.-N. (1989). Problems in assessment of learning. The social construction of questions and answers in the scholastic context. *International Journal of Educational Issues*, *13*(6), 671–684.

Trognon, A. & Rétornaz, A. (1989). Clinique du rationnel. Psychologie cognitive et analyse conversationnelle. [Clinic of rationality. Cognitive psychology and conversation analysis]. *Connexions*, 53, 69–90.

Valsiner, J. (1994). Bidirectional cultural transmission and constructive sociogenesis. In W. de Graaf & R Maier (Eds.). *Sociogenesis Reexamined* (pp. 47–70). New York: Springer.

Vygotsky, L.S. (1978). *Mind in Society: The Development of Higher Psychological Processes*. Cambridge, MA: Harvard University Press.

Wertsch, J. V. (1991). *Voices of the Mind: A Sociocultural Approach to Mediated Action*. London: Harvester Wheatsheaf.

Chapter 5

Burkhardt, H. (1994). Mathematical applications in school curriculum. In T. Husen & T.N. Postlethwaite (Eds.). *The International Encyclopedia of Education* (2nd ed.) (pp. 3621–3624). Oxford & New York: Pergamon Press.

Cai, J. & Silver, E.A. (1994). Solution processes and interpretations of solutions in solving a division-with-remainder story problem: do Chinese and US students have similar difficulties? *Journal for Research in Mathematics Education, 26*, 491–497.

Cobb, P., Yackel, E. & Wood, T. (1992). A constructivist alternative to the representational view of mind in mathematics education. *Journal for Research in Mathematics Education, 23*, 2–33.

De Corte, E. (1995a). Designing powerful teaching–learning environments conducive to the acquisition of cognitive skills. In R. Olechowski & G. Khan-Svik (Eds.). *Experimental Research on Teaching and Learning* (pp. 67–82). Frankfurt am Main: Peter Lang.

De Corte, E. (1995b). Fostering cognitive growth: a perspective from research on mathematics learning and instruction. *Educational Psychologist, 30*, 37–46.

De Corte, E. & Verschaffel, L. (1985). Beginning first graders' initial representation of arithmetic word problems. *Journal of Mathematical Behavior, 4*, 3–21.

De Corte, E., Greer, B. & Verschaffel, L. (1996). Psychology of mathematics teaching and learning. In D.C. Berliner & R.C. Calfee (Eds.). *Handbook of Educational Psychology* (pp. 491–549). New York: Macmillan.

Freudenthal, H. (1991). *Revisiting Mathematics Education*. Dordrecht, The Netherlands: Kluwer.

Gravemeijer, K.P.E. (1995). *Developing Realistic Mathematics Education*. Utrecht: Freudenthal Institute, University of Utrecht.

Greer, B. (1993). The modeling perspective on wor(l)d problems. *Journal of Mathematical Behavior, 12*, 239–250.

Greer, B. (1997). Modelling reality in mathematics classrooms: the case of word problems. *Learning and Instruction, 7*(4), 293–307.

Hamers, J.H.M., Sijtsma, K. & Ruijssenaars, A.J.J.M. (1993). *Learning Potential Assessment: Theoretical, Methodological and Practical Issues*. Amsterdam/Lisse: Swets & Zeitlinger.

Kilpatrick, J. (1987). Problem formulating: where do good problems come from? In A.H. Schoenfeld (Ed.). *Cognitive Science and Mathematics Education* (pp. 123–147). Hillsdale, NJ: Erlbaum.

McLeod, D.B. & Adams, V.M. (1989). *Affect and Mathematical Problem Solving*. New York: Springer.

Reusser, K. (1988). Problem solving beyond the logic of things: contextual effects on understanding and solving word problems. *Instructional Science, 17*, 309–338.

Reusser, K. & Stebler, R. (1997). Every word problem has a solution — the social rationality of mathematical modeling in schools. *Learning and Instruction, 7*(4), 309–327.

Säljö, R. (1991). Learning and mediation. Fitting reality into a table. *Learning and Instruction, 1*(3), 261–273.

Schoenfeld, A.H. (1985). *Mathematical Problem Solving*. New York: Academic Press.

Schoenfeld, A.H. (1988). When good teaching leads to bad results: the disasters of "well-taught" mathematics courses. *Educational Psychologist, 23*, 145–166.

Selter, C. (1994). How old is the captain? *Strategies, 5*(1), 34–37.

Silver, E.A., Shapiro, L.J. & Deutsch, A. (1993). Sense making and the solution of division problems involving remainders: an examination of middle school students' solution processes and their interpretations of solutions. *Journal for Research in Mathematics Education, 24*, 117–135.

Treffers, A. (1987). *Three Dimensions. A Model of Goals and Theory Description in Mathematics Education. The Wiskobas Project*. Dordrecht, The Netherlands: Reidel.

Verschaffel, L. & De Corte, E. (1997). Teaching realistic mathematical modeling in the elementary school. A teaching experiment with fifth graders. *Journal for Research in Mathematics Education*, 5, 577–601.

Verschaffel, L., De Corte, E. & Borghart, I. (1997). Pre-service teachers' conceptions and beliefs about the role of real-world knowledge in mathematical modeling of school word problems. *Learning and Instruction*, 7(4), 339–359.

Verschaffel, L., De Corte, E. & Lasure, S. (1994). Realistic considerations in mathematical modeling of school arithmetic word problems. *Learning and Instruction*, 4, 273–294.

Verschaffel, L., De Corte, E. & Lasure, S. (in press). Children's conceptions about the role of real-world knowledge in mathematical modeling: analysis and improvement. In W. Schnotz, S. Vosniadou & M. Carretero (Eds.). *New Perspectives on Conceptual Change*. Oxford: Elsevier Science.

Vygotsky, L.S. (1978). *Mind in Society. The Development of Higher Psychological Processes*. Cambridge, MA : Harvard University Press.

Wyndhamn, J. & Säljö, R. (1997).Word problems and mathematical reasoning — a study of children's mastery of reference and meaning in textual realities. *Learning and Instruction*, 7(4), 361–382.

Yoshida, H., Verschaffel, L. & De Corte, E. (1997). Realistic considerations in solving word problems: do Japanese and Belgian children have the same difficulties? *Learning and Instruction*, 7(4), 329–338.

Chapter 6

Bruner, J. (1986). *Actual Minds, Possible Worlds*. Cambridge, MA: Harvard University Press.

Bruner, J. (1990). *Acts of Meaning*. Cambridge, MA: Harvard University Press.

Edwards, D. (1997). *Discourse and Cognition*. London: Sage.

Hagström, G. & Sani, N. (1991). Elever och matematiken. En undersökning om elevernas matematiska kunskapsutveckling [Students and mathematics. A study of the development of mathematical knowledge among students]. University of Linköping, unpublished report.

Hannerz, U. (1992). *Cultural Complexity: Studies in the Social Organization of Meaning*. New York: Columbia University Press.

Harré, R. & Gillett, G. (1994). *The Discursive Mind*. London: Sage.

Hutchins, E. (1995). *Cognition in the Wild*. Cambridge, MA: MIT Press.

Keller, C. & Keller, J.D. (1993). Thinking and acting with iron. In S. Chaiklin & J. Lave (Eds.). *Understanding Practice. Perspectives on Activity and Context* (pp. 125–143). Cambridge: Cambridge University Press.

Laborde, C. (1990). Language and mathematics. In P. Nesher & J. Kilpatrick (Eds.). *Mathematics and Cognition: A Research Synthesis by the International Group for the Psychology of Mathematics Education* (pp. 53–69) Cambridge: Cambridge University Press.

Lave, J. (1988). *Cognition in Practice: Mind, Mathematics, and Culture in Everyday Life*. Cambridge, MA: Cambridge University Press.

Linell, P. (1992). The embeddedness of decontextualization in the context of social practices. In A. Heen-Wold (Ed.). *The Dialogical Alternative: Towards a Theory of Language and Mind* (pp. 253–271). Oslo: Scandinavian University Press.

Pettersson, A. (1990). *Att utvecklas i matematik. En studie av elever med olika prestationsutveckling* [To develop in mathematics. A study of pupils with differing progress rates]. Stockholm: Almqvist & Wiksell.

Reuterberg, S-E. (1989). *Utvärdering genom uppföljning. Analys av matinstrument använda i ärskurs 3* [Evaluating through following-up. Analysis of measurement instruments used in grade 3]. Department of Education and Educational Research, University of Göteborg.

Säljö, R. & Wyndhamn, J. (1988). A week has seven days. Or does it? On bridging linguistic openness and mathematical precision. *For the Learning of Mathematics*, 8(3), 16–19.

Säljö, R. & Wyndhamn, J. (1990). Problem-solving, academic performance and situated reasoning. A study of joint cognitive activity in the formal setting. *British Journal of Educational Psychology*, 60, 245–254.

Säljö, R. & Wyndhamn, J. (1993). Solving everyday problems in the formal setting. An empirical study of the school as context for thought. In S. Chaiklin & J. Lave (Eds.). *Understanding Practice. Perspectives on Activity and Context* (pp. 327–342). Cambridge: Cambridge University Press.

Voloshinov, V.N. (1973). *Marxism and the Philosophy of Language*. [L. Matejka & I.R. Titunik Trans.] New York: Seminar Press. (Original work published in 1930)

Vygotsky, L.S. (1986). *Thought and Language*. [A. Kozulin Trans.]. Cambridge, MA: MIT Press. (Original work published in 1934).

Wertsch, J.V. (1991). *Voices of the Mind: A Socio-cultural Approach to Mediated Action.* Cambridge, MA: Harvard University Press.

Wertsch, J.V. (1998). *Mind as Action*. Oxford, New York: Oxford University Press.

Winograd, T. & Flores, F. (1986). *Understanding Computers and Cognition*. Norwood, NJ: Ablex.

Wistedt, I. (1994). Reflection, communication, and learning mathematics: a case study. *Learning and Instruction*, 4(2), 123–138.

Wistedt, I. & Martinsson, M. (1996). Orchestrating a mathematical theme: eleven-year olds discuss the problem of infinity. *Learning and Instruction*, 6(2), 173–185.

Wyndhamn, J. (1993). *Problem-solving Revisited. On School Mathematics as a Situated Practice*. Linköping Studies in Arts and Science 98. Linköping: Linköping University.

Chapter 7

Anderson, J.R., Reder, L.M., & Simon, H.A. (1996). Situated learning and education. *Educational Researcher*, 25 (4), 5–11.

Bransford, J.D., Goldman, S.R. & Vye, N.J. (1991). Making a difference in people's ability to think: Reflections on a decade of work and some hopes for the future. In R.J. Sternberg & L. Okagaki (Eds.). *Influences on Children* (pp. 147–180). Hillsdale, NJ: Erlbaum.

Cognition and Technology Group at Vanderbilt (1992). The Jasper series as an example of anchored instruction: theory, program, description, and assessment data. *Educational Psychologist*, 27, 291–315.

Cognition and Technology Group at Vanderbilt (1993). Designing learning environments that support thinking: the Jasper series as a case study. In T.M. Duffy, J. Lowyck & D.H. Jonassen (Eds.). *Designing Environments for Constructive Learning* (pp. 9–36). Berlin: Springer.

Cognition and Technology Group at Vanderbilt (1996). Anchored instruction and situated learning revisited. In H. McLellan (Ed.). *Situated Learning Perspectives* (pp. 123–153). Englewood Cliffs, NJ: Educational Technology Publications.

Collins, A., Brown, J.S. & Newman, S.E. (1989). Cognitive apprenticeship: Teaching the crafts of reading, writing, and mathematics. In L.B. Resnick (Ed.). *Knowing, Learning, and Instruction* (pp. 453–494). Hillsdale, NJ: Erlbaum.

Gardner, H. (1991). *The Unschooled Mind. How Children Think and How Schools Should Teach*. New York: Basic Books.

Gräsel, C. & Mandl, H. (1993). Förderung des Erwerbs diagnostischer Strategien in fallbasierten Lernumgebungen [Fostering the acquisition of diagnostic strategies in cased-based learning environments]. *Unterrichtswissenschaft, 21*, 355–369.

Gruber, H., Renkl, A., Mandl, H. & Reiter, W. (1994). Exploration strategies in an economics simulation game. In D. M. Towne, T. de Jong, & H. Spada (Eds.). *Simulation-based Experiential Learning* (pp. 225–233). Berlin: Springer.

Hiebert, J., Carpenter, T.P., Fennenma, E., Fuson, K., Human, P., Murray, H., Olivier, A. & Wearne, D. (1996). Problem-solving as a basis for reform in curriculum and instruction: The case of mathematics. *Educational Researcher, 25* (4), 12–21.

Hofer, M., Niegemann, H.M., Eckert, A. & Rinn, U. (1996). Pädagogische Hilfen für interaktive selbstgesteuerte Lernprozesse und Konstruktion eines neuen Verfahrens zur Wissensdiagnose [Instructional support for interactive self-directed learning environments and construction of a new procedure for knowledge diagnosis]. *Zeitschrift für Berufs- und Wirtschaftspädagogik, Beiheft, 13*, 53–67.

Huber, G.L. (1993). Ungewißheits- vs. Gewißheitsorientierung im interkulturellen Vergleich [Uncertainty- vs. certainty-orientation in intercultural comparisons]. In H. Mandl, M. Dreher & H.J. Kornadt (Eds.). *Entwicklung und Denken im kulturellen Kontext* (pp. 75–98). Göttingen: Hogrefe.

Jacobson, M.J. & Spiro, R.J. (1994). Hypertext learning environments, epistemic beliefs, and the transfer of knowledge. In S. Vosniadou, E. De Corte & H. Mandl (Eds.). *The Psychological and Educational Foundations of Technology-based Learning Environments* (pp. 290–295). Berlin: Springer.

Lamon, M., Secules, T., Petrosino, A.J., Hackett, R., Bransford, J. D. & Goldman, S.R. (1996). Schools for thought: Overview and lessons learned from one of the sites. In L. Schauble & R. Glaser (Eds.). *Innovation in Learning. New Environments for Education* (pp. 243–288). Mahwah, NJ: Erlbaum.

Mandl, H., Gruber, H. & Renkl, A. (1993). Neue Lernkonzepte für die Hochschule [New learning concepts for higher education]. *Das Hochschulwesen, 41*, 126–130.

Marcus, N., Cooper, M. & Sweller, J. (1996). Understanding instructions. *Journal of Educational Psychology, 88*, 49–63.

Njoo, M.K.H., & De Jong, T. (1993). Exploratory learning with a computer simulation for control theory: Learning processes and instructional support. *Journal of Research in Science Education, 30*, 821–844.

O'Donnell, A.M. & Dansereau, D.F. (1992). Scripted cooperation in student dyads: A method for analyzing and enhancing academic learning and performance. In R. Hertz-Lazarowitz & N. Miller (Eds.). *Interactions in Cooperative Groups. The Theoretical Anatomy of Group Learning* (pp. 120–141). Cambridge, MA: Cambridge University Press.

Renkl, A. (1996). Lernen durch Erklären — oder besser doch durch Zuhören [Learning by explaining — or better still by listening]. *Zeitschrift für Entwicklungspsychologie und Pädagogische Psychologie, 28*, 148–168.

Renkl, A. (1997). Learning from worked-out examples: a study on individual differences. *Cognitive Science, 21* (1), 1–29.
Renkl, A., Gruber, H. & Mandl, H. (1996). Kooperatives problemorientiertes Lernen in der Hochschule [Cooperative problem-oriented learning in higher education]. In J. Lompscher & H. Mandl (Eds.). *Lehr-Lernprobleme im Studium — Bedingungen und Änderungs möglichkeiten* (pp. 131–147). Bern: Huber.
Renkl, A., Mandl, H. & Gruber, H. (1996). Inert knowledge: analyses and remedies. *Educational Psychologist, 31,* 115–121.
Resnick, L.B. (1987). Learning in school and out. *Educational Researcher, 16*(9), 13-20.
Spiro, R.J., Feltovich, P.J., Jacobson, M.J. & Coulson, R.L. (1991). Cognitive flexibility, constructivism, and hypertext: random access instruction for advanced knowledge acquisition in ill-structured domains. *Educational Technology, 31*(5), 24–33.
Stark, R., Graf, M., Renkl, A., Gruber, H. & Mandl, H. (1995). Förderung von Handlungskompetenz durch geleitetes Problemlösen und multiple Lernkontexte [Fostering action competence by guided problem-solving and multiple learning contexts]. *Zeitschrift für Entwicklungspsychologie und Pädagogische Psychologie, 27,* 289–312.
Stark, R., Gruber, H., Renkl, A. & Mandl, H. (1997). "Wenn um mich herum alles drunter und drüber geht, fühle ich mich so richtig wohl" — Ambiguitätstoleranz und Transfererfolg ["If everything is topsy-turvy, I feel at home" — Tolerance of ambiguity and transfer]. *Psychologie in Erziehung und Unterricht, 44,* 204–215.
Sweller, J. (1994). Cognitive load, learning difficulty, and instructional design. *Learning and Instruction, 4,* 295–312.
VanLehn, K. (1996). Cognitive skill acquisition. *Annual Review of Psychology, 47,* 513–539.
Van Merrienboer, J.J.G. & De Croock, M.B.M. (1992). Strategies for computer-based programming instruction: program completion vs. program generation. *Journal of Educational Computing Research, 8,* 365–394.
Vollmeyer, R., Burns, B.D. & Holyoak, K.J. (1996). The impact of goal specificity on strategy use and the acquisition of problem structure. *Cognitive Science, 20,* 75–100.
Webb, N.M. (1991). Task-related verbal interaction and mathematics learning in small groups. *Journal for Research in Mathematics Education, 22,* 366–389.
Weinert, F.E. (1996). Für und Wider die "neuen Lerntheorien" als Grundlage pädagogisch-psychologischer Forschung. *Zeitschrift für Pädagogische Psychologie, 10,* 1–12.
Whitehead, A.N. (1929). *The Aims of Education.* New York: Macmillan.
Wood, D., Wood, H., Ainsworth, S. & O'Malley, C. (1995). On becoming a tutor: Toward an ontogenetic model. *Cognition and Instruction, 13,* 565–581.

Chapter 8

Ackerman, M.S. (1994). Answer garden: a tool for growing organisational memory. PhD thesis, Massachusetts Institute of Technology, Boston, MA.
Ackerman, M. & Malone, T.W. (1990). Answer garden: a tool for growing organizational memory. *Proceedings of ACM Conference on Office Information Systems.* Amsterdam: ACM Press, pp. 31–39.
Adler, P.A. & Winograd, T.A. (1992). *Usability: Turning Technology into Tools.* New York: Oxford University Press.
Bauer, M.I. & Johnson-Laird, P.N. (1993). How diagrams can improve reasoning. *Psychological Science, 4*(6), 372–378.

Berlin, L.M., Jeffries, R., O'Day, V.L., Paepcke, A. & Wharton, C. (1993). Where did you put it? Issues in the design and use of a group memory. *Proceedings of INTERCHI '93*, Amsterdam. ACM Press, pp. 23–30.

Cheng, P.W. & Holyoak, K.J. (1985). Pragmatic reasoning schema. *Cognitive Psychology*, 17, 55–81.

Cohen, M.D. (1991). Individual learning and organisational routine: emerging connection. *Organizational Science*, 2(1), 135–139.

Conklin, J. (1996). Designing organizational memory: Preserving intellectual assets in a knowledge economy. http://www.cmsi.com/.

Evans, J.S.B.T. (1989). *Bias in Human Reasoning. Causes and Consequences.* Hillsdale, NJ: Erlbaum.

Garnham, A. & Oakhill, J. (1994). *Thinking and Reasoning.* Oxford: Blackwell.

Gorman, M.E., Gorman, M.E., Latta, R.M. & Cunningham, G. (1984). How disconfirmatory, confirmatory and combined strategies affect group problem solving. *British Journal of Psychology*, 75, 65–79.

Hutchins, E. (1995). *Cognition in the Wild.* Cambridge, MA: MIT Press.

Johnson-Laird, P.N. & Byrne, R. (1991). *Deduction.* Hove, UK: Lawrence Erlbaum Associates.

Lave, J. (1988). *Cognition in Practice: Mind, Mathematics and Culture in Everyday Life.* Cambridge, MA: Cambridge University Press.

Levitt, B. & March, J.G. (1988). Organizational learning. *Annual Review of Sociology*, 14, 319–340.

Light, P., Littleton, K., Messer, D. & Joiner, R. (1994). Social and communicative processes in computer-based problem solving. *European Journal of Psychology of Education*, 9, 93–110.

Maltzahn, C. & Vollmar, D. (1994). *ToolBox: A Living Directory for Unix Tools Owned by the Community* (CU-CS-747-94). Boulder, CO.: University of Colorado.

Marti, P., Rizzo, A., Bagnara, S., Lomagistro, P. & Tanzini, L. (1997). From system evaluation to service redesign: a case study. In M.J. Smith, G. Salvendy, R.J. Koubeck (Eds.). *Proceedings of International Conference on Human–Computer Interaction* (pp. 779–782). San Francisco, CA: Elsevier.

Mynatt, C.R., Doherty, M.E. & Tweney, R.D. (1977). Confirmation bias in a simulated research environment: an experimental study of scientific inference. *Quarterly Journal of Experimental Psychology*, 29, 85–95.

Norman, D.A. (1993). *Things That Make us Smart.* New York: Addison-Wesley.

Owen, D. (1986). Answer first, then questions. In D. A. Norman & S. W. Draper (Eds.). *User Centered System Design.* Hillsdale, NJ: Lawrence Erlbaum Associates.

Resnick, LB. Levine, R.A. & Teasley, S.D. (1991). *Perspectives on Socially Shared Cognition.* Washington: APA Press.

Schatz, B. & Chen, H. (1996). *Building large-scale digital libraries.* Special issue of *Computer*, 29 (5).

Slater, A. (1993). Demonstration of Answer Web: X11R5 and the interact communication facility. *Proceedings of First International WWW Conference, Geneva.* http://neptune.cee.hw.ac.uk/~afs/.

Sperber, D., Cara, F. & Girotto, V. (1995). Relevance theory explains the selection task. *Cognition*, 57, 31–95.

Sperber, D. & Wilson, D. (1986). *Relevance, Communication and Cognition.* Oxford: Blackwell.

Twidale, M. (1995). How to study and design for collaborative browsing in the DigitalLibrary. http://www.comp.lancs.ac.uk/computing/research/cseg/projects/ariadne/.

Vygotsky, L.S. (1929). The problem of cultural development of the child. *Journal of Genetic Psychology*, 36, 415–432.

Walsh, J.P. & Ungson, G.R. (1991). Organizational memory. *The Academy of Management Review*, *16*(1), 57–91.

Wason, P.C. (1977). On the failure to eliminate hypotheses. A second look. In P.N. Johnson-Laird & Wason, P.C. (Eds.). *Thinking* (pp. 307–314). Cambridge: Cambridge University Press.

Zhang, J. & Norman, D.A. (1994). Representations in distributed cognitive tasks. *Cognitive Science*, *18*, 87–122.

Chapter 9

Ackermann, E. (1991). From decontextualized to situated knowledge: revisiting Piaget's water-level experiment. In I. Harel & S. Papert (Eds.). *Constructionism*. (pp. 269–295). Norwood, NJ: Ablex Publishing Company.

Austin, J.L. (1962). *How to do Things With Words*. New York: Oxford University Press.

Bideaud, J. & Lautrey, J. (1983). De la résolution empirique à la résolution logique du problème d'inclusion: evolution des réponses en fonction de l'âge et des situations expérimentales. [From empirical resolution to logical resolution of the problem of inclusion: the evolution of responses as a function of age and experimental situation]. *Cahiers de Psychologie Cognitive*, *3*, 295–326.

Blaye, A. & Bernard-Peyron, V. (1996). Une étude développementale des activités de catégorisation: variabilité inter- et intra-individuelle des conduites. [A developmental study of categorisation: inter- and intra-individual variability]. Communication presented to the XIemes Journées de Psychologie Différentielle. Lorient (France): 8–10 octobre 1996.

Bouzigue, S., Chamorey, A. & Delcenserie, C. (1994). Etude sur l'effet du champ de la négation sur la compréhension des phrases négatives chez les enfants de CE1. [Effect of the scope of negation on the understanding of negative sentences by 2nd graders: a study]. Dissertation for Masters Degree, unpublished manuscript, University of Provence, France.

Bruner, J.S., Goodnow, J. & Austin, G. (1966). *A Study of Thinking*. New York: Wiley.

Cheng, P.W. & Holyoak, K. (1985). Pragmatic reasoning schemas. *Cognitive Psychology*, *17*, 391–416.

Deak, G.O. & Bauer, P.J. (1996). The dynamics of preschoolers' categorization choices. *Child Development*, *67*, 741–767.

Ducrot, O. (1980). *Les Échelles Argumentatives*. [Argumentative Scales]. Paris: Minuit.

Dunn, J. (1988). *The Beginning of Social Understanding*. Oxford: Blackwell.

Evans, J.St.B.T. (1982). *The Psychology of Deductive Reasoning*. London: Routledge.

Evans, J.St.B.T. (1989). *Bias in Human Reasoning: Causes and Consequences*. Hillsdale, NJ: Erlbaum.

Evans, J.St.B.T., Newstead, S. & Byrne, R. (1993). *Human Reasoning: The Psychology of Deduction*. Hove: Erlbaum.

Evans, J.St.B.T., Over, D.E. & Manktelow, K.I. (1993). Reasoning, decision making and rationality. *Cognition*, *49*, 165–187.

Frydman, O., Light, P. & Alegria, J. (1999). Pragmatic determinants of children's responses to the Wason Selection Task. *Psychologia*, *42*, 59–68.

Girotto, V. & Light, P. (1992). The pragmatic bases of children's reasoning. In P. Light & G. Butterworth (Eds.). *Context and Cognition: Ways of Learning and Knowing*. New York: Harvester.

Girotto, V., Gilly, M., Blaye, A. & Light, P. (1989). Children's performance in the selection task: plausibility and familiarity. *British Journal of Psychology*, *80*, 79–95.

Girotto, V., Light, P. & Colbourn, C. (1988). Pragmatic schemas and conditional reasoning in children. *Quarterly Journal of Experimental Psychology, 40A*, 469–482.

Harris, P. & Nunez, M. (1996). Understanding of permission rules by preschool children. *Child Development, 67*, 1572–1591.

Houdé, O. (1990). Logical categorization: schematic knowledge, categorical knowledge, and image versus linguistic format. A study in six-to-eleven-year-olds. *European Bulletin of Cognitive Psychology, 10*(4), 343–384.

Houdé, O. (1995). *Rationalité, Développement et Inhibition. Un Nouveau Cadre d'Analyse* [Rationality, Development and Inhibition. A New Framework of Analysis]. Paris: Presses Universitaires de France.

Inhelder, B. & Piaget, J. (1958). *The Growth of Logical Thinking from Childhood to Adolescence: An Essay on the Construction of Formal Operational Structures.* London: Routledge.

Inhelder, B. & Piaget, J. (1964). *The Early Growth of Logic in Children.* New York: Norton.

Jakubowicz, C. (1971). *La Compréhension des Phrases Négatives.* [Understanding of negative sentences]. Thèse de doctorat, unpublished manuscript, University of Paris V, France.

Johnson-Laird, P.N., Legrenzi, P. & Legrenzi, M. (1972). Reasoning and a sense of reality. *British Journal of Psychology, 63*, 395–400.

Lecacheur, M., Desprels-Fraysse, A. & Blaye, A. (1997). Context and categorization in preschool children. A study of flexibility in children's free sorting strategies. Communication to the VIIIth European Conference on Developmental Psychology. Rennes, 3–5 September. To be published (in French) in *Enfance*, 1999.

Light, P., Blaye, A., Gilly, M. & Girotto, V. (1989). Pragmatic schemas and logical reasoning in 6- to 8-year-old children. *Cognitive Development, 4*, 49–64.

Mandler, J.M. (1983). Representation. In J.H. Flavell & E.M. Markman (Eds.). *Cognitive Development*, Vol. 3 of P. Mussen (Ed.). *Manual of Child Psychology* (pp. 420–494). New York: Wiley.

Markman, E.M. (1978). Empirical vs logical solutions to part-whole comparison problems concerning classes and collections. *Child Development, 49*, 168–177.

Markman, E.M., Cox, B. & Machida, S. (1981). The standard object sorting task as a measure of conceptual organization. *Developmental Psychology, 17*, 115–117.

Marti, E. (1991). La négation de la ressemblance chez l'enfant: Différence ou altérité?. [Negation of similarity in children: Difference or alterity?]. *Archives de Psychologie, 49*, 25–45.

Moeschler, J. (1990). Les aspects pragmatiques de la négation linguistique: actes de langage, argumentation et inférence pragmatique. [Pragmatic aspects of linuistic negation: speech acts, argumentation, and pragmatic inference]. In *La Négation. Le Rôle de la Négation dans l'Argumentation et le Raisonnement.* [Negation: the Role of Negation in Argument and Reasoning]. Actes du colloque de Neuchâtel. Neuchâtel: Centre de Recherches.

Nelson, K. (1986). *Event Knowledge: Structure and Function in Development.* Hillsdale, NJ: Erlbaum.

Ninio, A. & Snow, C.E. (1996). *Pragmatic Development.* Oxford: Westview Press.

Olson, D. (1994). *The World on Paper.* Cambridge, MA: Cambridge University Press.

Politzer, G. (1993). *La psychologie du raisonnement: Lois de la pragmatique et de la logique formelle.* [Psychology of reasoning: pragmatic rules and formal logic]. Thèse de Doctorat d'Etat, University of Paris VIII, France.

Rosch, E.H., Mervis, C.B., Gray, W.D., Johnson, D.M. & Boyes-Braem, P. (1976). Basic objects in natural categories. *Cognitive Psychology, 8*, 382–439.

Rumain, B. (1988). Syntactics of interpretation of negation: a developmental study. *Journal of Experimental Child Psychology, 45*, 119–140.

Schank, R.C. & Abelson, R.P. (1977). *Scripts, Plans, Goals, and Understanding*. Hillsdale, NJ: Erlbaum.
Searle, J.R. (1969). *Speech Acts*. Cambridge: Cambridge University Press.
Siegal, M. (1991). *Knowing Children: Experiments in Conservation and Cognition*. Hillsdale, NJ: Erlbaum.
Sperber, D. & Wilson, D. (1986). *Relevance*. Oxford: Basic Blackwell.
Sperber, D., Cara, F. & Girotto, V. (1995). Relevance theory explains the selection task. *Cognition, 57*, 31–95.
Vygotsky, L. (1986). *Thought and Language* (A. Kozulin, Trans.). Cambridge, MA: Cambridge University Press. (Original work published in 1934.)
Ward, S.L. & Overton, W.F. (1990). Semantic familiarity, relevance and the development of deductive reasoning. *Developmental Psychology, 26*, 488–493.
Wason, P.C. (1961). Response to affirmative and negative binary statements. *British Journal of Psychology, 52*, 133–142.
Wason, P.C. (1966). Reasoning. In B.M. Foss (Ed.). *New Horizons in Psychology*. Harmondsworth: Penguin.

Chapter 10

Berger, P. & Luckman T. (1966). *The Social Construction of Reality*. London: Penguin University Books.
Bliss, J. (1994a). Knowledge acquisition as an active process. In R. Lewis & P. Mendelsson (Eds.). *Lessons from Learning* (pp. 23–37). Berlin: Springer.
Bliss, J. (1994b). Children learning science. In B. Jennison & J. Ogborn (Eds.). *Wonder and Delight* (pp. 45–61). Bristol: Institute of Physics Publishing.
Bliss, J. (1995). Piaget and after: the case of learning science. *Studies in Science Education, 25*, 139–172.
Bliss, J. & Ogborn J. (1994). Force and motion from the beginning. *International Journal of Instruction and Learning, 4*, 7–25.
Brown, A. (1990). Domain specific principles affect learning and transfer in children. *Cognitive Science, 14*, 107–133.
Giddens, A. (1997). *The Constitution of Society* (Reprint, first published 1984). Cambridge: Polity Press.
Laboratory of Comparative Human Cognition (1984). Culture and cognitve development. In W. Kessen & P. Mussen (Eds.). *History, Theory and Methods, Handbook of Child Psychology* (4th ed.).
Longeot, P. (1978). *Les Stades Opératoires de Piaget et les Facteurs de l'Intelligence* [Piaget's Operational Stages and Factors of Intelligence]. Grenoble: Presses Universitaires de Grenoble.
Piaget, J. (1964). *La Memoire et l'Intelligence* [Memory and Intelligence]. Geneva: Delachaux & Niestle.
Piaget, J. (1971). *Les Explications Causales* [Caused Explanations], *EEGXXVI*. Paris: Presses Universitaires de France. (English translation: Piaget, J. (1974). *Understanding Causality*. New York: W.W. Norton.)
Piaget, J. & Garcia, R. (1987). *Vers une Logique des Significations* [Toward a Logic of Meaning]. New Jersey: Lawrence Erlbaum.

Rogoff, B. (1990). *Apprenticeship in Thinking; Cognitive Development in Social Context.* New York: Oxford University Press.

Chapter 11

Ackermann, E. (1996). Perspective-taking and object construction. In Y. Kafai, & M. Resnick (Eds.). *Constructionism in Practice* (pp. 25–37). Mahwah, NJ: Lawrence Erlbaum Associates.
Ackermann, E. (1995). Construction and transference of meaning through form. In L. Steffe & J. Gale (Eds.). *Constructionism in Education* (pp. 341–355). Hillsdale, NJ: Lawrence Erlbaum Associates,
Ackermann, E. (1991). The agency model of transaction.. In I. Harel & S. Papert (Eds.). *Constructionism.* (pp. 367-379). Norwood, NJ: Ablex Publishing Company.
Astington, J., Harris, P. & Olson, D. (Eds.) (1988). *Developing Theories of Mind.* Cambridge: Cambridge University Press.
Bachelard, G. (1971). *On Poetic Imagination and Reverie* (Trans. C. Gaudin). Dallas, TX: Spring Publications, Inc.
Bettelheim, B. (1977). *The Uses of Enchantment.* New York: Vintage Books.
Brandes, A. (1992). *Children's Ideas about Machines.* Paper presented at the annual meeting of the American Educational Research Association.
Brown, J. S., Collins, A. & Duguid, P. (1989). Situated knowledge and the culture of learning. *Educational Researcher, 18* (1), 32–42.
Carey, S. (1985). *Conceptual Change in Childhood.* Cambridge, MA: MIT Press.
De Loache, J., Uttal, D.H. & Pierroutsakos, S.L. (1998). The development of early symbolization: educational implications. In A. Demetriou (Ed.) *Cognitive Development: Steps en route to Developmental Cognitive Science.* Special Issue of *Learning and Instruction*: The Journal of the European Association for Research on Learning and Instruction, 8(4), 325–341.
Gordon, R. (1996). Radical simulationism. In P. Carruthers & P. Smith (Eds.). *Theories of Theories of Mind.* Cambridge: Cambridge University Press.
Grannott, N. (1991). Puzzled minds and weird creatures. In I. Harel & S. Papert (Eds.). *Constructionism* (pp. 295–311), Norwood, NJ: Ablex Publishing Company.
Hatano, G. & Inagaki, K. (1987). Everyday biology and school biology: how do they interact? *Quarterly Newsletter of the Laboratory of Comparative Human Cognition, 9,* 120–128.
Jordan, J., Kaplan, A., Miller, J.B., Stiver, I. & Surrey, J. (1991). *Women's Growth in Connection: Writings from the Stone Center.* New York, London: Guildford Press.
Karmiloff-Smith, A. (1995). *Beyond Modularity.* Cambridge, MA: MIT Press.
Kegan, R. (1982). *The Evolving Self.* Cambridge, MA: Harvard University Press.
Keller, E.F. (1985). *Reflections on Gender and Science.* New Haven, CT: Yale University Press.
Lakoff, G. & Johnson, M. (1981). *Metaphors We Live By.* Chicago: Chicago University Press.
Laurel, B. (1991). *Computers as Theater.* New York: Addison-Wesley Publishing Company.
Leslie, A.M. (1984). Spatio-temporal contiguity and the perception of causality in infants. *Perception, 13,* 287–305.
Martin, F. (1996). Ideal and real systems: a study of notions of control in undergraduates who design robots. In Y. Kafai & M. Resnick (Eds.). *Constructionism in Practice* (pp. 297–325). Mahwah, NJ: Lawrence Erlbaum Associates.

Masciotra, D. (1996). La genèse de la relation à l'autre: de la symbiose, à l'extériorité et à l'intériorité. [Development of self-other relations: from symbiosis to exteriority to interiority] *Archives de Psychologie, 64,* 207–225.

Ogborn, J. & Miller, R. (1994). Computational issues in modeling. In H. Mellar, J. Bliss, R. Boohan, J. Ogborn and C. Tompsett (Eds). *Learning with Artificial Worlds: Computer-based Modelling in the Curriculum.* London: Falmer Press.

Papert, S. (1980). *Mindstorms: Children, Computers and Powerful Ideas.* New York: Basic Books.

Perner, J. (1996). Simulation as explicitation of prediction-implicit knowledge about the mind: arguments for a simulation-theory mix. In P. Carruthers & P. Smith (Eds.). *Theories of Theories of Mind* (pp. 96–104). Cambridge: Cambridge University Press.

Piaget, J. (1962). *Plays, Dreams and Imitation in Childhood.* New York: W.W. Norton & Company, Inc.

Quéau, P. (1986). *Eloge de la Simulation.* [On the virtues of simulation]. Seyssel, France: Collection Milieux. Edition du Champ Vallon.

Resnick, M. (1990). *Turtles, Termites, and Traffic Jams: Explorations in Massively Parallel Microworlds.* Cambridge, MA: MIT Press.

Rogoff, B. & Lave, L. (1984). *Everyday Cognition: Its Development in Social Context.* Cambridge, MA: Harvard University Press.

Sayeki, Y. (1989). Anthropomorphic epistemology. Unpublished paper, Laboratory of Comparative Human Cognition, University of California, San Diego.

Stamback, M. & Sinclair, H. (1990). *Les Jeux de Fiction entre Enfants de 3 ans.* [Pretense play among three year old children]. Paris: PUF.

Steward, J. (1982). Perception of animacy. Doctoral Thesis. University of Pennsylvania.

Suchman, L. A. (1987). *Plans and Situated Actions: the Problem of Human–Machine Communication.* Cambridge: Cambridge University Press.

Turkle, S. (1984). *The Second Self.* New York: Simon & Schuster.

Turkle, S. (1995). *Life on the Screen.* New York: Simon & Schuster.

Winnicott, D.W. (1989) *Playing and Reality.* London, New York: Routledge.

Chapter 12

Achtenhagen, F. (1995). Fusing experience and theory. *Learning and Instruction, 5,* 409–417.

Akin, O. (1990). Necessary conditions for design expertise and creativity. *Design Studies, 11,* 107–113.

Ball, L. & Ormerod, T. (1995). Structured and opportunistic processing in design. *International Journal of Human–Computer Studies, 43,* 131–151.

Batra, D. & Davis, J. (1992). Conceptual data modelling in database design. *International Journal of Man–Machine Studies, 37,* 83–101.

Benner, P. (1984). *From Novice to Expert.* Menlo Park, CA: Addison-Wesley.

Bereiter, C. & Scardamalia, M. (1993). *Surpassing Ourselves: An Inquiry into the Nature and Implications of Expertise.* Chicago: Open Court.

Berliner, D. (1992). The nature of expertise in teaching. In F. Oser, A. Dick & J. Patry (Eds.). *Effective and Responsible Teaching* (pp. 227–248). San Francisco: Jossey-Bass.

Boshuizen, H., Smith, H., Custer, E. & van de Viel, M. (1995). Knowledge development and restructuring in the domain of medicine. *Learning and Instruction, 5,* 269–289.

Bromme, R. & Tillema, H. (1995). Fusing experience and theory: the structure of professional knowledge. *Learning and Instruction, 5*, 261–268.

Brown, J.S. & Duguid, P. (1994a). Borderline issues: social and material aspects of design. *Human–Computer Interaction, 9*, 3–35.

Brown, J. S. & Duguid, P. (1994b). Patrolling the border: a reply. *Human–Computer Interaction, 9*, 137–143.

Chi, M., Glaser, R. & Farr, M. (Eds.). (1988). *The Nature of Expertise*. Hillsdale, NJ: Erlbaum.

Christiaans, H. (1992). *Creativity in Design*. Utrecht: Lemma.

Davies, S. (1991). Characterizing the program design activity. *Behaviour and Information Technology, 10*, 173–190.

Dreyfus, H. & Dreyfus, S. (1986). *Mind over Machine*. Oxford: Blackwell.

Eraut, M. (1994). *Developing Professional Knowledge and Competence*. London: Falmer Press.

Ericsson, K. & Smith, J. (1991). *Toward a General Theory of Expertise*. New York: Cambridge University Press.

Eteläpelto, A. (1993). Metacognition and the expertise of computer program comprehension. *Scandinavian Journal of Educational Research, 37*, 243–254.

Eteläpelto, A. (1994a). Learning process in the control theory. *Applied Psychology: An International Review, 43*, 370–376.

Eteläpelto, A. (1994b). Work experience and the development of expertise. In W. Nijhof & J. Streumer (Eds.). *Flexibility and Training in Vocational Education* (pp. 319–341). Utrecht: Lemma.

Eteläpelto, A. (1998). *Learning of expertise in the design of information systems*. Manuscript submitted for publication.

Goel, V. & Pirolli, P. (1992). The structure of design problem spaces. *Cognitive Science, 16*, 395–429.

Gruber, H.L. Law, L.-C., Mandl, H. & Renkl. A. (1995). Situated learning and transfer. In P. Reiman & H. Spada (Eds.). *Learning in Humans and Machines: Towards an Interdisciplinary Learning Science* (pp. 168–188). Oxford: Elsevier Science.

Guberman, S.R. & Greenfield, P.M. (1991). Learning and transfer in everyday cognition. *Cognitive Development, 6*, 233–260.

Guindon, R. (1990). Designing the design process. *Human–Computer Interaction, 5*, 305–344.

Hatano, G. & Inagaki, K. (1992). Desituating cognition through the construction of conceptual knowledge. In P. Light & G. Butterworth (Eds.). *Context and Cognition* (pp. 115–133). London: Harvester.

Hoc, J.-M. (1988). *Cognitive Psychology of Planning*. San Diego: Academic Press.

Jeffries, R, Turner, A., Polson, P. & Atwood, M. (1981). The processes involved in designing software. In J. Anderson (Ed.). *Cognitive Skills and their Acquisition* (pp. 255 283). Hillsdale, NJ: Erlbaum.

King, J. (1994). Life and death in the borderlands. *Human–Computer Interaction, 9*, 82–86.

Lave, J. (1988). *Cognition in Practice*. Cambridge: Cambridge University Press.

Lave, J. & Wenger, E. (1991). *Situated Learning*. Cambridge: Cambridge University Press.

Lawson, B. (1986). *How Designers Think*. London: Architectural Press.

Light, P. & Butterworth, G. (Eds.). (1992). *Context and Cognition*. London: Harvester.

Lyytinen, K. (1987). Different perspectives on information systems. *ACM Computing Surveys, 19*, 5–46.

Mayer, R. & Wittrock, M. (1996). Problem-solving transfer. In D. Berliner & R. Calfee (Eds.). *Handbook of Educational Psychology* (pp. 47–62). New York: Macmillan.

Moran, T. (1994). Introduction to special issue on context in design. *Human–Computer Interaction, 9*, 37–135.

Nardi, B. (1996). Studying context. In B. Nardi (Ed.). *Context and Consciousness* (pp. 69–102). Cambridge, MA: MIT Press.

Newman, S. (1994). Interpretation, negotiation, and practice in system design. *Human-Computer Interaction, 9*, 94–98.

Norman, D.A. (1993). Cognition in the head and in the world. *Cognitive Science, 17*, 1–6.

Parnas, D. & Clements, P. (1986). A rational design process. *IEEE Transactions on Software Engineering, 12*, 251–257.

Rambow, R. & Bromme, R. (1995). Implicit psychological concepts in architects' knowledge. *Learning and Instruction, 5*, 337–356.

Radziszewska, B. & Rogoff, B. (1991). Children's guided participation in planning imaginary errands with skilled adult or peer partners. *Developmental Psychology, 27*, 381–389.

Rogoff, B. (1990). *Apprenticeship in Thinking*. New York: Oxford University Press.

Saariluoma, P. (1995). *Chess Players' Thinking*. New York: Routledge.

Salomon, G. & Perkins, D. (1989). Rocky roads to transfer: rethinking mechanisms of a neglected phenomenon. *Educational Psychologist, 24*, 113–142.

Sarvimaki, A. (1988). *Knowledge in Interactive Practice Disciplines*. Research Bulletin 68, University of Helsinki, Department of Education. Helsinki.

Simon, H.A. (1973). The structure of ill-structured problems. *Artificial Intelligence, 4*, 145–180.

Simon, H.A. (1981). *Sciences of the Artificial* (2nd ed.). Cambridge, MA: MIT Press.

Sonnentag, S. (1995). Excellent software professionals. *Behaviour & Information Technology, 14*, 289–299.

Sternberg, R. & Horvath, J. (1995). A prototype view of expert teaching. *Educational Researcher*, 24, 9–17.

Sternberg, R. & Wagner, R. (1994). *Mind in Context*. New York: Cambridge University Press.

Visser, W. & Hoc, J.-M. (1990). Expert software design strategies. In J.-M. Hoc, T. Green, R. Samurcay & D. Gilmore (Eds.). *Psychology of Programming* (pp. 235–249). London: Academic Press.

Walz, D., Elam, J. & Curtis, B. (1993). Inside a software design team. *Communication of the ACM, 36*, 63–77.

Part 3: Introduction

Bourdieu, P. (1991). *Language and Symbolic Power*. Cambridge: Polity Press.

Giddens, A. (1997). *The Constitution of Society* (Reprint, first published 1984). Cambridge: Polity Press.

Chapter 13

Anderson, R.E. (1987). Females surpass males in computer problem solving: Findings from the Minnesota computer literacy assessment. *Journal of Educational Computing Research, 3*, 39–51.

Arbinger, R. & Bannert, M. (1993). Computerwissen von Schülern der Sekundarstufe I. [Computer literacy of secondary school students]. *Empirische Pädagogik, 7*, 103–124.

Artman, H. (1995). *Situated Cognition*. Paper presented at the Swedish Workshop on a Situated Cognitive View on the Design of Tutoring, Help, and Explanation Systems, 12–13 June, Barnens OE, Vaeddoe, Sweden.

Bamossy, G.J. & Jansen, P.G.W. (1994). Children's apprehension and comprehension: Gender influences on computer literacy and attitude structures toward personal computers. In J.A. Costa (Ed.). *Gender Issues and Consumer Behavior* (pp. 142–163). London: Sage.

Bannert, M. & Arbinger, R. (1996). Gender related differences in the exposure to and use of computers. *European Journal of Psychology of Education, 11*, 269–282.

Barbieri, M.S. & Light, P. (1992). Interaction, gender and performance on a computer-based problem solving task. *Learning and Instruction, 2*, 199–213.

Bergin, D.A., Ford, M.E. & Hess, R.D. (1993). Patterns of motivation and social behavior associated with microcomputer use of young children. *Journal of Educational Psychology, 85*, 437-445.

Beynon, J. (1993). Computers, dominant boys and invisible girls: Or, 'Hannah, it's not a toaster, it's a computer!'. In J. Beynon & H. Mackay (Eds.). *Computers into Classrooms. More Questions than Answers* (pp. 160–189). London, Washington: Falmer Press.

Beynon, J. & Mackay, H. (Eds.). (1993). *Computers into Classrooms. More Questions than Answers*. London, Washington: Falmer Press.

Brown, J.S., Collins, A. & Duguid, P. (1989). Situated cognition and the culture of learning. *Educational Researcher, 18*, 32–42.

Chen, M. (1986). Gender and computers: the beneficial effects of experience on attitudes. *Journal of Educational Computing Research, 2*, 265–282.

Clark, M. (1989). *The Great Divide*. Commonwealth Department of Employment, Education and Training, Canberra, Australia.

Coolican, H. (1997). *Computer Anxiety: Research and Practical Experience*. Paper presented at the Gender and Technology Seminar, February, The Open University, Milton Keynes, UK.

Cooper, J., Hall, J. & Huff, Ch. (1990). Situational stress as a consequence of sex-stereotyped software. *Personality and Social Psychology Bulletin, 16*, 419–429.

Culley, L. (1988). Girls, boys and computers. *Educational Studies, 14*, 3–8.

Culley, L. (1993). Gender equity and computing in secondary schools: Issues and strategies for teachers. In J. Beynon & H. Mackay (Eds.). *Computers into Classrooms. More Questions than Answers* (pp. 147–159). London, Washington: Falmer Press.

Cummings, R. (1985). Small group discussions and the microcomputer. *Journal of Computer Assisted Learning, 1*, 149–58.

Duffy, T.M. & Jonassen, D.H. (Eds.). (1992). *Constructivism and the Technology of Instruction. A Conversation*. Hillsdale, NJ: Erlbaum.

Eastman, S. & Krendl, K. (1987). Computers and gender: differential effects of electronic search on students' achievement and attitudes. *Journal of Research and Development in Education, 20*(3), 41–48.

Ebach, J. (1994). Der Rückgang des Frauenanteils in der Informatik — Überlegungen zu möglichen Ursachen aus psychologischer Sicht. [A drop in the proportion of females in computer science — reflections on possible reasons from a psychological perspective]. *Zeitschrift für Frauenforschung, 3*, 16–27.

Fetler, M. (1985). Sex differences on the California statewide assessment of computer literacy. *Sex Roles, 13*, 181–191.

Fife Schaw, C.R., Breakwell, G.M., Lee, T. & Spencer, J. (1986). Patterns of teenage computer usage. *Journal of Computer Assisted Learning, 2*, 152–161.

Finlayson, H. (1984). *The Transfer of Mathematical Problem Solving Skills from LOGO Experience*. Research Paper Number 238. Edinburgh: University of Edinburgh, Department of Artificial Intelligence.

Forman, E.A., Minick, N. & Stone, C.A. (1993). *Contexts for Learning: Sociocultural Dynamics in Children's Development*. Oxford: Oxford University Press.

Funken, C. (1993). Ist die Koedukation ein Fortschritt? Empirische Befunde zu Geschlechterunterschieden im Informatikunterricht. [Is co-education an improvement? Empirical results of gender-related differences in computer classes]. *Computer und Unterricht*, *3*, 43–48.

Gruber, H.L., Law, L.-C., Mandl, H. & Renkl, A. (1995). Situated learning and transfer. In P. Reimann & H. Spada (Eds.). *Learning in Humans and Machines. Towards an Interdisciplinary Learning Science* (pp. 168–188). Oxford: Elsevier Science.

Hargreaves, D., Bates, H. & Foot, J. (1985). Sex-typed labelling affects task performance. *British Journal of Social Psychology*, *25*, 153–155.

Hattie, J. & Fitzgerald, D. (1988). Sex differences in attitudes, achievement and use of computers. *Australian Journal of Education*, *31*(1), 3–26.

Hoffmann, U. (1987). *Computerfrauen: Welchen Anteil haben Frauen an Computergeschichte und -arbeit?* [Computer women: How much have females contributed to computer development and computer work?]. Munich: Hampp.

Hoyles, C. (1988). (Ed.). *Girls and Computers*. Bedfore Way Papers No. 34, (pp. 40–63). London: University of London, Institute of Education.

Hoyles, C., Sutherland, R. & Healy, L. (1991). Children talking in computer environments: new insights on the role of discussion in mathematics learning. In K. Durkin & B. Shire (Eds.). *Language in Mathematical Education: Research and Practice* (pp. 162–176). Milton Keynes: Open University Press.

Huff, C. & Cooper, J. (1987). Sex bias in educational software: the effect of designers' stereotypes on the software they design. *Journal of Applied Psychology*, *17*, 519–532.

Hughes, M., Brackenridge, A. & Macleod, H. (1987). Children's ideas about computers. In J. Rutkowska & C. Crook (Eds.). *Computers, Cognition and Development* (pp. 9–34). London: Wiley.

Huguet, P., Chambres, P. & Blaye, A. (1994). Interactive learning: does social presence explain the results? In H.C. Foot, C.J. Howe, A. Anderson, A.K. Tolmie & D.A. Warden (Eds.). *Group and Interactive Learning* (pp. 445-51), Southampton, Boston: Computational Mechanics Publications.

Issing, L.J. & Klimsa, P. (Eds.). (1995). *Information und Lernen mit Multimedia*. [Information and learning with multimedia]. Weinheim: Psychologie Verlags Union.

Issroff, K. (1994). Gender and cognitive and affective aspects of co-operative learning. In H.C. Foot, C.J. Howe, A. Anderson, A.K. Tolmie & D.A. Warden (Eds.). *Group and Interactive Learning*, pp. 67–7). Southampton, Boston, Computational Mechanics.

Jäger, R.S., Arbinger, R., Bannert, M. & Lissmann, U. (1993). Ergebnisse der wissenschaftlichen Begleitung. [Results of the evaluation study]. In A. Rissberger (Ed.). *Computerunterstütztes Lernen an allgemeinbildenden Schulen Teil II. Abschlussbericht des Modellversuchs CULAS.* [Computer-assisted learning in general schools. Report of the pilot study CULAS] (pp. 193–251). (Ministerium für Bildung und Kultur Rheinland-Pfalz, Schulversuche und Bildungsforschung, Berichte und Materialien, Bd. 72). Mainz: v. Hase & Koehler.

Jakobsdottir, S., Krey, C.L. & Sales, G.C. (1994). Computer graphics: preferences by gender in grades 2, 4, and 6. *Journal of Educational Research*, *88*, 91–99.

Johnson, R., Johnson, D. & Stanne, M. (1985). Effects of co-operative, competitive and individualistic goal structures on computer-assisted instruction. *Journal of Educational Psychology*, *77*, 668–677.

Jones, T. & Clarke, V. A. (1995). Diversity as a determinant of attitudes: a possible explanation of the apparent advantage of single-sex settings. *Journal of Educational Computing Research*, *12*, 51–64.

Kay, R. (1989). Gender differences in computer attitudes, literacy, locus of control and commitment. *Journal of Research on Computing in Education, 21*, 307–316.

Kay, R. (1992). An analysis of methods used to examine gender differences in computer-related behavior. *Journal of Educational Computing Research, 8*, 277–290.

Kirkup, G. (1992). The social construction of computers. In G. Kirkup & S. Keller (Eds.). *Inventing Women: Science, Gender and Technology* (pp. 267–281). Cambridge: Polity/Open University Press.

Kreienbaum, M.A. & Metz-Goeckel, S. (1992). *Koedukation und Technikkompetenz von Mädchen*. [Coeducation and technical competence of girls]. Weinheim/Munich: Juventa Verlag.

Lage, E. (1991). Boys, girls, and microcomputing. *European Journal of Psychology of Education, 6*, 29–44.

Law, L.C. & Wong, K.M. (1996). Expertise und instructional design. [Expertise and instructional design]. In H. Gruber & A. Ziegler (Eds.). *Expertiseforschung. Theoretische und methodische Grundlagen*. [Expertise research. Theoretical and methodological foundations] (pp. 115–147). Opladen: Westdeutscher Verlag.

Light, P. & Colbourn, C. (1987). The role of social processes in children's microcomputer use. In W. Kent & R. Lewis (Eds.). *Computer Assisted Learning in the Social Sciences and Humanities* (pp. 109–114). Oxford: Basil Blackwell.

Light, P., Littleton, K. & Bale, S. (1994) *Children Solving Problems in Mixed Gender Pairs*. Paper presented at the British Psychological Society Developmental Section Annual Conference, September 1994, University of Portsmouth.

Linn, M.C. (1985). Fostering equitable consequences from computer learning environments. *Sex Roles, 13*, 229-240.

Linnakylä, P. (1996). *High Expectations — High Achievement on Literacy*. Paper presented at 1996 World Conference on Literacy, March 12–15, Philadelphia.

Littleton, K. (1996). Girls and information technology. In P. Murphy & C. Gipps (Eds.). *Equity in the Classroom*. (pp. 81–96) London: Falmer Press.

Littleton, K. & Light, P. (1996). *Understanding Gender and Software Effects in Children's Computer-based Problem Solving*. Paper presented at the British Psychological Society Educational Section Annual Conference, November 1996, Wokingham, UK.

Littleton, K., Light, P., Joiner, R., Messer, D. & Barnes, P. (1992). Pairing and gender effects on children's computer-based learning. *European Journal of Psychology of Education, 7*, 311-324.

Littleton, K., Light, P., Barnes, P., Messer, D. & Joiner, R. (1993). Gender and software effects on children's computer-based problem-solving. Poster presented at the Society for Research in Child Development Annual Convention, March 1993, New Orleans, USA.

Lockheed, M. E., Nielsen, A. & Stone, M. K. (1985). Determinants of microcomputer literacy in high school students. *Journal of Educational Computing Research, 1*, 81-96.

Martin, R. (1991). School children's attitudes towards computers as a function of gender, course subjects and availability of home computers. *Journal of Computer Assisted Learning, 7*, 187–194.

Monteil, J.M. (1993). *Soi & le Contexte: Constructions Autobiographiques, Insertions Sociales, Performances Cognitives*. [Self and context: the autobiographical self, social positioning and cognitive performance] Paris: Armand Colin.

Newton, P. & Beck, E. (1993). Computing: an ideal occupation for women? In J. Beynon & H. Mackay (Eds.). *Computers into Classrooms. More Questions than Answers* (pp. 130–146). London, Washington: Falmer Press.

Pelgrum, W. J. & Plomp, T. (1991). Gender equity in relation to computers. In W.J. Pelgrum & T. Plomp (Eds.). *The Use of Computers in Education Worldwide. Results From the IEA 'Computers in Education' Survey in 19 Education Systems* (pp. 85–99). Oxford: Pergamon Press.

Podmore, V. (1991). Four-year-olds, 6-year-olds, and microcomputers: a study of perceptions and social behaviours. *Journal of Applied Developmental Psychology, 12*, 87–101.

Reisman, J. (1990). Gender inequality in computing. *Computers in Human Services, 7*, 45–63.

Robertson, I., Calder, J., Fung, P., Jones, A. & O'Shea, T. (1995). Computer attitudes in an English secondary school. *Computers and Education, 24*, 73–81.

Robinson-Staveley, K. & Cooper, J. (1990). Mere presence, gender, and reactions to computers: studying human-computer interaction in the social context. *Journal of Experimental Social Psychology, 26*, 168–183.

Säljö, R. (1995). Mental and physical artefacts in cognitive practices. In P. Reimann & H. Spada (Eds.). *Learning in Humans and Machines. Towards an Interdisciplinary Learning Science* (pp. 83–96). Oxford: Pergamon Press.

Schiersmann, C. (1992). Geschlechtstypische Unterschiede beim Zugang zum Computer — Problemstellung und Stand der Forschung. [Gender-related differences in exposure to computers — state of the art]. In Bundesminister für Bildung und Wissenschaft (Ed.). *Mädchen und Computer. Ergebnisse und Modelle zur Mädchenförderung in Computerkursen*. [Girls and Computers. Results and Models of Supporting Girls in Computer Classes] (pp. 7–21). Bad Honnef: Bock.

Schofield, J.W. (1995). *Computers and Classroom Culture*. Cambridge: Press Syndicate of the University of Cambridge.

Sewell, D.F. (1990). *New Tools for New Minds. A Cognitive Perspective on the Use of Computers with Young Children*. New York: Harvester Wheatsheaf.

Sternberg, R.J. & Wagner, R.K. (Eds.). (1994). *Mind in Context — Interactionist Perspectives on Human Intelligence*. Cambridge: Press Syndicate of the University of Cambridge.

Straker, A. (1989). *Children Using Computers*. Oxford: Blackwell.

Sutherland, R. & Hoyles, C. (1988). Gender perspectives on Logo programming in the mathematics curriculum. In C. Hoyles (Ed.). *Girls and Computers*. Bedford Way Papers, No. 34. London: University of London, Institute of Education.

Sutton, R.E. (1991). Equity and computers in the schools: a decade of research. *Review of Educational Research, 61*, 475–503.

Todman, J. & Dick, G. (1993). Primary children and teachers' attitudes to computers. *Computer Education, 20*, 199–203.

Turkle, S. (1984). *The Second Self. Computers and the Human Spirit*. New York: Simon & Schuster.

Turkle, S. & Papert, S. (1990). Epistemological pluralism: styles and voices within the computer culture. *Signs: Journal of Women in Culture and Society, 16*, 128–157.

Vosniadou, S., De Corte, E., Glaser, R. & Mandl, H. (Eds.). (1996). *International Perspectives on the Design of Technology-supported Learning Environments*. Mahwah, NJ: Lawrence Erlbaum Associates.

Webb, N. (1984). Microcomputer learning in small groups: cognitive requirements and group processes. *Journal of Educational Psychology, 76*, 1076–1088.

Wilder, G., Mackie, D. & Cooper, J. (1985). Gender and computers: two surveys of computer-related attitudes. *Sex Roles, 13*, (3/4), 215–228.

Williams, S.W. & Ogletree, S.M. (1992). Preschool children's computer interest and competence: Effects of sex and gender role. *Early Childhood Research Quarterly, 7*, 135–143.

Willis, S. & Kenway, J. (1986). On overcoming sexism in schooling: to marginalise or mainstream. *Australian Journal of Education*, *30*, 2.

Wosnitza, M. (1993). Mikroevaluation eines computerunterstützten Lernprogramms mit der Methode des lauten Denkens am Beispiel des Geometrieprogramms Capri Géomètre. [Microevaluation of the computer-based training program "Cabri Géométre" with the method of thinking-aloud]. *Empirische Pädagogik*, *7*, 391–420.

Chapter 14

Becker, H.S., Geer, B., Hughes, E.C. & Strauss, A.L. (1961). *Boys in White*. New Brunswick: Transaction Publishers.

Bruner, J. (1996). *The Culture of Education*. Cambridge, MA: Harvard University Press.

Crook, C.K. (1994a). *Computers and the Collaborative Experience of Learning*. London: Routledge.

Crook, C.K. (1994b). Computer networking and collaborative learning within a departmentally focused undergraduate course. In H. Foot, C. Howe, A. Tolmie, A. Anderson, & D. Warden (Eds.). *Group and Interactive Learning* (pp. 25–30). Southampton: Computational Mechanics.

Crook, C.K. (1996). *Resourcing Collaboration Among Undergraduates — the Place of Information Technology*. Paper presented at Conference of Learning Technology Dissemination Initiative, Stirling University, February.

Cuban, L. (1986). *Teachers and Machines*. New York: Teachers College.

Edwards, D. & Mercer, N. (1987) *Common Knowledge*. London: Methuen.

Grint, K. (1992). Sniffers, lurkers, actor networkers: computer mediated communications as a technical fix. In J. Beynon & H. Mackay (Eds.). *Technological Literacy and the Curriculum* (pp. 87–102). London; Falmer Press.

Hague, D. (1991). *Beyond Universities*. London: Institute of Economic Affairs.

I&T Magazine (1995). *Catchphrase Culture*. June. No. 17.

Kaye, A. (1995). Computer supported collaborative learning in a multi-media distance education environment. In C. O'Malley (Ed.). *Computer Supported Collaborative Learning* (pp. 125–143). Berlin: Springer.

Light, P. (1996). *Influences of New Technology on Informal Learning Processes Amongst Students*. Paper presented at Biennial Conference of the International Society for the Study of Behavioural Development, Quebec City, August 1996.

Light, P. & Light, V. (1999). Analysing asynchronous learning interactions: computer mediated communication in a conventional undergraduate setting. In K. Littleton & P. Light (Eds.). *Learning with Computers: Analysing Productive Interaction*. London: Routledge.

Mayes, T. & Neilson, I. (1995) Learning from other people's dialogues: questions about computer-based answers. In B. Collis & G. Davies (Eds.). *Innovative Adult Learning with Innovative Technologies*. North Holland: Elsevier Science.

Oldenburg, R. (1989). *The Great Good Place*. New York: Paragon House.

Ramsden, P. (1992). *Learning to Teach in Higher Education*. London: Routledge.

Rogoff, B. (1990). *Apprenticeship in Thinking: Cognitive Development in Social Context*. New York: Oxford University Press.

Säljö, R. (1984). Learning from reading. In F. Marton, D. Hounsell & N. Entwistle (Eds.). *The Experience of Learning* (pp. 230–245). Edinburgh: Scottish Academic Press.

Wallace, W.L. (1966). *Student Culture*. Chicago: Aldine.

Chapter 15

Baker, M.J. & Lund, K. (1996). Flexibly structuring the interaction in a CSCL environment. *Proceedings of the European Conference on Artificial Intelligence in Education* (pp. 401–407). Lisbon: Colibri — Artes Graficas, Lda.

Becker, J. & Varelas, M. (1995). Assisting construction: The role of the teacher in assisting the learner's construction of pre-existing cultural knowledge. In L.P. Steffe & J. Gale (Eds.). *Constructivism in Education* (pp. 433–446). Hillsdale. NJ: Erlbaum.

Bliss, J., Askew, M. & Macrae, S. (1996). Effective teaching and learning: Scaffolding revisited. *Oxford Review of Education, 22*(1), 37–61

Bloom, B.S. (1984). The 2 sigma problem: the search for methods of group instruction as effective as one-to-one tutoring. *Educational Researcher, 13*(6), 4–16.

Brecht, B. (1988). Student models: the genetic graph approach. *International Journal of Man-Machine Studies, 28*, 483–504.

Burton, M. & Brna, P. (1996). Clarissa: an exploration of collaboration through agent-based dialogue games. *Proceedings of the European Conference on Artificial Intelligence in Education* (pp. 393–400). Lisbon: Colibri — Artes Graficas, Lda.

Chan, T.W. & Baskin, A. B. (1990). Learning companion systems. In C. Frasson & G. Gauthier (Eds.). *Intelligent Tutoring Systems: at the Crossroads* (pp. 7–33). Norwood, NJ: Ablex.

Clancey, W.J. (1992). Representations of knowing: in defence of cognitive apprenticeship. *Journal of Artificial Intelligence in Education, 3*(2), 139–168.

Corbett, A.T., Anderson, J.R. & Patterson, E.J. (1988). Problem compilation and tutoring flexibility in a lisp tutor. In C. Frasson, G. Gauthier & G. McCalla (Eds.). *Proceedings of Intelligent Tutoring Systems* (pp. 423–429). Berlin: Springer.

Dillenbourg, P., Traum, D.R. & Sneider, D.K. (1996). Grounding in multi-modal task oriented collaboration. *Proceedings of the European Conference on Artificial Intelligence in Education* (pp. 415–425). Lisbon: Colibri — Artes Graficas, Lda.

Department for Education (1995). *The National Curriculum*. London: HMSO.

Evans, P. (1993). Some implications of work for special education. In H. Daniels (Ed.). *Charting the Agenda: Educational Activity after Vygotsky* (pp. 30–45). London: Routledge.

Gegg-Harrison, T.S. (1991). Learning prolog in a schema based environment. *Instructional Science, 20*(2/3), 173–192.

Gegg-Harrison, T.S. (1992). Adapting instruction to the student's capabilities. *Journal of Artificial Intelligence in Education, 3*, 169–181.

Goldstein, P. (1982). The genetic graph: a representation for the evolution of procedural knowledge. In D. Sleeman & J.S. Brown (Eds.). *Intelligent Tutoring Systems* (pp. 51–77). New York. Academic Press.

Griffiths, A.K. & Grant, A.C. (1985). High school students' understanding of food webs. *Journal of Research in Science Teaching, 22*(5), 421–436.

Jensen, F.V. (1996). Bayesian networks basics. *Society for the Study of Artificial Intelligence and Simulation of Behaviour Quarterly Newsletter, 94*, 9–23.

Laurillard, D. (1993). *Rethinking University Teaching: a Framework for the Use of Educational Technology*. London: Routledge.

Luckin, R. (1996a). TRIVAR: exploring the zone of proximal development. *Proceedings of the European Conference on Artificial Intelligence in Education* (pp. 16–21). Lisbon: Colibri — Artes Graficas, Lda.

Luckin, R. (1996b). You'll never walk alone in Vygotsky's zone. *The 9th White House Papers Technical Report CSRP No 440*. School of Cognitive and Computing Sciences, University of Sussex, UK.
Luckin, R. (1998). 'ECOLAB': Explorations in the Zone of Proximal Development. *CSRP Technical Report No. 486*. School of Cognitive and Computing Sciences, University of Sussex, UK.
Lumpe, A.T. & Staver, J.R. (1995). Peer collaboration and concept development: Learning about photosynthesis. *Journal of Research in Science Teaching, 32*(1), 71–98.
Mercer, N. (1992). Culture, context and the construction of knowledge in the classroom. In P. Light & G. Butterworth (Eds.). *Context and Cognition: Ways of Learning and Knowing* (pp. 28–46). Hillsdale, NJ: Erlbaum.
Murphey, T. (1996). *Proactive adjusting to the zone of proximal development: learner and teacher strategies*. Paper presented at the IInd Conference for socio-cultural research 1896–1996. September 1996, University of Geneva.
Norman, D.A. & Spohrer, J.C. (1996). Learner centred education. *Communications of the ACM, 39*, (4), 24–27.
Palthepu, S., Greer, J. & McCalla, G. (1995). Granularity in reverse engineering. *Proceedings of the Third Workshop on AI and Software Engineering: Breaking the Toy Mould*. (In conjunction with Fourteenth International Joint Conference on Artificial Intelligence). August 1995, Montreal, Canada.
Rosch, E.H. (1978). Principles of categorization. In E. Rosch & B. Lloyd (Eds.). *Cognition and Categorization* (pp. 312–322). Hillsdale, NJ: Erlbaum.
Self, J. (1990). Bypassing the intractable problem of student modelling. In C. Frasson & G. Gauthier (Eds.). *Proceedings of ITS* (pp. 107–123). Berlin: Springer.
Vygotsky, L.S. (1978). *Mind in Society: The Development of Higher Psychological Processes*, (M. Cole, V. John-Steiner, S. Scribner & E. Souberman, Eds. and Trans.). Cambridge, MA: Harvard University Press.
Vygotsky, L.S. (1986). *Thought and Language* (Original work published in 1934). Cambridge, MA: MIT Press.
Vygotsky, L.S. (1987). *The Collected Works of L.S. Vygotsky, Vol 1: Problems of General Psychology*. New York: Plenum Press.
Wood, D., Shadbolt, N., Reichgelt, H., Wood, H. & Paskiewitz, T. (1992). EXPLAIN: experiments in planning and instruction. *Society for the Study of Artificial Intelligence and Simulation of Behaviour Quarterly Newsletter, 81*, 13–16
Wood, D. & Wood, H. (1996). Vygotsky, tutoring and learning. *Oxford Review of Education, 22*(1), 5–16.
Wood, D.J., Wood, H.A. & Middleton, D.J. (1978). An experimental evaluation of four face-to-face teaching strategies. *International Journal of Behavioural Development, 1*, 131–147.

Chapter 16

Bliss, J. (1994). Reasoning with a semi-quantitative tool. In H. Mellar, J. Bliss, R. Boohan, J. Ogborn, & C. Tompsett (Eds.). *Learning with Artificial Worlds: Computer Based Modelling in the Curriculum* (pp. 128–141). Lewes, UK: Falmer Press.
Bliss, J. (1996). Externalising thinking through modelling. In S. Vosniadou, E. De Corte, R. Glaser & H. Mandl (Eds), *International Perspectives on the Design of Technology-Supported Learning Environments* (pp. 25–40). Hillsdale, N.J: Lawrence Erlbaum Associates.

Bliss, J. (1997). Making models and reasoning with them. *Journal of Computing in Higher Education*, *8*(2), 3–28.

Bliss, J. & Ogborn, J. (1992). *Summary Report. ESRC Information Technology in Education Initiative, Tools for Exploratory Learning Programme*. End-of-award report.

Carpenter, K.J. (1986). *The History of Scurvy and Vitamin C*. Cambridge: Cambridge University Press.

Corruble, V. (1996). Une approche inductive de la découverte en médecine: les cas de la lèpre et du scorbut. (An inductive approach to discovery in medicine: the cases of leprosy and scurvy) Thèse de Doctorat de l'Université Paris 6, spécialit: Informatique.

Corruble, V. & Ganascia, J.G. (1996). The discovery of the causes of leprosy: a rational reconstruction. *Proceedings of the National Conference on Artificial Intelligence*. July 1996, Portland, Oregon.

Corruble, V. & Ganascia, J.G. (1997). Induction and the discovery of the causes of scurvy: a computational reconstruction. *Artificial Intelligence*, *91*(2), 205–223

Ganascia, J.G. (1987). AGAPE et CHARADE: deux techniques d'apprentissage symbolique appliquées à la construction de base de connaissances. [AGAPE and CHARADE: two techniques of symbolic apprenticeship applied to the basic construction of knowledge] Doctorat d'Etat, Université Paris-Sud: Centre d'Orsay.

Lind, J. (1753). *A Treatise of the Scurvy*. Edinburgh: Millar.

Mellar, H., Bliss, J., Boohan, R., Ogborn, J. & Tompsett, C. (Eds). (1994). *Learning with Artificial Worlds: Computer Based Modelling in the Curriculum*. Lewes, UK: Falmer Press

Utgoff, P. (1986). Shift of bias for inductive concept learning. In R.S. Michalski, J.G. Carbonell & T.M. Mitchell (Eds.). *Machine Learning: An Artificial Intelligence Approach* Vol. 2. (pp. 107–149). Palo Alto, CA: Springer.

Chapter 17

Agre, P.E. (1993). The symbolic worldview: Reply to Vera and Simon. *Cognitive Science*, *17*(1), 61–69, special issue on situated action.

Agre, P.E. & Chapman, D. (1990). What are plans for? In P. Maes (Ed.). *Designing Autonomous Agents: Theory and Practice from Biology to Engineering and Back* (pp. 17–34). Cambridge, MA: MIT Press.

Breuker, J. & Van de Velde, W. (Eds.). (1994). *The Common KADS Library*. Amsterdam: IOS Press.

Brooks, R.A. (1986). A robust layered control system for a mobile robot. *IEEE Journal of Robotics and Automation*, *2*(1), 14–23.

Brooks, R.A. (1990). Elephants don't play chess. *Robotics and Autonomous Systems*, *6*(1), 3–15.

Brooks, R.A. (1991) Intelligence without reason. In J. Mylopoulos & R. Reiter (Eds.). *Proceedings of the 12th International Joint Conference on Artificial Intelligence* (pp. 569–595). San Mateo, CA: Morgan Kaufmann.

Brooks, R.A. (1997). From earwigs to humans. *Robotics and Autonomous Systems*, *20*(2–4), 291–304.

Brooks, R.A. & Stein, L.A. (1994). Building brains for bodies. *Autonomous Robots*, *1*(1), 7–25.

Cañamero, D. (1994). Modeling plan recognition for decision support. In L. Steels, G. Schreiber & W. Van de Velde (Eds.). *A Future for Knowledge Acquisition* (pp. 158–177). Berlin-Heidelberg: Springer, LNAI 867.

Cañamero, D. (1997). Modeling motivation and emotions as a basis for intelligent behavior. In W. Lewis Johnson (Ed.). *Proceedings of the First International Conference on Autonomous Agents* (pp. 148–155). New York: ACM Press.

Chandrasekaran, B. & Johnson, T.R. (1993). Generic tasks and task structures: History, critique and new directions. In J.-M. David, J.-P. Krivine & R. Simmons (Eds.) *Second Generation Expert Systems* (pp. 232–272). Berlin-Heidelberg: Springer.

Clancey, W.J. (1985). Heuristic classification. *Artificial Intelligence*, 27(3), 289–350.

Clancey, W.J. (1989). The knowledge level reinterpreted: modeling how systems interact. *Machine Learning*, 4(3/4), 285–291.

Clancey W. (1993). Situated action: a neuropsychological interpretation: response to Vera and Simon. *Cognitive Science*, 17(1), 87–116, special issue on situated action.

Clancey, W. (1995). A boy scout, toto, and a bird. In L. Steels & R. Brooks (Eds.). *The Artificial Life Route to Artificial Intelligence* (pp. 227–236). Hillsdale, NJ: Lawrence Erlbaum Associates.

Dennett, D. (1978). *Brainstorms*. Cambridge, MA: MIT Press.

Harnad, S. (1995). Grounding symbolic capacity in robotic capacity. In L. Steels & R. Brooks (Eds.). *The Artificial Life Route to Artificial Intelligence* (pp. 277–286). Hillsdale, NJ: Lawrence Erlbaum Associates.

Lakatos I. (1978). The methodology of scientific research programmes. In J. Worral & G. Currie (Eds.). *Philosophical Papers* (Vol. 1). Cambridge: Cambridge University Press.

Maes, P. (1989). How to do the right thing. *Connection Science Journal*, 1(3), 291–323.

Mataric, M.J. (1991). Navigation with a rat brain: a neurobiologically-inspired model for robot spatial representation. In J.-A. Meyer & S.W. Wilson (Eds.). *From Animals to Animats: Proceedings of the First International Conference on Simulation of Adaptive Behavior* (pp. 169–175). Cambridge, MA: MIT Press.

Maturana, H.R. & Varela, F.J. (1987). *The Tree of Knowledge: The Biological Roots of Human Understanding*. Boston: New Science Library.

Minsky, M. (1985). *The Society of Mind*. New York, NY: Simon & Schuster.

Musen, M.A. (1992). Dimensions of sharing and reuse. *Computers and Biomedical Research*, 25, 345–367.

Newell, A. (1980). Physical symbol systems. *Cognitive Science*, 4(2), 135–183.

Newell, A. (1982). The knowledge level. *Artificial Intelligence*, 18(1), 87–127.

Newell, A. & Simon, H.A. (1972). *Human Problem Solving*. Englewood Cliffs, NJ: Prentice-Hall.

Newell, A. & Simon, H.A. (1976). Computer science as empirical enquiry: symbols and search. *Communications of the ACM* 19(March). Reprinted in M. Boden (Ed.). *The Philosophy of Artificial Intelligence* (pp. 105–132). New York: Oxford University Press.

Prem, E. (1996). Elements of an epistemology of embodied AI. In M. Mataric (Ed.). *Embodied Cognition and Action. Papers from the AAAI Fall Symposium* (pp. 97–101). Menlo Park, CA: The AAAI Press.

Rademakers, P. & Vanwelkenhuysen, J. (1993). Generic models and their support in modeling problem solving behavior. In J.-M. David, J.-P. Krivine & R. Simmons (Eds.). *Second generation Expert Systems* (pp. 350–375). Berlin-Heidelberg: Springer.

Simon, H.A. (1947). *Administrative Behavior*. New York: Macmillan.

Simon, H.A.(1982). *The Sciences of the Artificial*. Cambridge, MA: MIT Press.

Smithers, T. (1995). Are autonomous agents information processing systems? In L. Steels & R. Brooks (Eds.). *The Artificial Life Route to Artificial Intelligence* (pp. 123–162). Hillsdale, NJ: Lawrence Erlbaum Associates.

Steels, L. (1990). Components of expertise. *AI Magazine*, 11(2), 28–49.

Steels, L. (1991). Towards a theory of emergent functionality. In J.-A. Meyer & S.W. Wilson (Eds.). *From Animals to Animats. Proceedings of the First International Conference on Simulation of Adaptive Behavior* (pp. 451–461). Cambridge, MA: MIT Press.

Slezak, P. (1994). Situated cognition: Empirical issue, 'paradigm shift' or conceptual confusion?. In A. Ram & K. Gisett (Eds.). *Proceedings of the 16th Annual Conference of the Cognitive Science Society* (pp. 806–811). Hillsdale, NJ: Lawrence Erlbaum Associates.

Vanwelkenhuysen, J. (1993). Participative conceptual system design of industrial knowledge systems. PhD thesis, Artificial Intelligence Laboratory, Vrije Universiteit Brussel, Belgium, AI-Lab Memo 931.

Van de Velde, W. (1993). Issues in knowledge-level modelling. In J.-M. David, J.-P. Krivine & R. Simmons (Eds.). *Second Generation Expert Systems* (pp. 211–231). Berlin-Heidelberg: Springer.

Van de Velde, W. (1994). A constructivist view on knowledge engineering. In A.G. Cohn (Ed.). *Proceedings of the 11th European Conference on Artificial Intelligence* (pp. 727–731). Chichester: John Wiley & Sons.

Vera, A.H. & Simon, H.A. (1993). Situated action: a symbolic interpretation. *Cognitive Science*, *17*(1), 7–48, special issue on situated action.

Wielinga, B.J., Schreiber, A.Th. & Breuker, J.A. (1992). KADS: a modelling approach to knowledge engineering. *Knowledge Acquisition*, *4*(1), 5–53.

Chapter 18

Brooks, R.A. & Stein, L.A. (1994). Building brains for bodies. *Autonomous Robots*, *1*, 7–25.

Cliff, D., Harvey, I. & Husbands, P. (1993). Explorations in evolutionary robotics. *Adaptive Behavior*, *2*(1), 71–108.

Donnett, J. & Smithers, T. (1991). Lego vehicles: A technology for studying intelligent systems. In J.-A. Meyer & S.W. Wilson, (Eds.). *Proceedings of the First International Conference on Simulation of Adaptive Behavior* (pp. 540–549). Cambridge, MA: MIT Press.

Drescher, G.L. (1991). *Made-up Minds: a Constructivist Approach to Artificial Intelligence.* Cambridge, MA: MIT Press.

Foner, L.N. (1994). Paying attention to what's important: Using focus of attention to improve unsupervised learning. MS thesis. Massachusetts Institute of Technology, Media Laboratory.

Gaussier, P. & Zrehen, S.A. (1994). Topological neural map for on-line learning: Emergence of obstacle avoidance in a mobile robot. In D. Cliff, P. Husbands, J.-A. Meyer & S.W. Wilson (Eds.). *From Animals to Animats 3. Proceedings of the Third International Conference on Simulation of Adaptive Behavior* (pp. 282–290). Cambridge, MA: MIT Press.

Harnad, S. (1990). The symbol grounding problem. *Physica D 42*, 335–346.

Hebb, D.O. (1988). The organization of behavior.In J.A. Anderson & E. Rosenfeld (Eds.). *Neurocomputing* (pp. 45–56). Cambridge, MA: MIT Press.

Holland, J.H. & Reitman, J.S. (1978). Cognitive systems based on adaptive algorithms. In D.A. Waterman & F. Hayes-Roth (Eds.). *Pattern-directed Inference Systems* (pp. 313–329). New York: Academic Press.

Holland, J.H. (1992). *Adaptation in Natural and Artificial Systems.* Cambridge, MA: MIT Press.

Kaelbling, L.P., Littman, M.L. & Moore, A.W. (1996). Reinforcement learning: A survey. *Journal of Artificial Intelligence Research*, *4*, 237–258.

Lambrinos, D., Scheier C. & Pfeifer, R. (1995). Unsupervised classification of sensory-motor states in a real world artefact using a temporal Kohonen map. In F. Fogelman & J.C. Rault (Eds). *Proceedings of the International Conference on Artificial Neural Networks ICANN '95* (pp. 467–472). Maison de la Chimie, Paris, October 9–13, 1995.

Littman, M. (1993). An optimization-based categorization of reinforcement learning environments. In J.-A. Meyer, H.L. Roitblat & S.W. Wilson, (Eds.). *From Animals to Animats 2. Proceedings of the Second International Conference on Simulation of Adaptive Behavior* (pp. 262–270). Cambridge, MA: MIT Press.

Maes, P. (1991). A bottom-up mechanism for behavior selection in an artificial creature. In J.-A. Meyer & S.W. Wilson (Eds.). *From Animals to Animats. Proceedings of the First International Conference on Simulation of Adaptive Behavior* (pp. 238–246). Cambridge, MA: MIT Press.

Maes, P. (1992). Learning behaviour networks from experience. In F.J. Varela & P. Bourgine (Eds.). *Toward a Practice of Autonomous Systems. Proceedings of the First European Conference on Artificial Life* (pp. 48–57). Cambridge, MA: MIT Press.

Maes, P. (1994). Modeling adaptive autonomous agents. *Artificial Life*, $1(1/2)$, 135–162.

Mahadevan, S. & Connell, J. (1992). Automatic programming of behavior based robots using reinforcement learning. *Artificial Intelligence*, 55, 311–365.

Marjanovic, M., Scassellati, B. & Williamson, M. (1996). Self-taught visually-guided pointing for a humanoid robot. In P. Maes, M. Mataric, J. Pollack, J.-A. Meyer & S. Wilson (Eds.). *From Animals to Animats 4. Proceedings of the Fourth International Conference on Simulation of Adaptive Behavior* (pp. 35–44). Cambridge, MA: MIT Press.

Mondada, F., Franzi E. & Ienne, P. (1994). Mobile robot miniaturisation: A tool for investigation in control algorithms. In *Experimental Robotics III, Proceedings of the 3rd International Symposium on Experimental Robotics*, Kyoto, Japan, October 28–30, 1993 (pp. 510–513) London: Springer.

Newell, A. (1980). Physical symbol systems. *Cognitive Science*, 4, 135–183.

Riolo, R.L. (1991). Lookahead planning and latent learning in a classifier system. In J.-A. Meyer & S.W. Wilson, (Eds.). *From Animals to Animats. Proceedings of the First International Conference on Simulation of Adaptive Behavior* (pp. 316–326). Cambridge, MA: MIT Press.

Rutkowska, J.C. (1994). Scaling-up sensorimotor systems: constraints from human infancy. *Adaptive Behavior*, 2(4), 349–373.

Smithers, T. (1996). On what embodiment might have to do with cognition. In M. Mataric (Ed.). *Embodied Cognition and Action. Papers from the 1996 AAAI Fall Symposium*. Menlo Park, CA: AAAI Press, (Technical Report FS-95-05),

Todd, P. & Wilson, S. (1993). Environment structure and adaptive behavior from the ground up. In J.-A. Meyer, H.L. Roitblat & S.W. Wilson (Eds.). *From Animals to Animats 2. Proceedings of the Second International Conference on Simulation of Adaptive Behavior* (pp. 11–20). Cambridge, MA: MIT Press.

Wilson, S.W. (1985). Knowledge growth in an artificial animal. In J.J. Grefenstette (Ed.). *Proceedings of the First International Conference on Genetic Algorithms and their Applications* (pp 16–23). Hillsdale, NJ: Lawrence Erlbaum.

Wilson, S.W. (1996). Explore/exploit strategies in autonomy. In P. Maes, M. Mataric, J. Pollack, J.-A. Meyer & S. Wilson (Eds.). *From Animals to Animats 4: Proceedings of the Fourth International Conference on Simulation of Adaptive Behavior* (pp. 325–332). Cambridge, MA: MIT Press.

Chapter 19

Agre, P. (1997). *Computation and Human Experience*. Cambridge: Cambridge University Press.
Alterman, R. (in press). Everyday reasoning. *Mind, Culture, and Activity, 6*.
Bateson, G. (1972). *Steps to an Ecology of Mind*. New York: Ballantine Books.
Beach, K. (in press). Consequential transitions: A sociocultural expedition beyond transfer in education. In A. Iran-Nejad (Ed.). *Review of Research in Education, 24*.
Bloch, M. E. F. (1998). *How We Think They Think: Anthropological Approaches to Cognition, Memory, and Literacy*. Boulder: Westview Press.
Bourdieu, P. (1977). *Outline of a Theory of Practice*. Cambridge: Cambridge University Press.
Bronfenbrenner, U. (1979). *The Ecology of Human Development: Experiments by Nature and Design*. Cambridge, MA: Harvard University Press.
Bruner, J. (1996). *The Culture of Education*. Cambridge, MA: Harvard University Press.
Carraher, T. N., Carraher, D. W. & Schliemann, A. D. (1985). Mathematics in the streets and in schools. *British Journal of Developmental Psychology, 3*, 21–29.
Clancey, W. J. (1997). *Situated Cognition: On Human Knowledge and Computer Representations*. Cambridge: Cambridge University Press.
Cole, M. (1995). The supra-individual envelope of development: Activity and practice; situation and context. In J. Goodnow, P. Miller & F. Kessel (Eds.). *Cultural Practices as Contexts for Development*. San Francisco: Jossey-Bass.
Cole, M., Engeström, Y. & Vasquez, O. (Eds.) (1997). *Mind, Culture, and Activity: Seminal Papers from the Laboratory of Comparative Human Cognition*. Cambridge: Cambridge University Press.
D'Andrade, R. & Strauss, C. (Eds.) (1992). *Human Motives and Cultural Models*. Cambridge: Cambridge University Press.
Darrah, C. (1996). *Learning and Work: An Exploration in Industrial Ethnography*. New York: Garland.
Engeström, Y. (1987). *Learning by Expanding: An Activity-theoretical Approach to Developmental Research*. Helsinki: Orienta-Konsultit.
Engeström, Y. (1994). The working health center project: Materializing zones of proximal development in a network of organizational learning. In T. Kauppinen & M. Lahtonen (Eds.), *Action Research in Finland*. Helsinki: Ministry of Labour.
Engeström, Y. (1996). Development as breaking away and opening up: A challenge to Vygotsky and Piaget. *Swiss Journal of Psychology, 55*, 126–132.
Engeström, Y. (1999). Innovative learning in work teams: Analyzing cycles of knowledge creation in practice. In Y. Engeström, R. Miettinen & R.-L. Punamäki (Eds.). *Perspectives on Activity Theory*. Cambridge: Cambridge University Press.
Engeström, Y. & Cole, M. (1997). Situated cognition in search of an agenda. In D. Kirshner & J. A. Whitson (Eds.). *Situated Cognition: Social, Semiotic, and Psychological Perspectives*. Mahwah: Lawrence Erlbaum.
Engeström, Y., Engeström, R. & Kärkkäinen, M. (1995). Polycontextuality and boundary crossing in expert cognition: Learning and problem solving in complex work activities. *Learning and Instruction, 5*, 319–336.
Engeström, Y., Engeström, R. & Vähäaho, T. (1999). When the center does not hold: The importance of knotworking. In S. Chaiklin, M. Hedegaard & U. J. Jensen (Eds.). *Activity Theory and Social Practice: Cultural–historical Approaches*. Aarhus: Aarhus University Press.

Engeström, Y., Miettinen, R. & Punamäki, R.-L. (Eds.) (1999). *Perspectives on Activity Theory*. Cambridge: Cambridge University Press.

Garfinkel, H. (1967). *Studies in Ethnomethodology*. Englewood Cliffs: Prentice-Hall.

Giddens, A. (1979). *Central Problems in Social Theory: Action, Structure, and Contradiction in Social Analysis*. Berkeley: University of California Press.

Greeno, J. G. (1989). A perspective on thinking. *American Psychologist, 44*, 134–141.

Greeno, J. G. (1998). The situativity of knowing, learning, and research. *American Psychologist, 53*, 5–26.

Gutierrez, K. D., Baquedano-Lopez, P. & Turner, M. G. (1997). Putting language back into language arts: When the radical middle meets the third space. *Language Arts, 74*, 368–377.

Gutierrez, K. D., Larson, J. & Kreuter, B. (1995). Cultural tensions in the scripted classroom: The value of the subjugated perspective. *Urban Education, 29*, 410–442.

Gutierrez, K. D., Rymes, B. & Larson, J. (1995). Script, counter-script, and underlife in the classroom: James Brown versus Brown *vs.* Board of Education. *Harvard Educational Review, 54*, 445–471.

Henning, P. (1998). Ways of learning: An ethnographic study of the work and situated learning of a group of refrigeration service technicians. *Journal of Contemporary Ethnography, 27*, 85–136.

Holland, D., Lachicotte, W., Jr., Skinner, D. & Cain, C. (1998). *Identity and Agency in Cultural Worlds*. Cambridge, MA: Harvard University Press.

Keller, C. M. & Keller, J. D. (1996). *Cognition and Tool Use: The Blacksmith at Work*. Cambridge: Cambridge University Press.

Keller, C. M. & Keller, J. D. (in press). Imagery in cultural tradition and innovation. *Mind, Culture, and Activity, 6*.

Koschmann, T. (Ed.)(1996). *CSCL: Theory and Practice of an Emerging Paradigm*. Mahwah: Lawrence Erlbaum.

Laboratory of Comparative Human Cognition (1983). Culture and cognitive development. In W. Kessen (Ed.). *History, Theory, and Methods*. Vol. 1 of P. H. Mussen (Ed.). *Handbook of Child Psychology*. New York: Wiley.

Lakoff, G. & Johnson, M. (1999). *Philosophy in the Flesh: Embodied Mind and its Challenge to Western Thought*. New York: Basic Books.

Lave, J. (1988). *Cognition in Practice: Mind, Mathematics and Culture in Everyday Life*. Cambridge: Cambridge University Press.

Lave, J. & Wenger, E. (1991). *Situated Learning: Legitimate Peripheral Participation*. Cambridge: Cambridge University Press.

Leont'ev, A. N. (1978). *Activity, Consciousness, and Personality*. Englewood Cliffs: Prentice-Hall.

Leontyev (Leont'ev), A. N. (1981). *Problems of the Development of the Mind*. Moscow: Progress.

Pintrich, P. R., Marx, R. W. & Boyle, R. A. (1993). Beyond conceptual change: The role of motivational beliefs and classroom contextual factors in the process of conceptual change. *Review of Educational Research, 63*, 167–199.

Rogoff, B. & Lave, J. (Eds.)(1984). *Everyday Cognition: Its Development in Social Context*. Cambridge, MA: Harvard University Press.

Schatzki, T. R. (1996). *Social Practices: A Wittgensteinian Approach to Human Activity and the Social*. Cambridge: Cambridge University Press.

Scribner, S. (1984). Studying working intelligence. In B. Rogoff & J. Lave (Eds.). *Everyday Cognition: Its Development in Social Context*. Cambridge, MA: Harvard University Press.

de Solla Price, D. (1984). The science/technology relationship: The craft of experimental science and policy on the improvement of high technology innovation. *Research Policy, 13*, 3–20.

Strauss, A. L. (1993). *Continual Permutations of Action*. New York: Aldine de Gruyter.

Strauss, C. & Quinn, N. (1997). *A Cognitive Theory of Cultural Meaning*. Cambridge: Cambridge University Press.

Suchman, L. A. (1987). *Plans and Situated Actions: The Problem of Human–machine Communication*. Cambridge: Cambridge University Press.

Sutter, B. (in preparation). Without instruction there is no learning! Ronneby: Department of Human Work Science, University of Karlskrona-Ronneby (manuscript).

Vygotsky, L. S. (1978). *Mind in Society: The Development of Higher Psychological Processes*. Cambridge, MA: Harvard University Press.

Wenger, E. (1998). *Communities of practice: Learning, meaning, and identity*. Cambridge: Cambridge University Press.

Author Index

Note – Page numbers in *italic type* indicate where references are printed in full.

Abelson, R.P. 126, *274*
Abreu, G. de 18, 19, 20, 21, 25, 26, 28, 29, 52, *260, 261, 263*
Achtenhagen, F. 155, *276*
Ackerman, M.S. 110, *270*
Ackermann, E. 120, 144, 146, 147, 150, 151, 152, *272, 275*
Adams, V.M. 61, *266*
Adler, P.A. 114, *270*
Agre, P.E. 169, 229, 232, 254, 286, *290*
Ainsworth, S. 107, *270*
Akin, O. 157, *276*
Alegria, J. 124, *272*
Alterman, R. 255, *290*
Anderson, J.R. 102, 105, 203, *268, 284*
Anderson, R.E. 179, *278*
Arbinger, R. 171, 172, 176, *279, 280*
Artman, H. 174, *278*
Askew, M. 196, *284*
Astington, J. 147, *275*
Atwood, M. 157, *277*
Austin, G. 125, *272*
Austin, J.L. 122, *272*

Bachelard, G. 145, 147, *275*
Bagnara, S. 110, 111, *271*
Baker, M.J. 197, *284*
Bakhurst, D. 8, *259*
Bale, S. 177, 178, *281*
Ball, L. 157, *276*
Bamossy, G.J. 171, *279*
Bannert, M. 171, 172, 176, *279, 280*
Baquedano-Lopez, P. 255, *290*
Barbieri, M.S. 172, 174, *279*
Barnes, P. 174, 175, 178, *281*

Barnett, M. 8, *259*
Baskin, A.B. 197, *284*
Bates, E. 33, 34, *262*
Bates, H. 175, *280*
Bateson, G. 46, *262*, 257, *289*
Batra, D. 160, *276*
Bauer, M.I. 112, *270*
Bauer, P.J. 126, *272*
Beach, K. *289*
Beck, E. 172, *281*
Becker, H.S. 187, 192, *283*
Becker, J. 196, *284*
Bell, N. 48, 56, *265*
Benner, P. 160
Bereiter, C. 155, 156, 159, 160, 161, *276*
Berger, P. 142, *274*
Bergin, D.A. 171, *279*
Bergqvist, K. 56, *265*
Berlin, L.M. 112, *271*
Berliner, D. 160, *276*
Bernard-Peyron, V. 126, *272*
Bettelheim, B. 146, *275*
Beynon, J. 171, 172, *279*
Bideaud, J. 122, 125, *272*
Blaye, A. 120, 123, 124, 126, 127, 179, *272, 273, 280*
Bliss, J. 1, 132, 134, 165, 196, 210, 211, 213, *274, 284, 286*
Bloch, M.E.F. 256, *289*
Bloom, B.S. 198, *284*
Boohan, R. 211, *286*
Borghart, I. 61, 71, *267*
Boshuizen, H. 156, *276*
Bourdieu, P. 52, 165, 249, *263, 278, 289*
Bourne, E. 40, *263*

Bouzigue, S. 128, *272*
Boyes-Braem, P. 126, *273*
Boyle, R.A. 254, *291*
Brackenridge, A. 172, *280*
Brandes, A. 151, *275*
Bransford, J.D. 101, 103, *268*, *269*
Breakwell, G.M. 172, *279*
Brecht, B. 200, *284*
Breuker, J. 227, *286*
Breuker, J.A. 226, *288*
Brna, P. 197, *284*
Bromme, R. 155, 160, *276*, *278*
Bronfenbrenner, U. 256, *289*
Brooks, R.A. 228, 230, 231, 233, 242, 247, *286*, *288*
Brown, A. 134, *274*
Brown, J.S. 101, 152, 158, 174, *269*, *275*, *277*, *279*
Brun, J. 56, *265*
Bruner, J. S. 13, 33, 34, 80, 125, 168, 169, 186, *260*, *262*, 257, *267*, *272*, *283*, *290*
Burge, B. 3, 14, *259*, *260*
Burkhardt, H. 62, *265*
Burns, B.D. 104, *270*
Burton, M. 197, *284*
Butterworth, G. 160, *277*
Byrne, R. 113, 121, 124, *271*, *272*

Cai, J. 68, *266*
Cain, C. 255, *291*
Calder, J. 172, *282*
Cañamero, D. 223, 227, 229, 236, *287*
Cara, F. 113, 118, 124, *271*, *274*
Carey, S. 150, *275*
Carpenter, K.J. 219, *286*
Carpenter, T.P. 105, *269*
Carraher, D.W. 17, 18, 19, 26, 29, 249, *261*, *290*
Carraher, T.N. 249, *290*
Chaiklin, S. 3, 14, *259*, *260*
Chambres, P. 179, *280*
Chamorey, A. 128, *272*
Chan, T.W. 197, *284*
Chandrasekaran, B. 226, *287*

Chapman, D. 229, 232, *286*
Chen, H. 111, *271*
Chen, M. 171, 172, *279*
Cheng, P.W. 113, 123, *271*, *272*
Chi, M. 158, 160, *277*
Childs, C.P. 18, *261*
Christiaans, H. 157, *277*
Clancey, W.J. 194, 226, 232, 234, 254, *284*, *287*, *290*
Clark, M. 181, *279*
Clarke, V.A. 180, *280*
Clements, P. 157, *278*
Cliff, D. 242, *288*
Cobb, P. 77, *266*
Cohen, M.D. 110, *271*
Colbourn, C. 124, 174, *273*, *281*
Cole, M. 1, 4, 8, 33, 47, 141, 249, *259*, 251, *262*, *263*, *274*, *290*
Collins, A. 101, 152, 174, *269*, *275*, *279*
Conklin, J. 110, 112, *271*
Connel, J. 242, *289*
Coolican, H. 172, *279*
Cooper, J. 167, 168, 172, 180, 181, *278*, *280*, *281*, *282*
Cooper, M. 107, *269*
Corbett, A.T. 203, *284*
Corruble, V. 210, 218, 223, *286*
Coulson, R.L. 102, *270*
Cox, B. 126, *273*
Cronin, O. 132
Crook, C.K. 183, 189, 190, *283*
Cuban, L. 185, *283*
Culley, L. 171, 172, 173, 180, 181, *279*
Cummings, R. 174, *279*
Cunningham, G. 114, 115, *271*
Curtis, B. 158, *278*
Custer, E. 156, *276*

D'Andrade, R. 255, *290*
Dansereau, D.F. 107, *269*
Darrah, C. 250, *290*
Davies, S. 157, *277*
Davis, J. 160, *276*
de Abreu, G. 17
Deak, G.O. 126, *272*

de Boer, B. 236
De Corte, E. 61, 62, 63, 64, 65, 66, 68, 70, 71, 72, 75, 76, 77, 171, *266, 267, 282*
De Croock, M.B.M. 109, *270*
De Góes, M.C.R. 42, 43, *263*
De Jong, T. 104, *269*
De Loache, J. 148, *275*
Delscenserie, C. 128, *272*
Dennett, D. 230, *287*
de Solla Price, D. 256–7, *292*
Desprels-Fraysse, A. 127, *273*
Deutsch, A. 65, *266*
Dick, G. 172, *282*
Dillenbourg, P. 197, *284*
Doherty, M.E. 114, *271*
Donaldson, M. 33, 47, 49, *262, 263*
Donnett, J. 242, *288*
Dreier, O. *263*
Drescher, G.L. 240, *288*
Dreyfus, H. 160, *276*
Dreyfus, S. 160, *277*
Ducrot, O. 122, *272*
Duffy, T.M. 171, *279*
Duguid, P. 152, 158, 174, *275, 277, 279*
Dunn, J. 123, *272*
Duveen, G. 18, *261*

Eastman, S. 174, *279*
Ebach, J. 172, *279*
Eckert, A. 104, *269*
Edwards, D. 81, 186, *267, 283*
Elam, J. 158, *278*
Elbers, E. 56, *263*
Elman, J. 34, *262*
Engeström, R. 256, *291*
Engeström, Y. 249, 251, 256, 257, *260, 290, 291*
Eraut, M. 155, *277*
Ericsson, K. 160, *277*
Etelapelto, A. 155, 160, 161, 162, *277*
Evans, J.St.B.T. 113, 121, 122, 124, 127, 129, 130, *271, 272*
Evans, P. 195, *284*

Fairclough, N. 2, *259*
Farr, M. 158, 160, *277*
Feltovich, P.J. 102, *270*
Fennenma, E. 105, *269*
Fetler, M. 171, 172, *279*
Fife-Schaw, C.R. 172, *279*
Finlayson, H. 174, *279*
Fitzgerald, D. 172, *280*
Flahault, F. 49, 56, *263*
Flores, F. 94, *268*
Foner, L.N. 241, *288*
Foot, J. 175, *280*
Ford, M.E. 171, *279*
Forman, E.A. 179, *279*
Franzi, E. 242, *289*
Freudenthal, H. 62, 70, *266*
Frydman, O. 124, *272*
Fung, P. 172, *282*
Funken, C. 180, *280*
Fuson, K. 105, *269*
Fuson, K.C. 17, *261*

Ganascia, J.G. 218, 219, *286*
Garcia, R. 133, *274*
Gardner, H. 102, *269*
Garfinkel, H. 50, 249, *263, 291*
Garnham, A. 113, *271*
Gaussier, P. 242, *288*
Geer, B. 187, 192, *283*
Geertz, C. 40, *262*
Gegg-Harrison, T.S. 197, *284*
Gergen, K. 33, 42, *262*
Giddens, A. 5, 142, 165, 249, *259, 274, 278, 291*
Gillett, G. 14, 33, 81, *260, 262, 267*
Gilly, M. 123, 124, *272, 273*
Girotto, V. 113, 118, 123, 124, *271, 272, 274*
Glaser, R. 158, 160, 171, *277, 282*
Glodstein, P. 200, *284*
Goel, V. 156, *277*
Goldman, S.R. 101, 103, *268, 269*
Goodnow, J. 125, *272*
Goodnow, J.J. 17, 18, *261*
Goody, J. 6, *259*

Gordon, R. 147, *275*
Gorman, M.E. 114, 115, *271*
Graf, M. 104, *270*
Grannott, N. 150, *275*
Grant, A.C. 200, *284*
Gräsel, C. 104, *269*
Gravemeijer, K.P.E. 77, *266*
Gray, W.D. 126, *273*
Greenfield, P.M. 18, 160, *261*, *277*
Greeno, J.G. 249, 255, *291*
Greer, B. 61, 62, 65, 68, 70, 72, 73, 75, 76, *266*
Greer, J. 197, *285*
Griffiths, A.K. 200, *284*
Grint, K. 191, *283*
Grossen, M. 47, 48, 56, *263*, *265*
Gruber, H. 101, 104, 105, 106, 108, 109, *269*, *270*
Gruber, H.L. 159, 174, *277*, *280*
Guberman, S.R. 160, *277*
Guindon, R. 157, 158, *277*
Gutierrez, K.D. 255, *290*, *291*

Hackett, R. 103, *269*
Hagström, G. 81, *267*
Hague, D. 183, *283*
Hall, J. 180, *279*
Hamers, J.H.M. 67, *266*
Hannerz, U. 80, *267*
Hargreaves, D. 175, *280*
Harnad, S. 228, 236, *287*, *288*
Harré, R. 14, 33, 81, *260*, *262*, *267*
Harris, P. 123, 125, 147, *273*, *275*
Harvey, I. 242, *288*
Hatano, G. 150, 160, *275*, *277*
Hattie, J. 172, *280*
Healy, L. 173, *280*
Hebb, D.O. 244, *288*
Heber-Suffrin, C. 50, *264*
Heber-Suffrin, M. 50, *264*
Henning, P. 250, *291*
Hess, R.D. 171, *279*
Hiebert, J. 105, *269*
Hoc, J.-M. 157, *277*, *278*
Hofer, M. 104, *269*

Hoffmann, U. 171, *280*
Holland, D. 255, *291*
Holland, J.H. 238, *288*
Holyoak, K. 123, *272*
Holyoak, K.J. 104, 113, *271*
Hood, L. 33, *262*
Horvath, J. 159, *278*
Houdé, O. 122, *273*
Hoyles, C. 172, 173, *280*, *282*
Huber, G.L. 106, *269*
Huff, C. 167, 168, 180, *278*, *280*
Hughes, E.C. 187, 192, *283*
Hughes, M. 172, *280*
Huguet, P. 179, *280*
Human, P. 105, *269*
Hundeide, K. 47, 56, *264*
Husbands, P. 242, *288*
Hutchins, E. 80, 114, 119, *267*, *271*

Iannaccone, A. 48, *264*
Ienne, P. 242, *289*
Inagaki, K. 150, 160, *275*, *277*
Inglis, B. 4, 5, *259*
Inhelder, B. 122, 123, 125, *273*
Issing, L.J. 171, *280*
Issroff, K. 174, *280*

Jacobi, D. 52, *264*
Jacobson, M.J. 102, 106, *270*
Jäger, R.S. 176, *280*
Jakobsdottir, S. 179, *280*
Jakubowicz, C. 128, *273*
Jansen, P.G.W. 171, *279*
Jeffries, R. 112, 157, *271*, *277*
Jensen, F.V. 204, *284*
Johnson, D. 174, *280*
Johnson, D.M. 126, *273*
Johnson, M. 34, 44, 45, 149, 153, *262*, 256, *275*, *291*
Johnson, M.H. 34, *262*
Johnson, R. 174, *280*
Johnson, T.R. 226, *287*
Johnson-Laird, P.N. 112, 113, 121, *270*, *271*, *273*

Joiner, R. 119, 174, 175, 178, *271*, *281*
Jonassen, D.H. 171, *279*
Jones, A. 172, *282*
Jones, T. 180, *280*
Jordan, J. 147, *275*
Jurdant, B. 50, 52, *264*

Kaelbling, L.P. 170, 237, *288*
Kaplan, A. 147, *275*
Kärkkäinen, M. 256, *290*
Karmiloff-Smith, A. 34, 147, 153, *262*, *275*
Kay, R. 171, 172, *281*
Kaye, A. 191, *283*
Kegan, R. 147, 153, *275*
Keller, C. 14, 81, *260*, *267*
Keller, C.M. 256, *291*
Keller, E.F. 147, *275*
Keller, J.D. 14, 81, *260*, 256, *267*, *291*
Kenway, J. 181, *283*
Kerbrat-Orecchioni, C. 49, 56, *264*
Kilpatrick, J. 70, *266*
King, J. 158, *277*
Kirkup, G. 173, *281*
Klimsa, P. 171, *280*
Koschmann, T. 253, *291*
Kreienbaum, M.A. 180, *281*
Krendl, K. 174, *279*
Kreuter, B. 255, *290*
Krey, C.L. 179, *280*
Kruger, A.C. 38, *263*
Kwon, Y. 17, *261*

Laboratory of Comparative Human Cognition 249, *291*
Laborde, C. 94, *267*
Lachicotte, W. Jr. 255, *291*
Lage, E. 171, *281*
Lakatos, I. 233, *287*
Lakoff, G. 44, 45, 149, 153, *262*, 256, *275*, *291*
Lambrinos, D. 242, *289*
Lamon, M. 103, *269*
Langacker, R. *262*

Larson, J. 255, *290*, *291*
Lasure, S. 61, 63, 64, 65, 66, 68, 71, *266*, *267*
Latta, R.M. 114, 115, *271*
Laurel, B. 145, *275*
Laurillard, D. 197, 198, *284*
Lautrey, J. 122, 125, *272*
Lave, J. 3, 4, 14, 18, 19, 25, 37, 47, 48, 80, 114, 158, 159, 249, *259*, 250, *260*, 251, *261*, 252, *262*, *264*, 256, *267*, *271*, *277*, *291*
Lave, L. 152, *276*
Law, L.C. 159, 174, *277*, *279*, *281*
Lawson, B. 157, *277*
Lecacheur, M. 127, *273*
Lee, T. 172, *279*
Legrenzi, M. 121, *273*
Legrenzi, P. 121, *273*
Leont'ev, A.N. 249, 255, *291*
Leslie, A.M. 146, *275*
Levine, J. 114, *271*
Levine, R.A. 47, *264*
Levitt, B. 110, *271*
Liengme Bessire, M.-J. 48, 56, *263*
Light, P. 47, 119, 120, 123, 124, 155, 160, 172, 174, 175, 177, 178, 183, 190, *264*, *271*, *272*, *273*, *277*, *279*, *280*, *281*, *283*
Light, V. 190, *283*
Lind, J. 219, *286*
Linell, P. 83, *267*
Linn, M.C. 171, 174, *281*
Linnakylä, P. 172, *281*
Lissmann, U. 176, *280*
Littleton, K. 119, 171, 174, 175, 177, 178, *271*, *280*, *281*
Littman, M.L. 237, 246, *289*
Lock, A. 33, *262*
Lockheed, M.E. 172, *281*
Lomagistro, P. 111, *271*
Longeot, P. 133, *274*
Luckin, R. 168, 194, 197, 207, 208, *281*, *285*
Luckman, T. 142, *274*
Lumpe, A.T. 200, *285*

Lund, K. 197, *284*
Lyytinen, K. 157, *277*

McCalla, G. 197, *285*
McDermott, R. 33, *262*
McGarrigle, J. 33, *262*
Machida, S. 126, *273*
Mackay, H. 171, *279*
Mackie, D. 172, *282*
McLeod, D.B. 61, *266*
Macleod, H. 172, *280*
Macrae, S. 196, *284*
Maes, P. 227, 229, 230, 239, 240, 247, *287, 289*
Mahadevan, S. 242, *289*
Malone, T.W. 110, *270*
Maltzahn, C. 110, *271*
Mandl, H. 101, 104, 105, 106, 108, 109, 159, 171, 174, *268, 269, 270, 277, 280, 282*
Mandler, J.M. 126, *273*
Manktelow, K.I. 129, 130, *272*
March, J.G. 110, *271*
Marcus, N. 107, *269*
Marjanovic, M. 242, *289*
Markman, E.M. 122, 125, 126, *273*
Marti, E. 128, *273*
Marti, P. 110, 111, *271*
Martin, F. 150, *275*
Martin, R. 172, *281*
Martinsson, M. 89, *268*
Marx, R.W. 254, *291*
Masciotra, D. 147, *276*
Mataric, M.J. 229, *287*
Maturana, H.R. 228, *287*
Mayer, R. 162, *277*
Mayes, T. 190, *283*
Mead, G.H. 47, *264*
Mehan, H. *264*
Mellar, H. 211, *286*
Mercer, N. 14, 186, 195, *260, 283, 285*
Mervis, C.B. 126, *273*
Messer, D. 119, 174, 175, 178, *271, 281*
Metz-Goeckel, S. 180, *281*
Middleton, D.J. 198, *285*

Miettinen, R. 249, *291*
Miller, J.B. 147, *275*
Miller, R. 145, *276*
Minick, N. 179, *279*
Minsky, M. 230, *287*
Moeschler, J. 122, *273*
Mondada, F. 242, *289*
Monteil, J.M. 179, *281*
Moore, A.W. 237, *288*
Moran, T. 158, *277*
Moscovici, S. 52, *264*
Muller, N. 47, 50, *264*
Murphey, T. 207, *285*
Murray, H. 105, *269*
Musen, M.A. 226, *287*
Mynatt, C.R. 114, *271*

Nardi, B. 158, 159, *278*
Neilson, I. 190, *283*
Nelson, K. 126, *273*
Newell, A. 224, 225, 233, 236, *287, 289*
Newman, S. 158, *278*
Newman, S.E. 101, *269*
Newstead 121, 124, *272*
Newton, P. 172, *281*
Nicolet, M. 48, 58, *264*
Niegemann, H.M. 104, *269*
Nielsen, A. 172, *281*
Ninio, A. 121, 128, *273*
Njoo, M.K.H. 104, *269*
Norman, D. 160, *278*
Norman, D.A. 112, 114, 196, *272, 285*
Ntamakiliro, L. 60, *265*
Nunes, T. 17, 19, 26, 29, *261*
Nunez, M. 123, 125, *273*

Oakhill, J. 113, *271*
O'Day, V.L. 112, *271*
O'Donnell, A.M. 107, *269*
Ogborn, J. 2, 132, 134, 145, 211, *259, 274, 276, 286*
Ogletree, S.M. 171, *282*
Oldenburg, R. 192, *283*
Olivier, A. 105, *269*
Olson, D. 6, 128, 147, *259, 273, 275*

O'Malley, C. 107, *270*
Ong, W.J. 6, *259*
Ormerod, T. 157, *276*
O'Shea, T. 172, *282*
Over, D.E. 129, 130, *272*
Overton, W.F. 123, *274*
Owen, D. 110, *271*

Paepcke, A. 112, *271*
Palthepu, S. 197, *285*
Papert, S. 149, 173, *276*, *282*
Parisi, D. 34, *262*
Parnas, D. 157, *278*
Paskiewitz, T. 196, 197, 202, *285*
Patterson, E.J. 203, *284*
Pelgrum, W.J. 171, *282*
Perkins, D. 163, *278*
Perner, J. 147, *276*
Perret-Clermont, A.-N. 47, 48, 56, 58, *263*, *264*, *265*
Petrosino, A.J. 103, *269*
Pettersson, A. 81, 82, 95, *268*
Pfeifer, R. 242, *288*
Piaget, J. 32, 122, 123, 125, 133, 142, 145, 146, *262*, *273*, *274*, *276*
Pierroutsakos, S.L. 148, *275*
Pino, A. 42, 43, *263*
Pintrich, P.R. 254, *291*
Pirolli, P. 156, *277*
Plomp, T. 171, *282*
Plunkett, K. 34, *262*
Podmore, V. 171, *282*
Politzer, G. 122, 125, *273*
Polson, P. 157, *277*
Pontecorvo, C. 3, 14, *259*, *260*
Posner, J. 17, *262*
Prem, E. 234, *287*
Punamäki, R.-L. 249, *291*

Quéau, P. 144, 146, *276*
Quinn, N. 255, *292*

Rademakers, P. 227, *287*
Radziszewska, B. 159, *278*
Rambow, R. 160, *278*

Ramsden, P. 186, *283*
Ratner, H.H. 38, *263*
Reader, W. 132
Reder, L.M. 102, 105, *268*
Reichgelt, H. 196, 197, 202, *285*
Reisman, J. 168, 181, *282*
Reiter, W. 104, *269*
Reitman, J.S. 238, *288*
Renkl, A. 101, 104, 105, 106, 107, 108, 109, 159, 174, *269*, *270*, *277*, *280*
Resnick, L.B. 3, 14, 47, 101, 114, *259*, *260*, *264*, *270*, *271*
Resnick, M. 149, 150, *276*
Rétornaz, A. 56, *265*
Reusser, K. 62, 65, 68, 69, 70, *266*
Reuterberg, S-E. 81, 82, 95, *268*
Rinn, U. 104, *269*
Riolo, R.L. 110, 239, *289*
Rizzo, A. 111, *271*
Robertson, I. 172, *282*
Robinson-Stavely, K. 181, *282*
Rogoff, B. 14, 17, 18, 47, 141, 152, 159, 187, 249, *260*, *261*, *265*, *274*, *276*, *278*, *283*, *291*
Rommetveit, R. 33, 49, *262*, *265*
Rosch, E. 201, *285*
Rosch, E.H. 126, *273*
Ross 168, 169, *282*
Ruijssenaars, A.J.J.M. 67, *266*
Rumain, B. 128, *273*
Rutkowska, J.C. 247, *289*
Rymes, B. 255, *291*

Saada, E.H. 56, *265*
Saariluoma, P. 160, *278*
Said, E. 36, *262*
Sales, G.C. 179, *280*
Säljö, R. 1, 3, 9, 10, 13, 14, 17, 29, 47, 48, 56, 60, 66, 70, 80, 81, 84, 174, 185, *259*, *260*, *261*, *265*, *266*, *267*, *268*, *282*, *283*
Salomon, G. 163, *278*
Sampson, E. 40, *263*
Sani, N. 81, *267*
Sarvimaki, A. 156, *278*

Saxe, G.B. 17, 19, 27, *262*
Sayeki, Y. 147, *276*
Scardamalia, M. 155, 156, 159, 160, 161, *276*
Scassellati, B. 242, *289*
Schank, R.C. 126, *274*
Schatz, B. 111, *271*
Schatzki, T.R. 249, *291*
Scheier, C. 242, *289*
Schiersmann, C. 172, *282*
Schliemann, A. 17, 19, 26, 29, *261*
Schliemann, A.D. 249, *290*
Schoenfeld, A.H. 61, 62, *266*
Schofield, J.W. 171, *282*
Schreiber, A.Th. 226, *288*
Schubauer-Leoni, M.-L. 47, 48, 49, 56, 60, *264*, *265*
Schweder, R. 40, *263*
Scribner, S. 17, 19, 29, 47, 141, 249, *262*, *263*, *274*, *291*
Searle, J.R. 122, *274*
Secules, T. 103, *269*
Self, J. 197, *285*
Selter, C. 62, *266*
Sewell, D.F. 171, *282*
Shadbolt, N. 196, 197, 202, *285*
Shapiro, L.J. 65, *266*
Shotter, J. 33, *263*
Siegal, M. 121, 122, 125, *274*
Sijtsma, K. 67, *266*
Silver, E.A. 65, 68, *265*, *266*
Simon, H.A. 102, 105, 156, 157, 224, 232, 233, *268*, *278*, *288*
Sinclair, H. 144, 146, *276*
Sinha, C. 32, 34, 35, 43, *262*, *263*
Skinner, D. 255, *291*
Slater, A. 110, *271*
Slezak, P. 232, *288*
Smith, H. 156, *276*
Smith, J. 160, *277*
Smithers, T. 228, 234, 241, 242, *287*, *288*, *289*
Smolka, A.L.B. 42, 43, *263*
Sneider, D.K. 197, *284*
Snow, C.E. 121, 128, *273*

Sonnentag, S. 161, *278*
Spencer, J. 172, *279*
Sperber, D. 113, 118, 122, 124, *271*, *274*
Spiro, R.J. 102, 106, *270*
Spohrer, J.C. 196, *285*
Stamback, M. 144, 146, *276*
Stanne, M. 174, *280*
Stark, R. 104, 106, *270*
Staver, J.R. 200, *285*
Stebler, R. 65, 68, 69, *266*
Steels, L. 226, 230, *288*
Stein, L.A. 231, 242, 247, *286*, *288*
Sternberg, R.J. 159, 160, 174, *278*, *282*
Steward, J. 150, *276*
Stiver, I. 147, *275*
Stone, C.A. 179, *279*
Stone, M.K. 172, *281*
Straker, A. 171, *282*
Strauss, A.L. 187, 192, 249, *283*, *292*
Strauss, C. 255, *290*, *292*
Suchman, L.A. 152, 249, *276*, *292*
Surrey, J. 147, *275*
Sutherland, R. 173, *280*, *282*
Sutter, B. *291*
Sutton, R.E. 171, *282*
Sweller, J. 107, 109, *269*, *270*

Tajfel, H. 25, *262*
Tanzini, L. 111, *271*
Taylor, C. 40, *263*
Teasley, S. 114, *271*
Teasley, S.D. 47, *264*
Tillema, H. 155, *277*
Todd, P. 246, *289*
Todman, J. 172, *282*
Tomasello, M. 38, *263*
Tompsett, C. 211, *286*
Traum, D.R. 197, *284*
Treffers, A. 62, *266*
Trognon, A. 56, *265*
Tsatsarelis, H. 132
Turkle, S. 146, 150, 173, *276*, *282*
Turner, A. 157, *277*
Turner, M.G. 255, *290*

Tweney, R.D. 114, *271*
Twidale, M. 111, *271*

Ungson, G.R. 110, *272*
Utgoff, P. 217, *286*
Uttal, D.H. 148, *275*

Vähäaho, T. 256, *290*
Valsiner, J. 49, *265*
Van de Velde, W. 227, *286, 288*
van de Viel, M. 156, *276*
VanLehn, K. 109, *270*
Van Merrienboer, J.J.G. 109, *270*
Vanwelkenhuysen, J. 227, *288*
Varela, F.J. 228, *286*
Varelas, M. 196, *284*
Vasquez, O. 249, *290*
Veneziano, V. 110
Vera, A.H. 232, *288*
Verschaffel, L. 61, 62, 63, 64, 65, 66, 68, 70, 71, 72, 75, 76, 77, *266, 267*
Visser, W. 157, *278*
Vollmar, D. 110, *271*
Vollmeyer, R. 104, *270*
Voloshinov, V.N. 80, *268*
Vosniadou, S. 171, *282*
Vye, N.J. 101, *268*
Vygotsky, L.S. 1, 2, 43, 47, 67, 96, 114, 125, 194, 195, 196, 202, 249, *259, 263,* 254, *265, 267, 268, 271, 274, 285, 291*

Waddington, C.H. 44, *263*
Wagner, R.K. 160, 174, *278, 282*
Walkerdine, V. 32, *263*
Wallace, W.L. 187, *283*
Walsh, J.P. 110, *272*
Walz, D. 158, *278*
Ward, S.L. 123, *274*
Wason, P.C. 114, 118, 121, 122, *272, 274*

Wearne, D. 105, *269*
Webb, N. 174, *282*
Webb, N.M. 107, *270*
Weinert, F.E. 102, *270*
Wenger, E. 25, 37, 158, 159, 249, 250, 251, *261,* 252, *262,* 256, *277, 291*
Wertsch, J. 14, 33, *260, 263*
Wertsch, J.V. 47, 81, *265, 268*
Wharton, C. 112, *271*
White, S.H. 4, 8, *259*
Whitehead, A.N. 101, *270*
Wielinga, B.J. 226, *288*
Wilder, G. 172, *282*
Williams, S.W. 171, *282*
Williamson, M. 242, *289*
Willis, S. 181, *283*
Wilson, D. 113, 122, *271, 274*
Wilson, S. 246, *289*
Wilson, S.W. 237, 238, *289*
Winnicott, D.W. 144, *276*
Winograd, T. 94, *268*
Winograd, T.A. 114, *270*
Wistedt, I. 84, 89, *268*
Wittrock, M. 162, *277*
Wong, K.M. 174, *281*
Wood 168, 169, *282*
Wood, D.J. 107, 196, 197, 198, 202, *270, 285*
Wood, H.A. 107, 196, 197, 198, 202, *270, 285*
Wood, T. 77, *266*
Wosnitza, M. 177, *283*
Wyndhamn, J. 17, 48, 60, 66, 80, 81, 84, *261, 265, 267, 268*

Yackel, E. 77, *266*
Yoshida, H. 61, 68, *267*

Zhang, J. 112, 114, *272*
Zlatev, J. 44, *263*
Zrehen, S.A. 242, *288*

Subject Index

achievement test 176–7
activity
  conceptual vocabulary 187
  notion 107
actuators 242–3
adaptation
  environment 236
  situatedness in learning as 245–7
adaptive agents 236
agenda
  attempts at pursuing 250
  for 1990s 249–50
  for new millenium 254–7
agricultural revolution 4
agro-chemical corporations 5
alphabet 6
ambiguity tolerance 106
analysis of variance (ANOVA) 117
anchored instruction 102
animats 238–9
  basic problem 238
apprenticeship 18, 37–9, 101–2
artefacts 44–5
  intelligent 150–1
  physical 2, 6, 8, 14
artificial intelligence (AI) 169–70
  and situation cognition 223–35
  behaviour-based 223, 227–34
  emergence 230
  engineering complete creatures 230–2
  intelligence activities 229–30
  physical embodiment 229
  situated movement 223
  situatedness 228–9
  symbolic 223, 228, 230, 232–3
  traditional 228, 232, 254

  world views in debate 232
artificial neural networks 34
artisanship 42
asynchronous collaboration 190
autistic learning 170
autonomous agents 227–8, 234
  definition 236
  environment 237
  learning behaviour 236–7
  operationalization of psychological theories 247
  situated learning 236–48
autonomous learning 241
autonomous robots 241–5

Bayes' theory 221
Bayesian Belief Network (BBN) 204
behaviour
  networks 239–40
  selection 239
behaviourism 246
bi-perspectism 34–5
book-indexing program 157
books 7
bounded rationality principle 169
braça 8, 20, 26–7

Cabri Geometre software package 176
calendar 8, 16, 89
case-based learning 104
categorisation 125–7
change and learning 256
CHARADE 169, 217–19, 221
child-centred discourses 39, 40
child-centred education 39
child-centred learning 39
child–computer collaboration 194–209

303

children
  and intelligent artefacts 150–1
  and miniature models 148
  and simple machines 151–2
  choices of best and worst pupil in school mathematics and unschooled people 29, 30
  cognition 120–31
  deductive reasoning in 123–5
  participation in school and farming practices 26–7
  reasoning 16
  use of mathematics in specific situations 28
  valorisation of school and farming mathematical practices 27–9
classifiers 238
Cog 231, 247
cognition
  children's 120–31
  social dimension of 47
Cognition and Technology Group 102, 103, 105, 108
cognitive activity 114
cognitive apprenticeship 101–2
cognitive development 152
cognitive discourses 40
cognitive growth 147
cognitive neurobiology 38
cognitive neuroscience 34
cognitive-optimist attitude 115
cognitive processes 47, 114–15
cognitive revolution 13
cognitive science 33, 163
  developmentalizing 34
cognitive science framework, design activity in 156–8
cognitivism 13
collaborative learning 188–91
communication
  as practical action 2–3
  computer-mediated (CMC) 190, 191

communication technologies *see* information and communication technologies (ICT)
communication technologies, development 4–6
community structure 186
complex learning 103–5, 108–9
complex problems 102
  engagement-inducing potential 105–6
computational aids 210–22
  case studies 211–21
computer-based activities 172
computer-based instruction 104
computer-based learning and programming, gender differences 174
computer-mediated communication (CMC) 190, 191
computer-mediated interaction 188–9
computer programs 210
computer rooms 188
computer-simulated micro-world 114
computer software 195
  design 164
  development activities 161
  implementations 237–41
computerised learning environment (CLE) design 197
computers 8, 9, 13, 45, 150, 167
  gender differences 171
  in teaching and learning system 188
  learning with 10
computing resources 188
conta 20, 26–7
context-free rules 160
context-independent models of design expertise 100
contexts 59, 125–7, 195
contextual approaches to learning and expertise 158–60
contextual constraints 158
contextual determinants, empirical evidence 174–9

contextual factors   162
contextual knowledge   160
  acquisition   162
  in design expertise   155–64
  integration with and strategic knowledge   161–3
continuous variables   83
conversational conventions   125–7
conversational reality   33
creative imagination   147
creativity   39, 40, 42
cubo   20, 26–7
cultural component   17
cultural–historical psychology   43
cultural practices   3
culture of student learning and information technology (IT)   183–93
customer constraints   157

databases   98
  RAI   111–13, 119
  searching   113–15
deductive reasoning in children   123–5
Department of Education   195
design activity   155
  in cognitive science framework   156–8
design expertise
  as integration of contextual and strategic knowledge   161–3
  context-independent models   100
  contextual knowledge in   155–64
  nature and development   163
design problems   157
design requirements   157
desk-top publishing   8
developmental theories   152
diagnostic competencies, acquisition   104
disclaimers   129
discursive classifications   88–95
discursive construction   33
distance education   191
double negatives   127

dynamic modelling   211–17
  and enactive representations   145–6

Economic and Social Research Council (ESRC)   135
economic revolutions   4
educational technology, gender differences in   171–2
electronic bulletin boards   190
Eleusis *see* New Eleusis
email   190
empirical abstraction   132–43
  critique   133
  research project   135–41
  use of term   133–5
empirical research   18
enactive representations   144–54
  and dynamic modelling   145–6
engagement-inducing potential of complex problems   105–6
environment
  adaptation   236
  complexity   246
  learning *see* learning environment
  social   177–9
épreuves operatoires   121
expert knowledge   155–6
expert–novice comparisons   160–1
expertise
  contextual and situational approaches   158–60
  defining   161
  development stage model   160
  social nature   159

face-to-face exchanges   192
face-to-face familiarity amongst students   191
face-to-face tutorial meetings   190
farmers   18–31
  cognition mediation by specific tools   20–1
  limitations of traditional mathematical tools   21–2
  measures used   19–20

farmers (*cont.*)
  perceived power of indigenous mathematics 22–4
  specific mathematics tools 19
farming 14–15
  children's participation 26–7
  children's valorisation of mathematical practices 27–9
  use of mathematics in specific situations 28
food production 5
food technologies 5
formal knowledge 155, 156, 162

gender differences
  and IT 171–82
    implications and future research 179–82
    in computer-based learning and programming 174
    in educational technology 171–2
  gender-related responses to technology 173–4
  gender-stereotyped software 180
general law of cultural development 194
genetic algorithm 239
genetically engineered food 5
German exchange 54–5

higher education, actual and virtual community 191–3
Honeybears 174–5, 177
human intelligence, symbolic nature of 153
human memory 7
hypothesis testing 113–15

images 174–5, 179
imaginative denial of rules task (topsy turvy task) 135–40
imaginative transformations tasks 135
individual and situational factors, integration 160–1

individually orientated learning culture 41
inductive reasoning 217–18
industrial revolution 4
information and communication technologies (ICT) 98, 166
  development 4–6
information management 112
information processing 13, 224–5
information systems development, strategic knowledge of 162
information technology (IT) 7, 98, 166, 195
  and culture of student learning 183–93
  and gender 171–82
    implications and future research 179–82
  psychological perspectives 185–6
  radical potentialities for higher education 192
  resources 188
  tools 167
innovation and learning 256
institutional culture 188, 192
instructional settings 101–9
  problems of situated learning in 103–7
instructional support 103–5
intellectual technologies 6
intelligent agents 227
intelligent artefacts 150–1
intelligent systems 225, 230
intelligent tutoring systems (ITS) 197
IQON 169, 211–12, 216

KEKADA system 232
King and Crown 174–5
knowledge 3, 195–6
  acquisition 112, 169, 224
  definition 58–9
  forms 18, 24
  object of 58–9
  practical 155, 156
  professional 155–6

transmission to new generations 23
valorisation 15, 25
see also contextual knowledge
knowledge-based system (KBS) 224
  engineering the knowledge
    level 226–7
knowledge base, VIS 199–202
knowledge exchange networks
  (KENs) 15
  analysis of interactions 53–4
  didactical interactions 60
  general background 50–3
  German exchange 54–5
  place assigned to object of
    knowledge 58–9
  representation of roles and
    negotiation of positions 56–7
  representation of situation and its
    objectives 54–6
  Strasbourg *see* Strasbourg
  writing exchange 55–6
knowledge level (KL)
  engineering 226–7
  hypothesis 224–6, 234
knowledge representation
  horizontal dimension 200–2
  vertical dimension 199
knowledge structures 160, 161
knowledge transmission, negotiating
  identities and meanings 47–60

Laboratory of Comparative Human
  Cognition (LCHC) 141
language 1, 2, 6
  acquisition 34
    Zapotec 36–7
language game 33
lateral transitions in learning 256
learners, situated perspective 32–46
learning 1, 3, 13–16
  as construction of social
    identities 25–9
  as internalisation of socio-cultural
    tools 19–24
  as mediated activity 187

  by explaining 107
  by listening 107
  by observation and emulation 41
  case-based 104
  child-centred 39
  collaborative 188–91
  complex 103–5, 108–9
  computer-based 10, 174
  contextual and situational
    approaches 158–60
  ethnography 35
  implications 15
  in context 97–100
  integration with innovation and
    change 256
  lateral transitions 256
  mathematics 17–31
  multiple instrumentalities in 256–7
  open 103, 104, 109
  organisational 110
  project-based experiences 162
  reasons for 255
  relationship with teaching 255
  romancing the subject 35–41
  situated perspective 32–46
  social psychology of 56
  sociocultural perspective 4
  theories 3, 33–4
  with computers *see* computers
  with machines *see* machines
learning environment 76–8, 102, 103,
  195
  VIS 198–9
learning results as function of
  complexity and guidance 105
learning self 41, 42
learning styles
  characterisation 185
  concept 186
legal protection of public persons
  against invasion of privacy 41
Lego/Logo creatures 150
logical inference in social context 99
logical reasoning 9, 130
logico-mathematical activity 132–3

Logo   173
Logo microworlds   150

machines   99–100, 145, 151–2
  children's fascination with   149
  identity   154
  in child's eyes   149–50
  learning with and by   165–70
macro-social structures   31
mass media   7
mathematical modelling, real-world knowledge in   70–5
mathematical problem-solving in upper primary school children   61–79
mathematical reasoning   15
  experimenting with pupils' positioning   84–95
mathematical tools, new   31
mathematics   9
  children's valorisation of mathematical practices   27–9
  choices of best and worst pupil in school mathematics and unschooled people   29, 30
  farmers' specific tools   19
  learning   17–31
  learning and teaching, cross-cultural research   68–70
  use in specific situations   28
meaning
  construction   13, 59
  creation   9, 14
  negotiation   13–16
  sciences of   34–5
metacognitive knowledge   155–6
metaphors   45, 174–5, 179
meta-problem   112
metonymic processes   45
micro-culture of praxis   186
micro-frame of teaching–learning by demonstration and emulation   41
microworlds   149
miniature models   148
Minnesota Computer Literacy Assessment   179

models   149
  context-independent, of design expertise   100
  dynamic   145–6, 211–17
  expertise development stage   160
  instructional   101–3, 107
  miniature   148
  situated learning   101–3, 107
  traffic   216
  VIS student   203
movable type   7
multi-cultural societies   18
multiple instrumentalities in learning   256–7
multiple meanings   45

National Curriculum   195
naturalistic stance   35
Navlab System   232
negations, form and function   127–9
negatives, understanding   128
New Eleusis   115–18
  experimental paradigm   115–18
new millenium
  agenda   254–7
  situated learning   249–81
non-linear systems   246
non-realistic reaction (NR)   64–8, 71, 74, 75

object of knowledge   58–9
on-line access   190
open learning   103, 104, 109
opportunistic decomposition   157
organisational learning   110
organisational memory   110–19
  design   112
over-confidence   105–6
overtraining   244

P-items   63–74, 76, 77
packages   140
  combinations   140
PASTEUR   169, 218
PC-based program   188

perspectival complementarity 35
Phoenix Project 232
physical artefacts 2, 6, 8, 14
physical symbol system (PSS) 224–6, 233, 234
physical tools 2
Piaget's Class Inclusion Task 122
Pirates problem-solving software 175
post-modernism 33
post-structuralism 33
practical knowledge 155, 156
pragmatic contextualization 130
pragmatic schema 124
pretense 146–7
printing 6–8
problem-solving 24, 104, 107, 109, 156, 157, 175, 224–7
professional expertise, re-conceptualisation 163
professional knowledge, components 155–6
project-based learning experiences 162
psychological development 99
psychological tools 2
psychology 13–14
publishing 7–8

quantifying time as discursive practice 80–96

radical situationism 159
RAI database 111–13, 119
random access instruction 102
reading 7
real-world knowledge
  in mathematical modelling 70–5
  in upper primary school children 61–79
realistic modelling, teaching to fifth graders 76–8
realistic reaction (RR) 64–9, 71, 73–5, 77, 78
reasoning 1
  and mastery of discursive classifications 88–95
  as practice 82–4
  children 16
  deductive 123–5
  forms 9
  in context 97–100
  inductive 217–18
  interactions of topic and type of 216–17
  logical 9, 130
  mathematical see mathematical reasoning
  social institutions as contexts for 3
  sociocultural perspective 4
  technologies as tools for 8–10
reflective abstraction 132–3
relevance
  judgements 123–5
  management 124
  of relevance 120–31
  theory 113, 122, 124
representations 149
robotic agents 170, 241–5
rules of conduct 188

scaffolding 66–8, 103, 168, 236
scale models 148
schema mechanism 240–1
schooling 3, 16
  children's participation 26–7
  children's valorisation of mathematical practices 27–9
  choices of best and worst pupil in school mathematics and unschooled people 29, 30
  farmers attitudes to 23–4
scientific knowledge, social representations attached to 53
scientistic reductionism 34
scurvy
  analysis of research results 221
  experiments on discovery of causes 218–21
  history of research 219
  induction from historical data 219–21

selection task
  deontic versions   123–4
  simplified (reduced array) version   123
self, Western idea   40
self-creation   41–4
self-regulatory knowledge   156
semiosis   44
semiotic activities   153
semiotic mediator   43
sensors   242–4
sentence-verification task   128
signs
  and tools   43–4
  as intellectual and practical resources   80–1
simulâcres   146
simulations   146, 149, 156
situated cognition   17, 141–2
  and artificial intelligence (AI)   223–35
situated knowledge   153
situated learning   17–46, 101–9, 158, 163, 194–7
  autonomous agents   236–48
  environment design   108
  in instructional settings, problems of   103–7
  instructional models   101–3, 107
  new millenium   249–81
  notion of   250
  theories   43
situatedness
  and knowledge level   224
  in learning as adaptation   245–7
situational and individual factors, integration   160–1
situational approaches to learning and expertise   158–60
skywriting   190–1
SMALL TALK   211
social action, technologies as tools for   8–10
social component   18
social constructionism   33, 42

social context, logical inference in   99
social dimension of cognition   47
social environment, working with others   177–9
social institutions as contexts for reasoning   3
social practices and technology   6–8
social psychology of learning   56
social relationships   18, 40
social representations attached to scientific knowledge   53
socio-cultural approach   17
socio-cultural perspective on learning and reasoning   4
socio-cultural tools, learning as internalisation of   19–24
socio-naturalistic approach to human development   35
software *see* computer software
stimulus–response (S–R) associations   238–9, 245
Strasbourg
  knowledge exchange networks (KENs)   48–9
  as structuring as well as constructed context   50–9
  basic principles   51
  general background   51–3
  methods of data collection   50
  reciprocity   53
  scientific exchanges   52–3
strategic knowledge
  acquisition   162
  integration with contextual knowledge   161–3
  of information systems development   162
STRIPS   239
student learning, researching in context   187–91
sugar-cane farmers *see* farmers
sugar-cane farming *see* farming
surplus meaning   44
symbol systems   149

symbolic nature of human
    intelligence   153
symbolic play   146–7
synthetic agents   170
system analysis   164
system design, technology-centred
    approach to   114

task
  assignment   157
  contextualisation   175–6
  development   135
  presentation   175
task-management skills   38
teacher role in VIS   202–4
teaching   196–7
  demonstration and emulation
    style   38
  relationship with learning   255
teaching–learning
  by demonstration and emulation,
    micro-frame of   41
  macro-context   39
  micro-context   39
teaching–learning units (TLU)   76–7
technical error (TE)   71
technological devices   2
technological revolution   6
technology
  and social practices   6–8
  as tool for reasoning and social
    action   8–10
  gender-related responses to   173–4
  in system design   114
  new   9, 165, 184, 188
thinking   1, 4, 14
THOG task   113
time measurement   81–2
  mistake in calculating   83
time period, strategies for establishing
    length   89
top-down decomposition   158
top-down strategies   157
topsy turvy task   135–40
Traffic models   216

transfer
  concept   3
  in autonomous agent learning   245
  issue   100
transforming practices   41–4
translation   7
Triplet game   117–18
TRIVAR   168, 197, 208
truth value   123–5
typewriter   8

universalism   32
unrealistic modelling in fifth
    graders   63–70

valorisation
  of knowledge   15, 25
  of school and farming mathematical
    practices   23, 27–9
virtual environments   192
VIS   168, 169, 194–209
  activities   206–8
  approach adopted by   197–204
  building mode   205
  collaboration   202, 207
  content and delivery of
    curriculum   203
  educational experience   204–8
  help statements   207
  knowledge base   199–202
  learning environment   198–9
  running mode   205–6
  structure of system   198–204
  student model   203
  teacher role in   202–4
  zone of available assistance
    (ZAA)   207

Wason's Selection Task   121
Wason's Sentence Verification
    Task   122
within-subjects design   129
working with others, social
    environment   177–9
workplace contexts   9

writing 6–8
writing exchange 55–6

zone of proximal adjustment, creation 207–8

zone of proximal development 168, 194–209
    collaboration within 202